Operator-Limit Distributions in Probability Theory

ZBIGNIEW J. JUREK

Institute of Mathematics
University of Wrocław
Wrocław, Poland

J. DAVID MASON

Department of Mathematics
The University of Utah
Salt Lake City, Utah

A Wiley-Interscience Publication
JOHN WILEY & SONS, INC.
New York · Chichester · Brisbane · Toronto · Singapore

This text is printed on acid-free paper.

Copyright © 1993 by John Wiley & Sons, Inc.

All rights reserved. Published simultaneously in Canada.

Reproduction or translation of any part of this work beyond
that permitted by Section 107 or 108 of the 1976 United
States Copyright Act without the permission of the copyright
owner is unlawful. Requests for permission or further
information should be addressed to the Permissions Department,
John Wiley & Sons, Inc., 605 Third Avenue, New York, NY
10158-0012.

Library of Congress Cataloging in Publication Data:
Jurek, Zbigniew J.
　Operator-limit distributions in probability theory/Zbigniew J.
Jurek and J. David Mason.
　　p.　cm.—(Wiley series in probability and mathematical
statistics. Probability and mathematical statistics)
　　"A Wiley-Interscience publication."
　　Includes bibliographical references and index.
　　ISBN 0-471-58595-5 (acid-free)
　　1. Central limit theorem.　　I. Mason, J. David (Jesse David),
1935-　.　II. Title.　III. Series.
QA273.67.J87　1993
519.2—dc20　　　　　　　　　　　　　　　　　　　　　　　93-7158

Printed in the United States of America

10 9 7 6 5 4 3 2 1

Probability and Mathematical Statistics (Continued)
  PURI, VILAPLANA, and WERTZ · New Perspectives in Theoretical and Applied Statistics
  RAO · Asymptotic Theory of Statistical Inference
  RAO · Linear Statistical Inference and Its Applications, *Second Edition*
  ROBERTSON, WRIGHT, and DYKSTRA · Order Restricted Statistical Inference
  ROGERS and WILLIAMS · Diffusions, Markov Processes, and Martingales, Volume II: Ito Calculus
  ROHATGI · A Introduction to Probability Theory and Mathematical Statistics
  ROSS · Stochastic Processes
  RUBINSTEIN · Simulation and the Monte Carlo Method
  RUZSA and SZEKELY · Algebraic Probability Theory
  SCHEFFE · The Analysis of Variance
  SEBER · Linear Regression Analysis
  SEBER · Multivariate Observations
  SEBER and WILD · Nonlinear Regression
  SERFLING · Approximation Theorems of Mathematical Statistics
  SHORACK and WELLNER · Empirical Processes with Applications to Statistics
  STAUDTE and SHEATHER · Robust Estimation and Testing
  STOYANOV · Counterexamples in Probability
  STYAN · The Collected Papers of T. W. Anderson: 1943–1985
  WHITTAKER · Graphical Models in Applied Multivariate Statistics
  YANG · The Construction Theory of Denumerable Markov Processes

Applied Probability and Statistics
  ABRAHAM and LEDOLTER · Statistical Methods for Forecasting
  AGRESTI · Analysis of Ordinal Categorical Data
  AGRESTI · Categorical Data Analysis
  ANDERSON and LOYNES · The Teaching of Practical Statistics
  ANDERSON, AUQUIER, HAUCK, OAKES, VANDAELE, and WEISBERG · Statistical Methods for Comparative Studies
  ASMUSSEN · Applied Probability and Queues
 *BAILEY · The Elements of Stochastic Processes with Applications to the Natural Sciences
  BARNETT · Interpreting Multivariate Data
  BARNETT and LEWIS · Outliers in Statistical Data, *Second Edition*
  BARTHOLOMEW, FORBES, and McLEAN · Statistical Techniques for Manpower Planning, *Second Edition*
  BATES and WATTS · Nonlinear Regression Analysis and Its Applications
  BELSLEY · Conditioning Diagnostics: Collinearity and Weak Data in Regression
  BELSLEY, KUH, and WELSCH · Regression Diagnostics: Identifying Influential Data and Sources of Collinearity
  BHAT · Elements of Applied Stochastic Processes, *Second Edition*
  BHATTACHARYA and WAYMIRE · Stochastic Processes with Applications
  BIEMER, GROVES, LYBERG, MATHIOWETZ, and SUDMAN · Measurement Errors in Surveys
  BIRKES and DODGE · Alternative Methods of Regression
  BLOOMFIELD · Fourier Analysis of Time Series: An Introduction
  BOLLEN · Structural Equations with Latent Variables
  BOX · R. A. Fisher, the Life of a Scientist
  BOX and DRAPER · Empirical Model-Building and Response Surfaces
  BOX and DRAPER · Evolutionary Operation: A Statistical Method for Process Improvement
  BOX, HUNTER, and HUNTER · Statistics for Experimenters: An Introduction to Design, Data Analysis, and Model Building
  BROWN and HOLLANDER · Statistics: A Biomedical Introduction

Continued on back end papers

*Now available in a lower priced paperback edition in the Wiley Classics Library.

Operator-Limit Distributions
in Probability Theory

Contents

Symbols and Notation vii

Preface ix

1. **Preliminaries,** 1
    1.1 Numakura Theorem, 1
    1.2 Linear Spaces, 3
    1.3 Integration and Differentiation of Vector-Valued Functions, 7
    1.4 One-Parameter Semigroups, 10
    1.5 Basic Facts from Lie Theory, 15
    1.6 Probability Measures on Metric Spaces, 20
    1.7 Probabilities on Banach Spaces, 23
    1.8 Infinitely Divisible Probabilities, 28
    1.9 The Skorohod Space, 39
    1.10 Bibliographic Comments, 44

2. **Convergence of Types Theorems, Symmetry Groups, and Decomposability Semigroups,** 45
    2.1 Full Measures, 46
    2.2 Convergence of Types Theorems, 48
    2.3 Urbanik Decomposability Semigroup, 53
    2.4 Compact Subgroups of $\mathscr{A}_l(\mathbb{R}^d)$ and $\mathbf{Aut}(\mathbb{R}^d)$, 59
    2.5 Examples in Infinite-Dimensional Spaces, 62
    2.6 The Case When $d = 1$; Comments, 65
    2.7 Standard Convergence of Types, 67
    2.8 Bibliographic Comments, 70

3. Operator-Selfdecomposable Measures — 71
- 3.1 Statement of the Problem, 71
- 3.2 Norming Sequences, 72
- 3.3 The Urbanik Semigroup of an Operator-Selfdecomposable Measure and Its Exponents, 76
- 3.4 Characteristic Functionals of $\exp(-tQ)$-Decomposable Measures, 89
- 3.5 Generators of the Class $L_0(Q)$, 108
- 3.6 Random Integral Representation, 116
- 3.7 Infinitesimal Generators, 144
- 3.8 The Absolute Continuity of $\exp(-tQ)$-Decomposable Measures, 162
- 3.9 Multivariate Selfdecomposable Measures, 177
- 3.10 Bibliographic Comments, 182

4. Operator-Stable Measures — 185
- 4.1 Statement of the Problem, 185
- 4.2 Sakovic–Sharpe Characterization, 186
- 4.3 A Class of t^B-Stable Measures, 193
- 4.4 The Class of $\mathscr{S}(B)$ and Fix Points of the Mapping $\mathscr{T}_B$, 198
- 4.5 Norming Sequences, 201
- 4.6 Structural Characterizations of Operator-Stable Measures and Their Exponents, 205
- 4.7 Commuting Exponents and Other Norms, 219
- 4.8 Elliptically Symmetric Operator-Stable Measures, 229
- 4.9 The Centering Function $b(t)$, 236
- 4.10 Absolute Continuity of Full Operator-Stable Measures, 237
- 4.11 Domains of Normal Attraction of Operator-Stable Measures, 239
- 4.12 Moments, 249
- 4.13 Independent Marginals, 254
- 4.14 Multivariate Stable Measures, 261
- 4.15 The Case When $d = 3$, 271
- 4.16 Bibliographic Comments, 274

Epilogue — 277

Bibliography — 281

Index — 289

Symbols and Notations

	Defined on Page		Defined on Page
$\langle \cdot, \cdot \rangle$	3	$\mathrm{ID}(X)$	28
$\|\cdot\|_\mu$	223	$\mathrm{ID}_{\log}(X)$	36
$\|\cdot\|_{\mu,B}$	223	$\mathrm{Inv}(\mu)$	50
$\|\cdot\|_Q$	92	$\ker(A)$	6
$\mathbf{A}(\mu)$	53	$\mathscr{K}_Q$	110
$\langle A; a \rangle$	24	$\mathrm{lin}(A)$	105
$\mathscr{A}_I$	60	$\mathrm{lin}_Q(A)$	105
$\mathbf{A}_J$	77	$\mathscr{L}(\xi)$	22
$[a, R, M]$	33	L_M	94
$\mathrm{Aut}(X)$	4	$L_0(Q)$	108
$\mathscr{A}(X)$	24	$L(X,Y)$	4
$\mathscr{B}(S)$	20	$\hat{\mu}$	25
$C_b(S)$	20	μ	26
$C_0(\mathbb{R}^d)$	4	μ^0	26
$\mathscr{D}(\Gamma)$	10	$\mu_n \Rightarrow \mu$	20
$\mathbf{D}(\mu)$	53	$\mu * \nu$	23
$\mathbf{D}_J(\mu)$	77	$\mathscr{O}$	18
$\mathrm{DONA}(\mu)$	240	$\mathrm{OL}(\mathbb{R}^d)$	72
$D(S; [0, \infty))$	42	Π_{x^*}	32
$D(S; [a, b])$	40	$\mathscr{P}(S)$	20
$\mathrm{End}(X)$	4	$\mathscr{Q}$	18
$e(m)$	30	$\mathrm{sem}\{a\}$	2
$\mathscr{E}_u(\mu)$	85	$\mathrm{sem}\ A$	2
$\mathscr{E}_{cu}(\mu)$	88	$\mathrm{sem}\{\mathbf{F}\}$	75
$\exp(A)$	15	S_Q	93
$[F; G]$	196	$\mathscr{T}(\mathbb{H})$	15
$\mathscr{F}(\mathbb{R}^d)$	47	$\mathscr{T}_Q$	127
$\mathrm{GDOA}(\mu)$	239	$\mathscr{T}_B$	198

Preface

The theory of limit distributions occupies a central place in probability and mathematical statistics. It describes limit phenomena of triangular series or sequences of independent random observations. Typically, one looks at statistics (functions) constructed from given random variables. Often these functions are assumed to be linear. Therefore, this leads to consideration of limit distributions of the sequences

$$a_n(\xi_1 + \cdots + \xi_n) + x_n, \quad n \geq 1, \tag{0.1.1}$$

where $\xi_1, \xi_2, \ldots$ are independent $\mathbb{R}$-valued random variables and a_n and x_n are real constants. But often when the observations $\xi_1, \xi_2, \ldots$ are random vectors ($\mathbb{R}^d$ or Banach space valued), the normalization in (0.1.1) is still done by *scalars*. In such a setting, we have coordinate-wise one-dimensional problems again. One should allow the interaction between coordinate normalization in (0.1.1) to be consistent with the structure of $\mathbb{R}^d$, that is, one should allow normalization by arbitrary *linear operators*. Thus, the main aims or principles for this book are:

1. Present a theory of limit distributions of sequences

$$A_n(\xi_1 + \cdots + \xi_n) + x_n, \quad n \geq 1, \tag{0.1.2}$$

   where $\xi_1, \xi_2, \ldots$ are $\mathbb{R}^d$-valued random vectors and $A_1, A_2, \ldots$ are matrices (operators) on $\mathbb{R}^d$. This explains the title, *Operator-Limit Distributions in Probability Theory*.
2. Present proofs which do not appeal to the one-dimensional results of (0.1.1). In other words, this exposition is much more "functional" or "coordinate-free."

3. Present complete exposition for $\mathbb{R}^d$ and indicate the essential differences for infinite-dimensional Banach space valued random variables.

It seems natural that the normalization of the random vectors be consistent with the algebraic structure of the space in which they take their values. Hence, for the real line one uses scalars, for a linear space one uses linear operators, and for a group one uses group endomorphisms. (As far as we know, the group case is completely open. For probabilities on groups, convolution powers are usually investigated and often the normalized Haar measure is the only limit measure.)

Limits of (0.1.1) are called *stable* distributions if $\xi_1, \xi_2, \ldots$ are independent and identically distributed. If ξ_n's are independent and the triangular array $\{a_n \xi_j : 1 \le j \le n\}$ is infinitesimal, then limits of (0.1.1) are called *selfdecomposable* distributions (or Lévy class L_0 distributions). Stable measures have been extensively investigated for the last 60 years or so [cf. Gnedenko and Kolmogorov (1954), Zolotarev (1986), and Linde (1986)]. The class L_0 of selfdecomposable measures contains the class of stable laws. It is related to autoregressive sequences, that is, a sequence $\{X_n\}$ such that $X_{n+1} \stackrel{d}{=} cX_n + \varepsilon_n$, whose $0 < c < 1$, X_n's are independent of ε_n's. Furthermore, L_0 distributions arise in limits of Ising models for ferromagnetism in statistical physics [cf. deConinck (1984)]. Elements from L_0 can be viewed as limits as $t \to \infty$ of some stochastic processes which are given by random integrals and are similar to Ornstein–Uhlenbeck processes (cf. Section 3.6). Finally, there is also the notion of self-similar processes. These processes satisfy $\{X(at): t \ge 0\}$ and $\{a^H X(t): t \ge 0\}$ have the same finite-dimensional distributions, where H is a scalar (or a matrix). In the case when the process has stationary independent increments, the distribution of $X(1)$ is (operator) stable, and in the case that the increments are only independent, the distribution of $X(1)$ is (operator) selfdecomposable [cf. Lamperti (1962, 1972), Hudson and Mason (1982), and Sato (1991)].

Limit distributions of sequences of the form (0.1.2) will be called *operator-stable* and *operator-selfdecomposable* when $\xi_1, \xi_2, \ldots$ are i.i.d. or only independent and infinitesimally small, respectively. Sakovic (1961, 1965), Ph.D. dissertation under B. V. Gnedenko, was the first to investigate operator-stable laws on $\mathbb{R}^d$. On the other hand, Fisz (1954) proved a theorem on the convergence of operator-types (normalization by matrices). Both, Fisz and Sakovic used a coordinate-wise approach which led to some computational difficulties and which perhaps explains why this direction was not continued later (cf. the above aim 2). Independently of Sakovic, Sharpe (1969), Ph.D. dissertation under S. Kakutani, gave a complete characterization of operator-stable measures on $\mathbb{R}^d$ and his proof used some algebraic methods. Similarly, Urbanik (1972) gave a "functional" proof for his description of operator-selfdecomposable measures. This was the beginning of a period of extensive study of operator-limit distributions. This functional approach also prompted the development of new purely algebraic methods (decomposability semigroups) in probability. This book attempts to summarize that period of investigation.

Chapter 1 is of an auxiliary character. It compiles well-known facts from the theory of limit distributions (infinitely divisible measures, weak convergence, Skorohod spaces) without proofs. On the other hand, algebraic facts (Numakura theorem, Lie groups, one-parameter semigroups of operators) are given with proofs. Bibliographic comments will always be given at the end of each chapter.

To be able to study the limits of (0.1.1) or (0.1.2), one needs theorems on the relationship between the limit distribution of the pair of sequences

$$\{\xi_n\} \quad \text{and} \quad \{A_n \xi_n + x_n\}. \tag{0.1.3}$$

We refer to these as convergence of *operator types* and these are proved in Chapter 2. Besides that, Chapter 2 contains theorems on decomposability and symmetry semigroups associated with probability measures. These semigroups will be the fundamental tools in Chapters 3 and 4, but they are of interest in themselves and therefore are presented in separate sections. The point is that some topological and algebraic properties of these semigroups are equivalent to measure theoretical properties. Examples in Section 2.5 indicate the "trouble" points when one deals with random variables having values in infinite-dimensional linear spaces. The last two sections specialize some results to one-dimensional and infinite-dimensional spaces. Bibliographic comments are given in Section 2.8.

Chapter 3 deals with operator-selfdecomposable measures, that is, limits of (0.1.2) with independent ξ_n's and infinitesimal triangular array $\{A_n \xi_j: 1 \le j \le n, n \ge 1\}$. First, properties of norming sequences $\{A_n\}$ are discussed, in particular, semigroups of operators generated by them. Then the main Urbanik decomposability theorem for operator-selfdecomposable measures, Theorem 3.3.5, is proved. Their characteristic functions are described in Section 3.4. Section 3.6 establishes random integral representations of operator-selfdecomposable probabilities, thus showing the connection with Ornstein–Uhlenbeck type processes. Subsequently, infinitesimal generators for such processes are found. Section 3.8 proves absolute continuity of full $\exp(-tQ)$-decomposable measures on $\mathbb{R}^d$, whereas Section 3.9 deals with selfdecomposable measures on arbitrary Banach spaces.

In Chapter 4, we investigate operator-stable distributions, that is, limits of (0.1.2) with i.i.d. sequences $\xi_1, \xi_2, \ldots$. The main characterization due to Sharpe (partially due to Sakovic) is proved in Theorem 4.2.12. Structural characterization of operator-stable distributions and operator exponents of such measures are proved in Section 4.6. Then commuting exponents and elliptically symmetric measures are discussed. Section 4.11 gives descriptions of the domain of normal attraction, that is, normalization is by operators of the form n^{-B} for some operator exponent B. In the case of the *generalized* domain of attraction, that is, using arbitrary norming operators A_n, we decided to omit the results. This general case is presently solved in the finite-dimensional setting, but it is dealt with by appealing to knowledge of one-dimensional results and applying them uniformly in every direction [cf. Hahn and Klass (1981, 1985) and Griffin (1986)]. This is not in the spirit of our aim 2 and therefore is not included here. The existence of moments of operator-stable laws are presented in Section 4.12, whereas the existence of a complete set of independent univariate marginals is given in Section 4.13. The special case of the usual multivariate stable laws is presented in Section 4.14. Chapter 4 closes by considering some special cases and with bibliographic comments.

The book ends with an epilogue in which we briefly discuss some areas of operator-limit distributions theory which are omitted, in particular, operator-semistability, $\mathscr{G}$-stability for a group $\mathscr{G}$ of bounded linear operators, and normalization by a one-parameter semigroup of operators.

This book is designed to be used by graduate students as a textbook either in a classroom or for directed studies. For that purpose, we added Chapter 1. Definitely, we do not suggest one starts with Chapter 1 since it is for reference purposes only and it should be consulted as needed. Because of aim 2, it is not necessary that readers be familiar with classical stability or decomposability (on $\mathbb{R}^1$ or a Banach space). As a by-product, those are discussed at the end of each chapter. We do not have sections with exercises, but many of the corollaries and lemmas can be assigned as homework. Because of aim 3, students can persue their own research in the area of operator-limit distributions on infinite-dimensional linear spaces and on topological groups. Those interested only in operator-stability can skip Chapter 3 since it is quite independent of Chapter 4.

We also see this book as the main reference for further research. It is the first monograph covering such a selection of limit laws. In some sense it can be viewed as a complement to the books by Araujó and Giné (1980), Linde (1986), and Zolotarev (1986).

It took a long time to complete this book due in large part to the "long-distance" communication between Poland and the United States. Thus, we divided the work and the responsibility. Of course, we share the responsibility for any errors and oversights. Much of the results are due to others and we point this out in the bibliographic comments at the end of each chapter.

We benefited from discussions with many of those working in the areas of operator-limit theorems. We thank them all for their encouraging, discouraging, or neutral comments. Those whom we particularly thank include (in alphabetical order): M. G. Hahn (Tufts University), W. Hazod (University of Dortmund), W. N. Hudson (Auburn University), R. Jajte (University of Łódz), M. Klass (University of California, Berkeley), W. Krakowiak, J. Kucharczak, B. Mincer, and T. Rajba (all from University of Wrocław), K. Sato (University of Nagoya), H. Tucker (University of California, Irvine), K. Urbanik (University of Wrocław), J. A. Veeh (Auburn University), and M. Yamazato (Nagoya Institute of Technology).

Last, but not least, we would like to thank our closest loved ones who have suffered through these years. Their constant reminding question: "When will you be ready?" was always motivating and encouraging on a road up to this line.

ZBIGNIEW J. JUREK
J. DAVID MASON

Operator-Limit Distributions
in Probability Theory

CHAPTER 1

Preliminaries

The purpose of this introductory chapter is to collect all the needed facts for future references. The theorems and tools from outside of probability theory are presented with complete proofs and are in the generality that covers our needs. However, facts from probability theory, mainly concerned with weak limit theorems, are presented mostly without proofs. References for them are given in the last section of this chapter.

1.1 NUMAKURA THEOREM

Let S be a Hausdorff topological space. If in S there is defined a single-valued product ab which is associative and continuous, then S is called a *topological semigroup*. By a *subsemigroup* of S we mean a nonempty subset A such that $A^2 \subset A$, that is, $ab \in A$ for all $a, b \in A$. Also, A is called a *subgroup* of S if $xA = Ax = A$ for all $x \in A$. By a *left* (*right*) *ideal* of S we mean a subset M such that $SM \subset M$ ($MS \subset M$). When M is both a left and a right ideal of S, M is called an *ideal* of S. Finally, an element $a \in S$ is called an *idempotent* of S provided $a^2 = a$. Obviously, the zero ($a \cdot 0 = 0 \cdot a = 0$ for all $a \in S$) and the identity ($e \cdot a = a \cdot e = a$ for all $a \in S$) elements of S are idempotent, whenever they exist.

Theorem 1.1.1. *If S is a compact semigroup, then S contains a compact subgroup, and hence at least one idempotent.*

Proof. Fix $a \in S$ and let $K(a)$ denote the set of all limit points of the sequence $\{a^n\}_{n \geq 1}$, that is,

$$K(a) := \bigcap_{n=1}^{\infty} \overline{\{a^i : i \geq n\}},$$

where A^- denotes the closure of $A \subset S$. We assert that $K(a)$ is a compact, commutative subsemigroup of S. The commutativity of $K(a)$ is given by: let $b_1 := \lim_{n \to \infty} a^{m_n}$, $b_2 := \lim_{n \to \infty} a^{k_n}$ with $\{m_n\}$ and $\{k_n\}$ strictly increasing. By continuity of products, $b_1 b_2 = \lim_{n \to \infty} a^{m_n + k_n} = b_2 b_1$. The subsemigroup property follows from a similar argument and the compactness is obvious.

To complete the proof, we show that $xK(a) = K(a)$ for all $x \in K(a)$. Clearly, we have $xK(a) \subset K(a)$. Suppose there is an $x \in K(a)$ for which $xK(a)$ is a proper subset of $K(a)$. Then there is a $z \in K(a)$ such that $z \notin xK(a)$. By continuity of products, there are open neighborhoods V, W, and U of x, $K(a)$, and z, respectively, such that $VW \cap U = \emptyset$. Since $x, z \in K(a)$, there are an integer m and a sequence of integers $\{n_i\}_{i \geq 1}$ such that $n_{i+1} > n_i > m$, $a^m \in V$, and $a^{n_i} \in U$ for all $i \geq 1$. Let b be a limit point of $\{a^{n_i - m}\}_{i \geq 1}$. Then $b \in K(a) \subset W$, so there is an integer j such that $a^{n_i - m} \in W$ for all $i \geq j$. Hence, $a^{n_i} = a^m a^{n_i - m} \in VW$ for $i \geq j$. But, $a^{n_i} \in U$, which contradicts $VW \cap U = \emptyset$. Thus, $xK(a) = K(a)x = K(a)$, so $K(a)$ is a subgroup of S. Obviously, the identify of $K(a)$ is an idempotent of S.
Q.E.D.

For $A \subset S$, let sem A denote the smallest closed subsemigroup in S containing A. In case $A = \{a\}$, sem$\{a\}$ is called a *monothetic* semigroup.

Theorem 1.1.2. *If the monothetic semigroup* sem$\{a\}$ *is compact, then the set of all limit points of* $\{a^n\}_{n \geq 1}$, $K(a)$, *is the minimal ideal in* sem$\{a\}$ *and the unit element*, e, *of the subgroup* $K(a)$ *is the only idempotent in* sem$\{a\}$.

Proof. Since sem$\{a\} = \{a^n : n \geq 1\} \cup K(a)$ and $K(a)$ is a commutative subgroup of sem$\{a\}$, we see that $K(a)$ is an ideal in sem$\{a\}$. Let b be an idempotent in sem$\{a\}$. When $b \in K(a)$, $b = e$ since $K(a)$ is a group. When $b \in \{a^n : n \geq 1\}$, then $b = a^m$ for some m, so for every $k \geq 1$, $b = b^k = a^{km}$, which implies $b \in K(a)$. This shows that the identity of $K(a)$ is the only idempotent of sem$\{a\}$.

It remains to show $K(a)$ is a minimal ideal. First, note that if H is a minimal ideal in sem$\{a\}$, then $xH = Hx = H$ for all $x \in$ sem$\{a\}$, because otherwise xH or Hx is a proper subideal of H in sem$\{a\}$. Hence, H is a subgroup with unit element $e \in K(a)$, since sem$\{a\}$ has only one idempotent. Second, note that $H = G(e)$, where H is a minimal ideal in sem$\{a\}$ and $G(e)$ is a maximal subgroup of sem$\{a\}$ containing e. We know that $G(e)$ exists by the Zorn lemma. To see this, we obviously have $H \subset G(e)$. Let $z \in G(e)$. Then $zH = H$, so there is $z^* \in H$ such that $zz^* = z^*z = e$. Hence, $z \in H$, so $G(e) \subset H$. Third, we show that $K(a) = G(e)$. Since $K(a)$ is a subgroup with identity e, we have $K(a) \subset G(e)$. Suppose there is $b \in G(e)$ and $b \notin K(a)$. Since $b \in \text{sem}\{a\}$, $b = a^m$ for some integer m. Since $a^m e = a^m$ and product is continuous, for every neighborhood W of b, there is a neighborhood V of e such that $bV \subset W$. Since $e \in K(a)$, $V \supset \{a^{n_k}: k \geq 1\}$ for some sequence $n_k < n_{k+1}$. Hence, $a^m a^{n_k} = a^{m+n_k} \in W$, so b must be in $K(a)$. This contradiction shows that $K(a) = G(e)$. These three steps show that $K(a)$ is the minimal ideal in $\text{sem}\{a\}$. Q.E.D.

Corollary 1.1.3 (Numakura Theorem). *Let S be a compact semigroup. For each $a \in S$, the monothetic semigroup $\text{sem}\{a\}$ is compact and the set of limit points of $\{a^n\}_{n \geq 1}$, $K(a)$, is a subgroup. Moreover, $K(a)$ is the minimal ideal of $\text{sem}\{a\}$ [hence, for x in $\text{sem}\{a\}$, $xK(a) = K(a)$] and the identity element, e, of the group $K(a)$ is the only idempotent in $\text{sem}\{a\}$.*

For future reference, we have the following corollary.

Corollary 1.1.4. *If the monothetic semigroup $\text{sem}\{a\}$ is compact, then there is $b \in K(a)$ such that $ab = ba = e$.*

Proof. Since $K(a)$ is commutative and $aK(a) = K(a)$, such $b \in K(a)$ exists. Q.E.D.

1.2 LINEAR SPACES

Let X be a real *Banach* space, that is, X is a real linear, normed, complete space, with norm $\|\cdot\|$. By X^* we denote its *topological dual Banach* space, that is, $x^* \in X^*$ are continuous linear functionals on X, and $\langle \cdot, \cdot \rangle$ is the dual pair between X^* and X. When the norm in X is given by a scalar product, X is called a *Hilbert* space. In that case, X^* is isomorphic to X and the dual pair is simply the scalar product. Furthermore, all real separable Hilbert spaces are isomorphic to l_2, the space of all real square-summable sequences with

$$\langle x, y \rangle := \sum_i x_i y_i, \qquad \|x\| := (\langle x, x \rangle)^{1/2}$$

for $x = \{x_i\}_{i \geq 1}$ and $y = \{y_i\}_{i \geq 1}$. Besides this example, we will deal with the following ones.

(a) $X = C_0(\mathbb{R}^d)$ is the set of all real-valued continuous functions on $\mathbb{R}^d$, d-dimensional Euclidean space, which vanish at infinity. The point ∞ can be used in the one-point compactification of $\mathbb{R}^d$. By the *Riesz representation theorem*, each $x^* \in [C_0(\mathbb{R}^d)]^*$ is uniquely determined by a finite Borel measure m on $\mathbb{R}^d$, not necessarily positive, such that, for $f \in C_0(\mathbb{R}^d)$,

$$\langle x^*, f \rangle = \int f(x) m(dx), \qquad \|x^*\| = m(\mathbb{R}^d).$$

(b) $X = \mathbb{R}^d$. Then $(\mathbb{R}^d)^* = \mathbb{R}^d$ and

$$\langle x, y \rangle = \sum_{i=1}^{d} x_i y_i.$$

Also, it is easy to see that in all finite-dimensional inner product spaces all norms are *equivalent*, that is, if $\|\cdot\|_1$ and $\|\cdot\|_2$ are two norms on X, then there are positive constants c_1 and c_2 such that for all $x \in X$, $c_1\|x\|_1 \leq \|x\|_2 \leq c_2\|x\|_1$.

Let X and Y be Banach spaces. By a *bounded linear operator* A from X into Y, we mean a function $A: X \to Y$ such that (1) A is linear, that is, $A(\alpha_1 x_1 + \alpha_2 x_2) = \alpha_1 A x_1 + \alpha_2 A x_2$ for all $x_1, x_2 \in X$ and all $\alpha_1, \alpha_2 \in \mathbb{R}^1$, and (2) there is a constant C such that $\|Ax\| \leq C\|x\|$ for all $x \in X$; the first norm is in Y and the second norm, possibly different, is in X. The infimum of all C in (2) is denoted by $\|A\|$, and is called the *norm of the operator* A. The assumption that A is bounded and linear is equivalent to A being continuous and linear from X to Y, where the topologies are given by the norms. The collection $L(X,Y)$ of all bounded linear operators from X into Y, using the operator norm, is also a Banach space. When $X = Y$, $L(X,Y)$ is denoted by $\text{End}(X)$; in which case, we also have that the product of two operators in $\text{End}(X)$ is a continuous linear operator: if $A, B \in \text{End}(X)$, then $AB: X \to X$ is given by $(AB)x = A(Bx)$ for $x \in X$. Moreover, $\|AB\| \leq \|A\|\|B\|$ for all $A, B \in \text{End}(X)$. With this multiplication of operators, $\text{End}(X)$ becomes a topological semigroup. By $\text{Aut}(X)$, we denote the set of all invertible operators in $\text{End}(X)$. These inverses are also continuous and linear, so $\text{Aut}(X)$ is a topological group.

LINEAR SPACES

Let $\mathscr{D}(A)$ be a linear subspace of a Banach space X, and let A be a linear operator from $\mathscr{D}(A)$ into a Banach space Y. By the *graph* of A is meant the set

$$\operatorname{graph} A := \{(x, Ax) : x \in \mathscr{D}(A)\} \subset X \times Y.$$

The product space $X \times Y$ can be treated as a Banach space; for example, $\|(x, y)\| := \|x\|_1 + \|y\|_2$, where $\|\cdot\|_1$ and $\|\cdot\|_2$ are the norms in X and Y, respectively. We say that the operator A is *closed* if its graph is a closed subset of $X \times Y$, that is, if $x_n \in \mathscr{D}(A)$, $x_n \to x_0$ in X, $Ax_n \to y_0$ in Y implies that $x_0 \in \mathscr{D}(A)$ and $y_0 = Ax_0$. An operator B defined on $\mathscr{D}(B) \subset X$ is called an *extension* of A with its domain $\mathscr{D}(A)$ whenever $\mathscr{D}(A) \subset \mathscr{D}(B)$ and $Ax = Bx$ for all $x \in \mathscr{D}(A)$.

The following is a criterion which determines when an operator A with domain $\mathscr{D}(A)$ has a closed extension.

Theorem 1.2.1. *Let A be a linear operator with domain $\mathscr{D}(A) \subset X$. Then A has a closed extension if and only if there is no $y \neq 0$ such that $(0, y)$ belongs to $(\operatorname{graph} A)^-$, the closure of $\operatorname{graph} A$. In this case, $(\operatorname{graph} A)^-$ is the $\operatorname{graph} A^-$, the smallest closed extension of A.*

Proof. Since $\operatorname{graph} A$ is a linear subspace of $X \times Y$, so is $(\operatorname{graph} A)^-$. Since $(x, y_1), (x, y_2) \in (\operatorname{graph} A)^-$ implies that $y_1 = y_2$ $[(x, y_1) - (x, y_2) = (0, y_1 - y_2) \in (\operatorname{graph} A)^-]$, the relation $(\operatorname{graph} A)^- \subset X \times Y$ determines a function, A^-. We see that the linearity and closeness of $(\operatorname{graph} A)^-$ implies that A^- is linear and closed. Obviously, $\operatorname{graph} A^- = (\operatorname{graph} A)^-$. It is also obvious that $(\operatorname{graph} A)^-$ is the graph of the smallest closed extension of A. The converse is trivial. Q.E.D.

We use A^- to denote the smallest closed extension of the linear operator A.

Corollary 1.2.2. *If a linear operator A has an extension which is a closed linear operator B, then A^- exists.*

Proof. Since $(\operatorname{graph} A)^- \subset \operatorname{graph} B$, there is no $y \neq 0$ such that $(0, y) \in (\operatorname{graph} A)^-$. Now, apply Theorem 1.2.1. Q.E.D.

Obviously, all bounded linear operators are closed. Other examples of closed operators are the infinitesimal generators of one-parameter semigroups [cf. Proposition 4.4.1(c)]. Also, note that the sum of a closed linear operator and a bounded linear operator is closed.

Now, we wish to define the trace of a linear operator on $\mathbb{R}^d$. When A is a $d \times d$ matrix, we define trace(A) to be the sum of the numbers on its main diagonal. Since similar matrices have the same trace, we may define the trace(A) for A a linear operator on $\mathbb{R}^d$ to be the trace of any matrix which represents A in an ordered basis.

For $A \in \text{End}(\mathbb{R}^d)$, we may define e^A by the series $e^A := \sum_{n=0}^{\infty} (n!)^{-1} A^n$. Since $\sum_{n=0}^{\infty} (n!)^{-1} \|A\|^n$ converges, $e^A \in \text{End}(\mathbb{R}^d)$. Actually, $e^A \in \text{Aut}(\mathbb{R}^d)$ since the inverse of e^A is given by e^{-A}.

It is well known how the determinant of a $d \times d$ matrix A is defined. For $A \in \text{End}(\mathbb{R}^d)$, det($A$) denotes the determinant of any matrix which represents A in a basis. Then $\det(e^A) = e^{\text{trace}(A)}$.

The final topic of this section is the primary decomposition theorem of $\mathbb{R}^d$. Let $A \in \text{End}(\mathbb{R}^d)$. The *minimal polynomial for A* is the unique polynomial $g(\cdot)$ over the reals with the properties that (1) the coefficient of its highest term is one, (2) $g(A) = 0$, and (3) if $h(A) = 0$, where $h(\cdot)$ is a polynomial over the reals, then the degree of $h(\cdot)$ is greater than or equal to the degree of $g(\cdot)$. A polynomial over the reals is said to be *irreducible* if it is not possible to factor it into the product of two polynomials, each having degree greater than or equal to one. Let ker(A) denote the set $\{x \in \mathbb{R}^d: Ax = 0\}$, and call it the *null space of A*. A subspace $W \subset \mathbb{R}^d$ is said to be *A-invariant* if $A(W) \subset W$. We write $\mathbb{R}^d = W_1 \oplus \cdots \oplus W_k$ if each W_i is a subspace of $\mathbb{R}^d$ and each $v \in \mathbb{R}^d$ has a unique representation of the form $v = v_1 + \cdots + v_k$ with $v_i \in W_i$ for all i. We also say that $\mathbb{R}^d$ is the *direct sum* of the subspaces $W_1, \ldots, W_k$. A nonzero polynomial $a_n x^n + a_{n-1} x^{n-1} + \cdots$ is called a *monic* polynomial if $a_n = 1$.

Theorem 1.2.3 (Primary Decomposition Theorem). *Let* $A \in \text{End}(\mathbb{R}^d)$, *and let* $g(\cdot)$ *be its minimal polynomial. Assume* $g = g_1^{r_1} \cdots g_k^{r_k}$, *where the* g_i *are distinct irreducible monic polynomials over* $\mathbb{R}^1$ *and the* r_i *are positive integers. Let* $W_i := \ker(g_i(A)^{r_i})$ *for* $1 \leq i \leq k$. *Then*

(i) $\mathbb{R}^d = W_1 \oplus \cdots \oplus W_k$;
(ii) *each* W_i *is A-invariant*;
(iii) *if* A_i *is the restriction of A to* W_i, *then the minimal polynomial of* A_i *is* $g_i^{r_i}$.

1.3 INTEGRATION AND DIFFERENTIATION OF VECTOR-VALUED FUNCTIONS

Let $(S, \mathscr{S}, \rho)$ be a measure space with ρ a finite measure on the σ-field $\mathscr{S}$. Let X be a Banach space with the Borel σ-field $\mathscr{B}(X)$ generated by the open sets in the metric on X given by a norm. We say that $f: S \to X$ is *measurable* if for each $B \in \mathscr{B}(X)$, $f^{-1}(B) \in \mathscr{S}$. This is equivalent to f being the limit ρ-a.e. of a sequence of simple functions, that is, of the form $\sum_{i=1}^{n} x_i I_{A_i}$ with each $x_i \in X$ and $A_i \in \mathscr{S}$.

A measurable function $f: S \to X$ is said to be (Bochner) *integrable* if there exists a sequence $\{f_n\}_{n \geq 1}$ of simple functions such that $f_n \to f$ ρ-a.e. and $\int_S \|f_n - f\| \, d\rho \to 0$ as $n \to \infty$. In this case, the sequence $\{\int_S f_n \, d\rho\}_{n \geq 1}$ is a Cauchy sequence in X, where $\int_S f_n \, d\rho := \sum_{i=1}^{k_n} x_i \rho(A_i)$ whenever $f_n = \sum_{i=1}^{k_n} x_i I_{A_i}$. To see this, note that

$$\left\| \int_S f_n \, d\rho - \int_S f_m \, d\rho \right\| = \left\| \int_S (f_n - f_m) \, d\rho \right\|$$

$$\leq \int_S \|f_n - f_m\| \, d\rho$$

$$\leq \int_S \|f_n - f\| \, d\rho + \int_S \|f_m - f\| \, d\rho \to 0$$

as $n, m \to \infty$. Hence, the sequence converges in X. We write

$$\int_S f \, d\rho := \lim_{n \to \infty} \int_S f_n \, d\rho$$

and call this the *integral of f over S with respect to the measure ρ*.

The following is a simple condition for a function to be (Bochner) integrable.

Proposition 1.3.1. *A measurable function $f: S \to X$ is (Bochner) integrable with integral $\int_S f \, d\rho$ if and only if $\int_S \|f\| \, d\rho < \infty$. In this case,*

$$\left\| \int_S f \, d\rho \right\| \leq \int_S \|f\| \, d\rho.$$

Proof. Assume that f is integrable. Since $\int_S \|f_n\| \, d\rho$ is a Cauchy sequence in $\mathbb{R}^1$,

$$\int_S \|f\| \, d\rho \leq \lim_{n \to \infty} \left(\int_S \|f_n\| \, d\rho + \int_S \|f - f_n\| \, d\rho \right) < \infty.$$

Now, assume $\int_S \|f\| \, d\rho < \infty$. Let f_n be simple functions such that $f_n \to f$ ρ-a.e. Define $g_n(x)$ to be $f_n(x)$ whenever $\|f_n(x)\| \leq \|f(x)\|(1 + 1/n)$, and to be 0 otherwise. Then $\|g_n(x)\| \leq \|f(x)\|(1 + 1/n)$ for all x and $\|g_n - f\| \to 0$ ρ-a.e. as $n \to \infty$. Also, $\|g_n - f\| \leq 2\|f\|(1 + 1/n)$, so by Fatou's lemma, $\int_S \|g_n - f\| \, d\rho \to 0$. Hence, f is (Bochner) integrable. Q.E.D.

Corollary 1.3.2

(a) *If f_1 and f_2 are integrable, then $af_1 + bf_2$ is integrable with $a, b \in \mathbb{R}^1$, and*

$$\int_S (af_1 + bf_2) \, d\rho = a \int_S f_1 \, d\rho + b \int_S f_2 \, d\rho.$$

(b) *If T is a bounded linear operator from X into a Banach space Y, and if $f \colon S \to X$ is integrable, then Tf is integrable and*

$$T\left(\int_S f \, d\rho \right) = \int_S Tf \, d\rho.$$

The proof is the standard argument. Show it first for simple functions and then take limits, noting that a linear combination and the image by a linear operator of simple functions is again a simple function.

Now, let $f \colon X \to Y$ with X and Y Banach spaces. We say that f is *differentiable at* $x \in X$ if there exists a bounded linear operator $Df(x)$ from X into Y, that is, $Df(x) \in L(X, Y)$, such that

$$\lim_{h \to 0} \frac{\|f(x + h) - f(x) - Df(x)h\|}{\|h\|} = 0.$$

The operator $Df(x)$ is unique whenever it exists. Furthermore, if $g \in L(X, Y)$, then $Dg(x) = g$ for all $x \in X$.

Now, say that $f \colon X \to Y$ is *continuously differentiable* if the mapping $x \to Df(x)$ is continuous. We illustrate these general notions in the particular case when $X = \mathbb{R}^n$ and $Y = \mathbb{R}^m$.

Proposition 1.3.3. *If $f = (f_1, \ldots, f_m): \mathbb{R}^n \to \mathbb{R}^m$ is differentiable at $x \in \mathbb{R}^n$, then the partial derivatives $\partial_j f_i(x)$, $1 \le i \le m$, $1 \le j \le n$, all exist and*

$$Df(x) = \left[\partial_j f_i(x)\right]_{1 \le i \le m, 1 \le j \le n}.$$

Conversely, if all partials $\partial_j f_i(x)$ exist in an open neighborhood of $a \in \mathbb{R}^n$, and if they are continuous at a, then $Df(a)$ exists.

For $n \ge 2$, we set $D^n f(x) := D(D^{n-1}f(x))(x)$ whenever they exist. If $g: \mathbb{R}^n \to \mathbb{R}^1$ is differentiable, then $Dg(x)$ is a vector in $\mathbb{R}^n$ with coordinates $\partial_j g(x)$, $1 \le j \le n$. Hence, if $D^2 g(x)$ exists, it is a matrix with entries $\partial^2_{ij} g(x)$, where ∂^2_{ij} denotes the second-order partial derivatives. We end this part on finite-dimensional spaces with an important fact.

Theorem 1.3.4 (Inverse Function Theorem). *Assume that $f: \mathbb{R}^m \to \mathbb{R}^m$ is continuously differentiable in an open neighborhood of $a \in \mathbb{R}^m$, and that $Df(a)$ is invertible, that is, $Df(a) \in \text{Aut}(\mathbb{R}^m)$. Then*

(a) *there exist open sets U and V in $\mathbb{R}^m$ such that $a \in U$, $f(a) \in V$, f is one-to-one on U, and $f(U) = V$;*

(b) *if g is the inverse of f defined on V, which exists by (a), then g is continuously differentiable on V and $Dg(y) = (Df(g(y)))^{-1}$.*

In many of our applications, we deal with functions f defined on subsets of $\mathbb{R}^1$ and with values in linear spaces. In this situation, we may define $\int_a^b f(t)\,dt$ as the limit of sums $\Sigma f(t_k)(t_k - t_{k-1})$, where $a = t_0 < t_1 < \cdots < t_n = b$ and $\max(t_k - t_{k-1}) \to 0$ as $n \to \infty$. For this Riemann-type integral, Corollary 1.3.2 holds. Furthermore, we may define the derivative of f at t to be the vector $f'(t)$ such that $\|(f(t+h) - f(t))/h - f'(t)\| \to 0$ as $h \to 0$. Properties of such differentiable and/or integrable functions are given in the following corollary.

Corollary 1.3.5

(a) *Let $T: X \to Y$ be a bounded linear operator, and let $f: [a,b] \to X$. If f is differentiable, then so is Tf, and $(Tf)'(t) = Tf'(t)$.*

(b) If f' is continuous on $[a, b]$, then

$$\int_a^b f'(t) \, dt = f(b) - f(a).$$

(c) If f is integrable on $[a, a + h]$, $h > 0$, and continuous from the right at a, then

$$\lim_{h \to 0+} \frac{1}{h} \int_a^{a+h} f(t) \, dt = f(a).$$

(d) If f is integrable on $[a, b]$, then $f(\cdot - h)$ is integrable on $[a + h, b + h]$ and

$$\int_{a+h}^{b+h} f(t - h) \, dt = \int_a^b f(t) \, dt.$$

Property (a) follows from the definition, and properties (b), (c), and (d) can be obtained by applying functionals $x^* \in X^*$ to both sides of the equalities and using the analogous facts for real-valued functions.

1.4 ONE-PARAMETER SEMIGROUPS

For each $t \geq 0$, let U_t be a bounded linear operator from a Banach space X into X, so $\mathbb{U} := \{U_t : t \geq 0\} \subset \text{End}(X)$. We say that $\mathbb{U}$ is a *one-parameter strongly continuous contraction semigroup* if

(i) for all $t, s \geq 0$, $U_t U_s = U_{t+s}$, with $U_0 = I$ (identity);
(ii) $\lim_{t \to 0+} U_t x = x$ for all $x \in X$;
(iii) $\|U_t\| \leq 1$ for all $t \geq 0$.

The *infinitesimal generator* Γ of the semigroup $\mathbb{U}$ is defined by

$$\Gamma x := \lim_{t \to 0+} \frac{U_t x - x}{t}. \tag{1.4.1}$$

Its domain $\mathscr{D}(\Gamma)$ consists of all $x \in X$ for which the limit in (1.4.1) exists. It is easy to see that $\mathscr{D}(\Gamma)$ is a linear subspace of X, that $U_t(\mathscr{D}(\Gamma)) \subset \mathscr{D}(\Gamma)$ for all $t \geq 0$, and that Γ is a linear operator.

Proposition 1.4.1. *For a one-parameter strongly continuous semigroup $\mathbb{U}$, its infinitesimal generator Γ has its domain $\mathscr{D}(\Gamma)$ dense on X, and*

(a) *for $x \in \mathscr{D}(\Gamma)$, $dU_t x/dt = \Gamma U_t x = U_t \Gamma x$;*
(b) *for $x \in \mathscr{D}(\Gamma)$, $U_t x - x = \int_0^t U_s \Gamma x\, ds$;*
(c) *Γ is a closed operator.*

Proof. Since $t \to U_t x$ is continuous from $[0, \infty)$ into X [cf. (i) and (ii)], it is integrable over every finite interval. Set $x(a) := \int_0^a U_t x\, dt$ for $a \geq 0$. From Corollaries 1.3.2(b) and 1.3.5(d), we obtain $U_h x(a) = \int_0^a U_{t+h} x\, dt = \int_h^{a+h} U_s x\, ds = x(a) + \int_a^{a+h} U_s x\, ds - x(h)$. Hence, by Corollary 1.3.5(c),

$$\lim_{h \to 0+} \frac{U_h x(a) - x(a)}{h} = \lim_{h \to 0+} \left[\frac{1}{h} \int_a^{a+h} U_s x\, ds - \frac{x(h)}{h} \right] = U_a x - x.$$

Consequently, $x(a) \in \mathscr{D}(\Gamma)$ for all $a > 0$. Since $x(a)/a \to x$ as $a \to 0+$, we see that $\mathscr{D}(\Gamma)$ is dense in X.

For (a), we use the equality

$$\frac{1}{h}(U_{t+h} x - U_t x) = U_t \left(\frac{U_h x - x}{h} \right) = \frac{1}{h}(U_h(U_t x) - U_t x)$$

for $x \in \mathscr{D}(\Gamma)$, to see that $d^+ U_t x/dt$ exists, the right derivative. Also, $U_t x \in \mathscr{D}(\Gamma)$ and

$$\frac{d^+ U_t x}{dt} = U_t \Gamma x = \Gamma U_t x.$$

For $0 < h < t$, we have

$$\left\| \frac{U_t x - U_{t-h} x}{h} - U_t \Gamma x \right\|$$

$$\leq \left\| U_{t-h} \left(\frac{U_h x - x}{h} - \Gamma x \right) \right\| + \| U_{t-h}(\Gamma x - U_h \Gamma x) \|$$

$$\leq \left\| \frac{U_h x - x}{h} - \Gamma x \right\| \leq \| \Gamma x - U_h \Gamma x \|.$$

Hence, as $h \to 0+$, we obtain $d^- U_t x/dt$ exists, the left derivative. Also,

$$\frac{d^- U_t x}{dt} = U_t \Gamma x = \frac{d^+ U_t x}{dt}.$$

For (b), we see that this easily follows from (a) and Corollary 1.3.5(b).

For (c), let $x_n \in \mathscr{D}(\Gamma)$ with $x_n \to x_0$ and $\Gamma x_n \to x_1$ for some $x_0, x_1 \in X$. From (b), we obtain

$$U_t x_n - x_n = \int_0^t U_s \Gamma x_n \, ds.$$

Letting $n \to \infty$, after dividing by t, we have

$$\frac{U_t x_0 - x_0}{t} = \frac{1}{t} \int_0^t U_s x_1 \, ds.$$

Now, let $t \to 0+$ to obtain $\Gamma x_0 = x_1$. Q.E.D.

Proposition 1.4.2. *Let $\mathbb{U}$ be a one-parameter strongly continuous contraction semigroup on X with infinitesimal generator Γ. Then for each $y \in X$ and for each $\lambda > 0$ there exists a unique $x \in \mathscr{D}(\Gamma)$ such that*

$$\lambda x - \Gamma x = y. \tag{1.4.2}$$

Moreover,

$$x = R_\lambda y := \int_0^\infty e^{-\lambda t} U_t y \, dt. \tag{1.4.3}$$

Proof. Since $t \to e^{-\lambda t} U_t y$ is a continuous function and $\|e^{-\lambda t} U_t y\| \le e^{-\lambda t} \|y\|$ which is integrable, the function R_λ in (1.4.3) is well defined, cf. Theorem 1.3.1. From Corollaries 1.3.2(b) and 1.3.5(d), we have, for x given by (1.4.3),

$$U_h x = \int_h^\infty e^{-\lambda(t-h)} U_t y \, dt = e^{\lambda h} \int_h^\infty e^{-\lambda t} U_t y \, dt$$

$$= e^{\lambda h} \left(x - \int_0^h e^{-\lambda t} U_t y \, dt \right).$$

Hence,

$$\frac{1}{h}(U_h x - x) = \frac{e^{\lambda h} - 1}{h} x - \frac{e^{\lambda h}}{h} \int_0^h e^{-\lambda t} U_t y\, dt$$

and as $h \to 0+$, the limit on the left exists by Corollary 1.3.5(c), we see that $\Gamma x = \lambda x - y$, that is, $x \in \mathscr{D}(\Gamma)$ and x given by (1.4.3) satisfies (1.4.2). It remains to show the uniqueness of x in (1.4.2). Suppose that for some $y \in X$, we have two different solutions of (1.4.2). Then there is a nonzero $x_0 \in \mathscr{D}(\Gamma)$ such that $\lambda x_0 - \Gamma x_0 = 0$. From Proposition 1.4.1(a), we obtain, for all $t > 0$,

$$\frac{dU_t x_0}{dt} = U_t \Gamma x_0 = \lambda U_t x_0.$$

Consequently, $d(e^{-\lambda t} U_t x_0)/dt = 0$, so from Corollary 1.3.5(b) we obtain $e^{-\lambda t} U_t x_0 - x_0 = 0$ for all $t > 0$. But, since $\mathbb{U}$ is a contraction semigroup, $\|x_0\| \le e^{-\lambda t}\|x_0\|$ for all $t > 0$. This implies that $x_0 = 0$, a contradiction. Q.E.D.

Corollary 1.4.3. *The infinitesimal generator Γ of a one-parameter strongly continuous contraction semigroup $\mathbb{U}$ uniquely determines $\mathbb{U}$.*

Proof. Suppose Γ with its domain $\mathscr{D}(\Gamma)$ is the infinitesimal generator of two such semigroups, $\mathbb{U} := \{U_t: t \ge 0\}$ and $\mathbb{U}' := \{U_t': t \ge 0\}$. Then from Proposition 1.4.2 we obtain, for all $y \in X$ and for all $\lambda > 0$,

$$\int_0^\infty e^{-\lambda t} U_t' y\, dt = \int_0^\infty e^{-\lambda t} U_t y\, dt.$$

Hence, for $x^* \in X^*$, by Corollary 1.3.2(b), we obtain, for all $\lambda > 0$,

$$\int_0^\infty e^{-\lambda t} \langle x^*, U_t' y \rangle\, dt = \int_0^\infty e^{-\lambda t} \langle x^*, U_t y \rangle\, dt. \tag{1.4.4}$$

In other words, (1.4.4) says that the two bounded continuous real-valued functions, $t \to \langle x^*, U_t' y \rangle$ and $t \to \langle x^*, U_t y \rangle$, on $[0, \infty)$ have the same Laplace transforms. Hence, $\langle x^*, U_t' y \rangle = \langle x^*, U_t y \rangle$ for all $x^* \in X^*$, which implies that $U_t' y = U_t y$ for all $y \in X$ and for all $t > 0$. Q.E.D.

The next lemma is helpful in the description of the infinitesimal generator Γ and its domain $\mathscr{D}(\Gamma)$. We need the following notion. For a closed linear operator A with domain $\mathscr{D}(A)$, a linear manifold $D \subset \mathscr{D}(A)$ is called a *core of A* provided A is the smallest closed extension of the restriction of A to D, that is, $(A|D)^- = A$ (cf. Theorem 1.2.1).

Lemma 1.4.4 (Watanabe Lemma). *Let $\mathbb{U} = \{U_t : t \geq 0\}$ be a one-parameter strongly continuous semigroup on a Banach space X. Assume that there exist $c > 0$ and $\beta > 0$ such that $\|U_t\| \leq c e^{\beta t}$ for all $t \geq 0$. Let Γ be the infinitesimal generator of $\mathbb{U}$ with domain $\mathscr{D}(\Gamma)$. If D is a linear manifold of X such that*

(i) $D \subset \mathscr{D}(\Gamma)$;
(ii) D is dense in X;
(iii) D is U_t-invariant for each $t > 0$, that is, $U_t D \subset D$, then D is a core for Γ.

Proof. First, we prove that for $\alpha > \beta$, $(\alpha I - \Gamma)(D)$ is dense in X. To this end, we show that for $x^* \in X^*$ the equality

$$\langle x^*, \alpha x - \Gamma x \rangle = 0 \quad \text{for all } x \in D \tag{1.4.5}$$

implies that $x^* = 0$. From Proposition 1.4.1(a), we have $\Gamma U_t x = dU_t x/dt$ for $x \in D$. Since D is U_t-invariant,

$$0 = \alpha \langle x^*, U_t x \rangle - \left\langle x^*, \frac{dU_t x}{dt} \right\rangle$$
$$= \alpha \langle x^*, U_t x \rangle - \frac{d}{dt} \langle x^*, U_t x \rangle$$

for all $x \in D$. Hence, $\langle x^*, U_t x \rangle = M e^{\alpha t}$ for some real M. On the other hand, for $t > 0$, $|\langle x^*, U_t x \rangle| \leq c \|x^*\| e^{\beta t}$ and $\beta < \alpha$. Thus, $M = 0$, and therefore, for all x in D, $\langle x^*, U_t x \rangle = 0$. Hence, $\langle x^*, x \rangle = 0$ for all $x \in D$, and (ii) implies $x^* = 0$. Thus, (1.4.5) implies $x^* = 0$. Therefore, for $\alpha > \beta$, $(\alpha I - \Gamma)(D)$ is a dense subset of X.

Second, note that for $\alpha > \beta$, $R_\alpha := \int_0^\infty e^{-\alpha t} U_t \, dt$ is in $\text{End}(X)$, that is, a bounded linear operator on X. Furthermore, from Propositions 1.4.2 and 1.4.1(a), we obtain

$$R_\alpha (\alpha I - \Gamma) y = y \quad \text{for all } y \in D. \tag{1.4.6}$$

Since $(\alpha I - \Gamma)(D)$ is dense in X, for each $u \in \mathscr{D}(\Gamma)$ there exists $x_n := (\alpha I - \Gamma)y_n$, with $y_n \in D$, such that $x_n \to (\alpha I - \Gamma)u$. By (1.4.6), we have $y_n = R_\alpha x_n \to R_\alpha(\alpha I - \Gamma)u = u$, and also $\Gamma y_n = \alpha y_n - (\alpha y_n - \Gamma y_n) = \alpha y_n - x_n \to \alpha u - (\alpha I - \Gamma)u = \Gamma u$, which shows that $(\Gamma|D)^- = \Gamma$. Q.E.D.

1.5 BASIC FACTS FROM LIE THEORY

This section contains some basic facts from Lie theory needed later. Mainly, we deal with subgroups of $\mathbf{Aut}(\mathbb{R}^d)$ and their Lie algebras. However, we begin with a semigroup of linear operators in a normed linear space, X.

Let $\mathbb{H} \subset \mathbf{End}(X)$. The *tangent space* $\mathscr{T}(\mathbb{H})$ at the identity consists of those $A \in \mathbf{End}(X)$ such that

$$A = \lim_{n \to \infty} \frac{H_n - I}{d_n} \tag{1.5.1}$$

for some sequence $\{H_n\}$ in $\mathbb{H}$ and for some $0 < d_n \to 0$, where the convergence is in the operator norm of $\mathbf{End}(X)$.

Lemma 1.5.1. *Let $\mathbb{H}$ be a subsemigroup of $\mathbf{End}(X)$. Then the exponential mapping takes $\mathscr{T}(\mathbb{H})$ into $\mathbb{H}^-$, the closure of $\mathbb{H}$ in $\mathbf{End}(X)$, where*

$$\exp(A) := \lim_{n \to \infty} \left(I + \frac{A}{n}\right)^n.$$

Proof. The usual proof that $(1 + x/n)^n$ converges may be used to see that $(I + A/n)^n$ converges. Let $A \in \mathscr{T}(\mathbb{H})$. Then $A = \lim_{n \to \infty}(H_n - I)/d_n$ for some sequence $\{H_n\}$ in $\mathbb{H}$ and $0 < d_n \to 0$. Set $n_k := [1/d_k]$, the integer part of $1/d_k$, and set $G_k := n_k(H_k - I) - A$. Then $n_k d_k \to 1$, $G_k \to 0$ as $k \to \infty$, and $n_k^{-1}(G_k + A) + I = H_k \in \mathbb{H}$. Hence, $H_k^{n_k} \in \mathbb{H}$ and $H_k^{n_k} \to \exp(A)$. Therefore, $\exp(A) \in \mathbb{H}^-$. Q.E.D.

Lemma 1.5.2

(a) *For $A \in \text{End}(X)$, we have*

$$\exp(A) = \sum_{n=0}^{\infty} (n!)^{-1} A^n,$$

$$\left\| \exp(A) - \sum_{n=0}^{k} (n!)^{-1} A^n \right\| \leq o(\|A\|^{k+1}).$$

(b) *If $f: [0, \infty) \to \mathbb{H}$ is right differentiable at zero with $f(0) = I$, then $f'(0) \in \mathcal{T}(\mathbb{H})$ for a semigroup $\mathbb{H} \subset \text{End}(X)$.*

(c) *If, additionally, $\mathbb{H}$ is closed in $\text{End}(X)$, then*
   (i) *for all $t \geq 0$ and for all $A \in \mathcal{T}(\mathbb{H})$, $\exp(tA) \in \mathbb{H}$;*
   (ii) *for all $A, B \in \mathcal{T}(\mathbb{H})$, $A + B \in \mathcal{T}(\mathbb{H})$.*

Proof

(a) Since $1 - \prod_{j=1}^{m} \alpha_j \leq \sum_{j=1}^{m} (1 - \alpha_j)$ for $0 < \alpha_j < 1$, we have

$$\left\| \sum_{k=0}^{n} (k!)^{-1} A^k - \left(I + \frac{A}{n} \right)^n \right\|$$

$$= \left\| \sum_{k=2}^{n} (k!)^{-1} A^k \left(1 - \prod_{j=1}^{k-1} \left(1 - \frac{j}{n} \right) \right) \right\|$$

$$\leq \sum_{k=2}^{n} (k!)^{-1} \|A\|^k k \frac{k-1}{n}$$

$$= \frac{\|A\|^2}{n} \sum_{k=0}^{n-2} (k!)^{-1} \|A\|^k \to 0 \quad \text{as } n \to \infty.$$

(b) This follows from the definitions of derivative and tangent space.

(c)(i) For $t > 0$, $tA \in \mathcal{T}(\mathbb{H})$ whenever $A \in \mathcal{T}(\mathbb{H})$ by (1.5.1). Since $\mathbb{H}$ is closed, Lemma 1.5.1 gives $\exp(tA) \in \mathbb{H}$.

(c)(ii) Let $A, B \in \mathcal{T}(\mathbb{H})$. Define $f: [0, \infty) \to \mathbb{H}$ by $f(t) := \exp(tA)\exp(tB)$. From (c)(i), $f(t) \in \mathbb{H}$. By (b), $f'(0) = A + B \in \mathcal{T}(\mathbb{H})$. Q.E.D.

BASIC FACTS FROM LIE THEORY 17

Note that if A and B commute, then $\exp(A)\exp(B) = \exp(A + B)$.

Lemma 1.5.3. *If $\mathbb{G}$ is a closed subgroup of* $\mathrm{Aut}(X)$, *then its tangent space is a Lie algebra in* $\mathrm{End}(X)$, *that is,* $\mathcal{T}(\mathbb{G})$ *is a linear space closed under the Lie bracket,* $[A, B] := AB - BA$.

Proof. For $t < 0$ and $A \in \mathcal{T}(\mathbb{G})$, by Lemma 1.5.2(c)(i), $(\exp(tA))^{-1} = \exp(-tA) \in \mathbb{G}$. Hence, $\exp(tA) \in \mathbb{G}$, so $tA \in \mathcal{T}(\mathbb{G})$. Therefore, $\mathcal{T}(\mathbb{G})$ is a linear space. Note that, for $A, B \in \mathcal{T}(\mathbb{G})$ and for $t > 0$,

$$f(t) := \exp(\sqrt{t}\,A)\exp(\sqrt{t}\,B)\exp(-\sqrt{t}\,A)\exp(-\sqrt{t}\,B) \in \mathbb{G},$$

and $f(t) = I + t(AB - BA) + o(t^{3/2})$. Since $f'(0) = AB - BA$, by Lemma 1.5.2(b), $[A, B] \in \mathcal{T}(\mathbb{G})$. Q.E.D.

Lemma 1.5.4. *Let $\mathbb{G}$ be a closed subgroup of* $\mathrm{Aut}(\mathbb{R}^d)$. *Then there are open sets U_1 and U_2 in* $\mathrm{End}(\mathbb{R}^d)$, *both containing zero, with* $U_2 \subset U_1$, *and*

(a) $\exp: U_1 \to \mathbb{G}$ *is invertible on* $\exp(U_1)$ *with a continuously differentiable inverse;*
(b) *for all $A, B \in U_2$,* $\exp(A)\exp(B) \in \exp(U_1)$.

Proof

(a) Since the derivative of the exponential mapping at zero is the identity, by the inverse function theorem 1.3.4, the required U_1 exists. Note that $I \in \exp(U_1)$.
(b) Since multiplication in $\mathrm{End}(X)$ is continuous, there is $U_2 \subset U_1$, containing zero, such that $U_2 U_2 \subset U_1$. Hence, from $\exp(U_2 U_2) \subset \exp(U_1)$ and continuity of exponentials, we obtain (b). Q.E.D.

The function $\log: \exp(U_1) \to \mathrm{End}(\mathbb{R}^d)$ is the continuously differentiable inverse of exp restricted to U_1 given by this lemma.

Lemma 1.5.5. *Let $\mathbb{G}$ be a closed subgroup of* $\mathrm{Aut}(\mathbb{R}^d)$. *For given $A, B \in \mathcal{T}(\mathbb{G})$ and for all t, s sufficiently near zero, the function*

$$f(t, s) := \log(\exp(tA)\exp(sB))$$

is well defined, analytic at $(0,0)$, and $\partial_{t,s} f(0,0) = [A, B]$, where $\partial_{t,s}$ denotes the second mixed partial derivative.

This lemma is a consequence of the previous material.

Theorem 1.5.6. *The identity I in the closed subgroup $\mathbb{G}$ of $\mathrm{Aut}(\mathbb{R}^d)$ belongs to the interior of $\exp(\mathscr{T}(\mathbb{G}))$.*

Proof. Suppose to the contrary that there exists a sequence $\{D_n\}$ such that $D_n \in \mathbb{G} \cap \exp(U_1)$, $D_n \notin \exp(\mathscr{T}(\mathbb{G}))$, for all $n \geq 1$, and $D_n \to I$, where U_1 is the open set in Lemma 1.5.4. For each $n \geq 1$, let $A_n \in \mathbb{G} \cap \exp(U_1)$ be such that $\exp(A_n) = D_n$, and let J be the orthogonal projection of $\mathrm{End}(\mathbb{R}^d)$ onto the linear space $\mathscr{T}(\mathbb{G})$. Since $A_n \notin \mathscr{T}(\mathbb{G})$, $A_n - JA_n \neq 0$ for all $n \geq 1$. However, $A_n = \log D_n \to 0$, so $A_n - JA_n \to 0$ as $n \to \infty$. Since the norms of the sequence $\{(\exp A_n - \exp JA_n)/\|A_n - JA_n\|\}$ converge to one, we may assume there is $D \in \mathrm{End}(\mathbb{R}^d)$ with $\|D\| = 1$ such that the sequence converges to D. From $(\exp)'(0) = I$, we obtain

$$\frac{\|\exp(A_n) - \exp(JA_n) - (A_n - JA_n)\|}{\|A_n - JA_n\|} \to 0 \quad \text{as } n \to \infty.$$

Thus, $(A_n - JA_n)/\|A_n - JA_n\| \to D$. Since $\exp(-JA_n) \to I$, $(\exp(A_n)\exp(-JA_n) - I)/\|A_n - JA_n\| \to D$ as $n \to \infty$. But, $\exp(A_n)\exp(-JA_n) \in \mathbb{G}$, so $D \in \mathscr{T}(\mathbb{G})$. Therefore,

$$D = JD = J\left(\lim_{n \to \infty} \frac{A_n - JA_n}{\|A_n - JA_n\|}\right)$$
$$= \lim_{n \to \infty} \frac{JA_n - JA_n}{\|A_n - JA_n\|} = 0,$$

which contradicts $\|D\| = 1$. Q.E.D.

Example 1.5.7. For an example of a tangent space, let $\mathscr{O}$ and $\mathscr{D}$ denote the groups of orthogonal and skew-symmetric linear transformations on $\mathbb{R}^d$, respectively. Then

$$\mathscr{T}(\mathscr{O}) = \mathscr{D}.$$

To see this, let $A \in \mathscr{T}(\mathscr{O})$. Then $\exp(tA) \in \mathscr{O}$ for all $t \in \mathbb{R}^1$. Therefore, $\exp(tA)\exp(tA^*) = I$ for all t. By differentiating this relation at $t = 0$, we obtain $A + A^* = 0$, that is, $A \in \mathscr{D}$. Conversely, let $A \in \mathscr{D}$. Then A and A^* commute, so for all t, $\exp(tA)\exp(tA^*) = \exp(t(A + A^*)) = I$. Thus, $\exp(tA) \in \mathscr{O}$ for all t. Since $A = \lim_{t \to 0+}(\exp(tA) - I)/t$, we have $A \in \mathscr{T}(\mathscr{O})$.

Example 1.5.8. Let $X = \mathbb{R}^3$, and let $\mathscr{O}$ be the group of orthogonal transformations on $\mathbb{R}^3$. Let $\mathscr{O}_0$ be any closed subgroup of $\mathscr{O}$. Then the dimension of $\mathscr{T}(\mathscr{O}_0)$ is either 0, 1, or 3. When $\dim \mathscr{T}(\mathscr{O}_0) = 0$, $\mathscr{O}_0$ is a discrete group; when $\dim \mathscr{T}(\mathscr{O}_0) = 3$, $\mathscr{O}_0 = \mathscr{O}$; and when $\dim \mathscr{T}(\mathscr{O}_0) = 1$, its tangent space $\mathscr{T}(\mathscr{O}_0)$ has the form

$$\left\{ \begin{bmatrix} 0 & -c & 0 \\ c & 0 & 0 \\ 0 & 0 & 0 \end{bmatrix} : c \in \mathbb{R}^1 \right\},$$

with respect to some orthonormal basis for $\mathbb{R}^3$.

Proof. By Example 1.5.7, $\mathscr{T}(\mathscr{O}) = \mathscr{D}$. Since $\mathscr{O}_0$ is a closed subgroup of $\mathscr{O}$, $\mathscr{T}(\mathscr{O}_0) \subset \mathscr{D}$. Also, $\mathscr{T}(\mathscr{O}_0)$ is closed under the Lie bracket, $[A, B] := AB - BA$ (cf. Lemma 1.5.3). Define $f: \mathbb{R}^3 \to \mathscr{D}$ by

$$f(a, b, c) := \begin{bmatrix} 0 & -a & -b \\ a & 0 & -c \\ b & c & 0 \end{bmatrix}.$$

Then f is linear and onto. Note that for $x, y \in \mathbb{R}^3$, $[f(x), f(y)] = f(x \times y)$, where $\times$ is the cross product on $\mathbb{R}^3$. Hence, $(\mathbb{R}^3, \times)$ is isomorphic to $\mathscr{D}$ and, therefore, the dimension of any subspace of $\mathscr{D}$ is either 0, 1, or 3.

Consider the case when $\dim \mathscr{T}(\mathscr{O}_0) = 1$. Let $Q \in \mathscr{D}$ form a basis for $\mathscr{T}(\mathscr{O}_0)$. Since $\det(Q) = \det(-Q^*) = (-1)^3 \det(Q)$, we have $\det(Q) = 0$, so Q is singular. Select $u \in \mathbb{R}^3$ with $\|u\| = 1$ and $Qu = 0$. Let this vector u be the third member of an orthonormal basis for $\mathbb{R}^3$. This is the basis needed in the lemma. By skew-symmetry, the third row and the third column are all zeros, and the diagonal is all zeros. Also, the (1, 2)-element is the negative of the (2, 1)-element. Since $\{Q\}$ is a basis for $\mathscr{T}(\mathscr{O}_0)$, we have the derived form. Q.E.D.

Remark. If $Q \in \mathscr{D}$, then $\exp(tQ)$ is a rotation for all t. This follows from: $\mathscr{D} = \mathscr{T}(\mathscr{O})$, so $\exp(\mathscr{D})$ is the connected component of $\mathscr{O}$ which contains the identity; hence, $\exp(tQ) \in \exp(\mathscr{D})$ cannot be a reflection.

1.6 PROBABILITY MEASURES ON METRIC SPACES

Let S be a metric space with the Borel σ-field $\mathscr{B}(S)$ generated by its open sets. All measures are defined on $\mathscr{B}(S)$ and are nonnegative and countably additive. A probability is a measure which gives S the measure of one. By $\mathscr{P}(S)$, we denote the class of all probabilities on S.

Theorem 1.6.1

(a) *Every $\mu \in \mathscr{P}(S)$ is regular, that is, for each $A \in \mathscr{B}(S)$ and for each $\varepsilon > 0$, there are a closed set F and an open set G such that $F \subset A \subset G$ and $\mu(G \setminus F) < \varepsilon$.*
(b) *If S is separable and complete, then each $\mu \in \mathscr{P}(S)$ is tight, that is, for each $\varepsilon > 0$, there is a compact set K such that $\mu(K) > 1 - \varepsilon$.*

Let $C_b(S)$ denote the set of all real-valued bounded continuous functions defined on S. For $\mu_n, \mu \in \mathscr{P}(S)$, we say that μ_n *converges weakly to* μ, $\mu_n \Rightarrow \mu$, if, for every $f \in C_b(S)$,

$$\lim_{n \to \infty} \int f(x) \mu_n(dx) = \int f(x) \mu(dx). \qquad (1.6.1)$$

This weak convergence in (1.6.1) is also used for finite measures, not necessarily probabilities, with the additional condition that $\mu_n(S) \to \mu(S)$. The following theorem gives some useful conditions which are equivalent to weak convergence. For $A \subset S$, ∂A denotes the boundary of A, that is, $\partial A := A^- \cap (S \setminus A)^-$.

Theorem 1.6.2 (Portmanteau Theorem). *Let $\mu_n, \mu \in \mathscr{P}(S)$. The following conditions are equivalent.*

(a) $\mu_n \Rightarrow \mu$.
(b) *For all real-valued bounded, uniformly continuous f on S,*

$$\lim_{n \to \infty} \int f(x) \mu_n(dx) = \int f(x) \mu(dx).$$

(c) *For all closed sets F in S,*

$$\limsup_{n \to \infty} \mu_n(F) \leq \mu(F).$$

(d) *For all open sets G in S,*

$$\liminf_{n \to \infty} \mu_n(G) \geq \mu(G).$$

(e) *For all μ-continuity sets A, that is, $\mu(\partial A) = 0$,*

$$\lim_{n \to \infty} \mu_n(A) = \mu(A).$$

The following simple fact is often useful.

Corollary 1.6.3. *For $\mu_n, \mu \in \mathscr{P}(S)$, $\mu_n \Rightarrow \mu$ if and only if each subsequence $\{\mu_{n'}\}$ of $\{\mu_n\}$ contains a further subsequence $\{\mu_{n''}\}$ such that $\mu_{n''} \Rightarrow \mu$ as $n'' \to \infty$.*

Let S' be a metric space with its Borel σ-field $\mathscr{B}(S')$, and let h_n, h be measurable mappings from S to S'. For $\mu \in \mathscr{P}(S)$, let $h\mu$ denote the probability in $\mathscr{P}(S')$ given by $(h\mu)(B') := \mu(h^{-1}B')$ for $B' \in \mathscr{B}(S')$. A relationship between the weak convergence of $\mu_n \in \mathscr{P}(S)$ and of $h_n\mu_n \in \mathscr{P}(S')$ is given in the following theorem.

Theorem 1.6.4 (Weak Convergence Mapping Theorem). *Let*

$$E := \{x \in S : \text{there exists } \{x_n\} \text{ in } S \text{ such that}$$
$$x_n \to x \text{ and } h_n(x_n) \nrightarrow h(x) \text{ in } S'\}.$$

Assume $E \in \mathscr{B}(S)$, which holds if S' is separable. If $\mu_n \Rightarrow \mu$ in $\mathscr{P}(S)$ and if $\mu(E) = 0$, then $h_n\mu_n \Rightarrow h\mu$ in $\mathscr{P}(S')$.

In case $h_n = h$ for all $n \geq 1$, $E = D_h$ = the set of all discontinuity points of h, and we obtain the following corollary.

Corollary 1.6.5. *For a real-valued bounded measurable function h on S, we have that $\mu_n \Rightarrow \mu$ in $\mathscr{P}(S)$ and $\mu(D_h) = 0$ imply $h\mu_n \Rightarrow h\mu$ in $\mathscr{P}(\mathbb{R}^1)$.*

The set $\mathcal{P}(S)$ with weak convergence is a topological space. There is a metric on $\mathcal{P}(S)$ which gives this topology. It is the Prohorov metric defined as follows. For $\mu, \nu \in \mathcal{P}(S)$,

$$\|\mu - \nu\|_p := \inf\{\varepsilon \geq 0 : \mu(F) \leq \nu(F_\varepsilon) + \varepsilon \text{ for all } F \text{ closed}\},$$

where $F_\varepsilon := \{x \in S : \text{distances from } x \text{ to } F \text{ is } < \varepsilon\}$.

Our next aim is to describe the conditionally compact sets in $\mathcal{P}(S)$. A subset Γ of $\mathcal{P}(S)$ is called *conditionally compact*, or *sequentially conditionally compact*, if every sequence $\{\mu_n\}$ in Γ contains a subsequence which is weakly convergent in $\mathcal{P}(S)$; the limit probability need not be in Γ. Also, a subset Γ of $\mathcal{P}(S)$ is called *tight* if for every $\varepsilon > 0$, there is a compact set K such that $\mu(K) > 1 - \varepsilon$ for all $\mu \in \Gamma$. When Γ consists of exactly one probability, this definition is the notion of a tight measure [cf. Theorem 1.6.1(b)]. The following fundamental theorem gives the relationship between conditional compactness and tightness.

Theorem 1.6.6 (Prohorov Theorem). *For a metric space S, every tight set Γ in $\mathcal{P}(S)$ is conditionally compact. When S is separable and complete, Γ being conditionally compact implies that Γ is tight.*

From Theorem 1.6.6 and Corollary 1.6.3, we have a method to establish weak convergence of a sequence in $\mathcal{P}(S)$. Show that the sequence is a tight family in $\mathcal{P}(S)$, and show that all the limits are the same.

The concept of weak convergence extends to S-valued random variables. By an S-valued random variable ξ, we mean a measurable function $\xi : \Omega \to S$, where $(\Omega, \mathcal{M}, P)$ is a probability space. The *distribution* of ξ, denoted by $\mathcal{L}(\xi)$, is the probability in $\mathcal{P}(S)$ given by

$$\mathcal{L}(\xi)(B) := (\xi P)(B) = P(\xi^{-1}B)$$

for $B \in \mathcal{B}(S)$. Then S-valued random variables ξ_n *converge in distribution* to ξ if $\mathcal{L}(\xi_n) \Rightarrow \mathcal{L}(\xi)$, that is, $\lim_{n \to \infty} \mathbf{E}(f(\xi_n)) = \mathbf{E}(f(\xi))$ for all $f \in C_b(S)$, where $\mathbf{E}(\cdot)$ denotes the expectation operator.

To have a simple example of a weakly convergent sequence in $\mathcal{P}(S)$, let $\delta(a)$ denote the probability with all its mass at $a \in S$. Then $\int f(x) \delta(a)(dx) = f(a)$.

Remark 1.6.7. For $a_n, a \in S$, $\delta(a_n) \Rightarrow \delta(a)$ if and only if $a_n \to a$.

The final result of this section gives a separable subset of $\mathcal{P}(S)$ whenever S is separable.

Proposition 1.6.8. *Let S be a separable metric space with a countable dense subset S_0. Then the set of all probabilities of the form of a finite sum $\Sigma c_i \delta(x_i)$, with $c_i > 0$, $\Sigma c_i = 1$, $x_i \in S_0$, is dense in $\mathcal{P}(S)$.*

This result will be useful in finding generators of some classes of distributions which are homeomorphic to $\mathcal{P}(S)$, or certain of its subsets (cf. Proposition 1.8.10 and Section 3.5).

1.7 PROBABILITIES ON BANACH SPACES

Let X be a real separable Banach space. The linear structure of X allows considering the sum of X-valued random variables. Then $\mathcal{L}(\xi_1 + \xi_2)$ gives the *convolution*, $\mu * \nu$, whenever ξ_1 and ξ_2 are independent with $\mathcal{L}(\xi_1) = \mu$ and $\mathcal{L}(\xi_2) = \nu$, that is, $\mu * \nu \in \mathcal{P}(X)$ is given by

$$(\mu * \nu)(E) := \int_X \mu(E - x)\nu(dx) = \int_X \nu(E - x)\mu(dx)$$

for $E \in \mathcal{B}(X)$. The set $\mathcal{P}(X)$ with convolution $*$ and with weak convergence is a topological semigroup with unit $\delta(0)$, where $\delta(x)$ denotes the point-mass probability at $x \in X$. Note that, for $A, B \in \text{End}(X)$ and for $\mu, \nu \in \mathcal{P}(X)$,

$$A(\mu * \nu) = A\mu * A\nu \quad \text{and} \quad (AB)\mu = A(B\mu).$$

To see this, let $\mu = \mathcal{L}(\xi_1)$, $\nu = \mathcal{L}(\xi_2)$ with ξ_1 and ξ_2 independent. Then $A(\mu * \nu) = \mathcal{L}(A(\xi_1 + \xi_2)) = \mathcal{L}(A\xi_1 + A\xi_2) = A\mu * A\nu$.

A set Γ in $\mathcal{P}(X)$ is said to be *shift conditionally compact* if for each sequence $\{\mu_n\}$ in Γ, there exists a sequence $\{x_n\}$ in X such that the sequence $\{\mu_n * \delta(x_n)\}$ contains a convergent subsequence. The next theorem is a consequence of the Prohorov theorem 1.6.6, and it gives an important property of the topological semigroup $\mathcal{P}(X)$.

Theorem 1.7.1. *Let $\{\lambda_n\}$, $\{\mu_n\}$, and $\{\nu_n\}$ be sequences in $\mathcal{P}(X)$ such that $\lambda_n = \mu_n * \nu_n$ for all $n \geq 1$.*

(a) *If the sequences $\{\lambda_n\}$ and $\{\mu_n\}$ are conditionally compact, then so is the sequence $\{\nu_n\}$.*

(b) *If the sequence $\{\lambda_n\}$ is conditionally compact, then the sequences $\{\mu_n\}$ and $\{\nu_n\}$ are shift conditionally compact.*

We often need the following proposition which is a consequence of the weak convergence mapping theorem 1.6.4.

Proposition 1.7.2. *Let A_n, A be bounded linear operators from a Banach space X into a Banach space Y. Assume $A_n x \to Ax$ for all $x \in X$, that is, $A_n \to A$ in the strong operator topology. Then $\mu_n \Rightarrow \mu$ in $\mathscr{P}(X)$ implies that $A_n \mu_n \Rightarrow A\mu$ in $\mathscr{P}(Y)$.*

Proof. By the Banach–Steinhaus theorem, we have that $\sup \|A_n\| < \infty$, [cf. Hille and Phillips (1957)]. Let $x_n \to x$ in X. Then $\|A_n x_n - Ax\| \le \|A_n\| \|x_n - x\| + \|A_n x - Ax\|$ implies that $A_n x_n \to Ax$, that is, the "exceptional" set E in Theorem 1.6.4 is empty. Q.E.D.

Corollary 1.7.3. *Let $x_n^*, x^* \in X^*$ be functionals such that $\langle x_n^*, x \rangle \to \langle x^*, x \rangle$ for all $x \in X$. Then $\mu_n \Rightarrow \mu$ in $\mathscr{P}(X)$ implies that $\Pi_{x^*}\mu_n \Rightarrow \Pi_{x^*}\mu$ in $\mathscr{P}(\mathbb{R}^1)$, where $\Pi_{x^*}: X \to \mathbb{R}^1$ is given by $\Pi_{x^*} y := \langle x^*, y \rangle$ for all $y \in X$.*

For the Banach space X, let $\mathscr{A}(X)$ denote the set of all *affine transformations* on X, that is, each $\alpha \in \mathscr{A}(X)$ is given by an operator $A \in \text{End}(X)$ and a vector $a \in X$, $\alpha := \langle A; a \rangle$, in the following way: $\alpha x := Ax + a$. By defining, for $\alpha = \langle A; a \rangle$ and $\beta = \langle B; b \rangle$, $\alpha + \beta := \langle A + B; a + b \rangle$, and for $r \in \mathbb{R}^1$, $r\alpha := \langle rA, ra \rangle$, the set $\mathscr{A}(X)$ becomes a Banach space with a norm

$$\|\alpha\| := \max\{\|A\|, \|a\|\}.$$

Since $\alpha_n x = A_n x + a_n \to 0$ for all $x \in X$ if and only if $A_n x \to 0$ and $a_n \to 0$ for all $x \in X$, and since convolution is a continuous operation (cf. Proposition 1.7.2 and Remark 1.6.7), we obtain the following corollary.

Corollary 1.7.4. *Let $\alpha_n, \alpha \in \mathscr{A}(X)$, and assume $\alpha_n x \to \alpha x$ for all $x \in X$. Then $\mu_n \Rightarrow \mu$ in $\mathscr{P}(X)$ implies that $\alpha_n \mu_n \Rightarrow \alpha\mu$.*

The study of the converse implication of this corollary is given in Section 2.2.

For $\mu \in \mathcal{P}(X)$, its *characteristic* functional, $\hat{\mu}$, the Fourier transform, is given by

$$\hat{\mu}(x^*) := \int_X \exp i\langle x^*, x\rangle \mu(dx)$$

for all $x^* \in X^*$, where $\langle \cdot, \cdot \rangle$ is the bilinear from between X^* and X (cf. Section 1.2). Basic properties of $\hat{\mu}$ are given in the following proposition.

Proposition 1.7.5

(a) $\hat{\mu}$ *uniquely determines* μ.
(b) $|\hat{\mu}(x^*)| \leq 1$ *for all* $x^* \in X^*$.
(c) $\hat{\mu}$ *is uniformly continuous in the norm topology of* X^*; *in fact, we have, for all* $x_1^*, x_2^* \in X^*$,

$$|\hat{\mu}(x_1^* + x_2^*) - \hat{\mu}(x_1^*)|^2 \leq 2(1 - \operatorname{Re} \hat{\mu}(x_2^*))$$

and

$$1 - \operatorname{Re} \hat{\mu}(x_1^* + x_2^*) \leq 2[1 - \operatorname{Re} \hat{\mu}(x_1^*) + 1 - \operatorname{Re} \hat{\mu}(x_2^*)].$$

(d) *For* $x^* \in X^*$ *and* $\mu, \nu \in \mathcal{P}(X)$, $(\mu * \nu)\hat{\,}(x^*) = \hat{\mu}(x^*)\hat{\nu}(x^*)$.
(e) *For* $A \in \operatorname{End}(X)$ *and* $x^* \in X^*$, *with* A^* *denoting the adjoint operator of* A, $(A\mu)\hat{\,}(x^*) = \hat{\mu}(A^*x^*)$.

We will use the characteristic functionals mainly to describe some measures [cf. property (a) of the above proposition]. Unfortunately, they are not quite the appropriate tool for the study of weak convergence in infinite-dimensional spaces. We have only the following proposition.

Proposition 1.7.6. *For a Banach space* X, *and* μ_n, μ *from* $\mathcal{P}(X)$, *we have*

(a) *if* $\mu_n \Rightarrow \mu$ *in* $\mathcal{P}(X)$, *then* $\hat{\mu}_n \to \hat{\mu}$ *uniformly over bounded balls in* X^*;
(b) *if* $\{\mu_n\}$ *is conditionally compact in* $\mathcal{P}(X)$ *and* $\hat{\mu}_n(x^*) \to \hat{\mu}(x^*)$ *for all* $x^* \in X^*$, *then* $\mu_n \Rightarrow \mu$;
(c) *if* $\{\mu_n\}$ *is shift conditionally compact in* $\mathcal{P}(X)$ *and* $\hat{\mu}_n \to \hat{\mu}$ *uniformly over bounded balls in* X^*, *then* $\mu_n \Rightarrow \mu$.

Example 1.7.7. Let $\mu_n := \delta(e_n)$ and $\mu := \delta(0)$, where $\{e_n\}$ is a complete orthonormal basis for an infinite-dimensional real separable Hilbert space H. Then $\hat{\mu}_n(y) \to \hat{\mu}(y)$ as $n \to \infty$ for every $y \in H$. But, $\|e_n - e_m\| = \sqrt{2}$ for $n \neq m$, so $\delta(e_n)$ cannot converge weakly to $\delta(0)$. Therefore, the continuity theorem 1.7.8 below does not hold.

In contrast to the infinite-dimensional case, we have the following extremely useful results in Euclidean spaces.

Theorem 1.7.8 (Cramér–Lévy Continuity Theorem)

(a) *If $\mu_n \Rightarrow \mu$ in $\mathscr{P}(\mathbb{R}^d)$, then $\hat{\mu}_n \to \hat{\mu}$ uniformly on compact sets in $\mathbb{R}^d$.*

(b) *Let $\mu_n \in \mathscr{P}(\mathbb{R}^d)$. If $\hat{\mu}_n \to \phi$, where ϕ is a complex-valued function on $\mathbb{R}^d$ and ϕ is continuous at zero, then there is $\mu \in \mathscr{P}(\mathbb{R}^d)$ such that $\phi = \hat{\mu}$ and $\mu_n \Rightarrow \mu$.*

(c) *Let $\mu_n \in \mathscr{P}(\mathbb{R}^d)$, and let ϕ be a complex-valued function on $\mathbb{R}^d$ which is continuous at zero. Assume there is a sequence $\{k_n\}$ of positive integers such that $k_n \to \infty$ and $\hat{\mu}_n^{k_n}(y) \to \phi(y)$ for all y in some open neighborhood of zero. Then $\hat{\mu}_n(y) \to 1$ for all $y \in \mathbb{R}^d$, that is, $\mu_n \Rightarrow \delta(0)$.*

Theorem 1.7.9 (Cramér–Wold Device). *Let ξ_n, ξ be $\mathbb{R}^d$-valued random variables. Then ξ_n converges in distribution to ξ in $\mathbb{R}^d$ if and only if for every $a \in \mathbb{R}^d$, $\langle a, \xi_n \rangle$ converges in distribution to $\langle a, \xi \rangle$ in $\mathbb{R}^1$.*

For $\mu \in \mathscr{P}(X)$, we define $\mu^- \in \mathscr{P}(X)$ by $\mu^-(E) := \mu(-E)$ for all $E \in \mathscr{F}$, where $-E := \{x: -x \in E\}$. Also, $\mu^0 \in \mathscr{P}(X)$ is defined by $\mu^0 := \mu * \mu^-$, and μ^0 is called the *symmetrization* of μ. When $\mu = \mu^-$, we say that μ is *symmetric*; in terms of characteristic functions, this is equivalent to $\hat{\mu}$ being real-valued. Note that $(\mu^0)\hat{\,}(x^*) = |\hat{\mu}(x^*)|^2$ and μ^0 is symmetric. For X-valued random variables, if $\mathscr{L}(\xi) = \mu$, then $\mathscr{L}(-\xi) = \mu^-$ and $\mu^0 = \mathscr{L}(\xi_1 - \xi_2)$, where ξ_1 and ξ_2 are independent copies of ξ.

By the *support* of a measure μ on X, we mean the smallest closed set whose complement has μ-measure zero; denote the support by $\operatorname{supp} \mu$. In the case of a separable metric space, $\operatorname{supp} \mu$ always exists, and $\operatorname{supp} \mu = \{x \in X: \text{for every open } G \text{ containing } x, \mu(G) \neq 0\}$.

Proposition 1.7.10. *Let $\mu, \nu \in \mathcal{P}(X)$. Then*

$$\mathrm{supp}(\mu * \nu) = (\mathrm{supp}\, \mu + \mathrm{supp}\, \nu)^-.$$

Proof. Let $z \in \mathrm{supp}\, \mu + \mathrm{supp}\, \nu$, so $z = x + y$ for some $x \in \mathrm{supp}\, \mu$ and some $y \in \mathrm{supp}\, \nu$. Then for each open neighborhood U_z of z, there are open neighborhoods U_x and U_y of x and y, respectively, such that $U_x + U_y \subset U_z$. For $u \in U_y$, $U_z - u \supset U_x + U_y - u \supset U_x$, so

$$(\mu * \nu)(U_z) \geq \int_{U_y} \mu(U_z - u)\nu(du) \geq \mu(U_x)\nu(U_y) > 0.$$

Hence, $\mathrm{supp}\, \mu + \mathrm{supp}\, \nu \subset \mathrm{supp}(\mu * \nu)$. Since supports are closed, $(\mathrm{supp}\, \mu + \mathrm{supp}\, \nu)^- \subset \mathrm{supp}(\mu * \nu)$.

Conversely, let $z \in \mathrm{supp}(\mu * \nu)$, and let U_z be an open neighborhood of z. Since $0 < (\mu * \nu)(U_z) = \int \mu(U_z - y)\nu(dy)$, there is $y \in \mathrm{supp}\, \nu$ such that $\mu(U_z - y) > 0$. Hence, $(U_z - y) \cap \mathrm{supp}\, \mu \neq \emptyset$. Therefore, there is $x \in \mathrm{supp}\, \mu$ such that $x + y \in U_z$. Hence, $(\mathrm{supp}\, \mu + \mathrm{supp}\, \nu) \cap U_z \neq \emptyset$. Thus, $z \in (\mathrm{supp}\, \mu + \mathrm{supp}\, \nu)^-$, so $\mathrm{supp}(\mu * \nu) \subset (\mathrm{supp}\, \mu + \mathrm{supp}\, \nu)^-$. Q.E.D.

Example. Let μ and ν be discrete probabilities on $\mathbb{R}^1$ with positive atoms on the set of integers, $\mathbb{Z}$, and on the set $a\mathbb{Z}$, with a irrational, respectively. Then

$$\mathrm{supp}(\mu * \nu) = \{m + an : m, n \in \mathbb{Z}\}^- = \mathbb{R}^1.$$

Proposition 1.7.11. *Let μ be a probability on the topological space S_1 and let $f: S_1 \to S_2$ be a continuous mapping into the topological space S_2. Then*

$$\mathrm{supp}(f\mu) = (f(\mathrm{supp}\, \mu))^-.$$

In particular, for Banach spaces X and Y, probability μ on X, and a bounded linear operator $A: X \to Y$, we obtain

$$\mathrm{supp}(A\mu) = A(\mathrm{supp}\, \mu).$$

Proof. Note that

$$(f\mu)\big[(f(\operatorname{supp}\mu))^-\big] \geq \mu\big[f^{-1}(f(\operatorname{supp}\mu))\big]$$
$$\geq \mu(\operatorname{supp}\mu) = 1.$$

Hence, $\operatorname{supp}(f\mu) \subset (f(\operatorname{supp}\mu))^-$ for any measurable mapping f. If $\operatorname{supp}(f\mu)$ is a *proper* subset of $(f(\operatorname{supp}\mu))^-$, then there are $s \in \operatorname{supp}\mu$ and an open set U in S_2 such that $f(s) \in U$ and $U \cap \operatorname{supp}(f\mu) = \emptyset$. Hence, $0 = (f\mu)(U) = \mu(f^{-1}(U))$ and s is in the open set $f^{-1}(U)$. This contradicts $s \in \operatorname{supp}\mu$. Therefore, $\operatorname{supp}(f\mu) = (f(\operatorname{supp}\mu))^-$. The second part of this proposition follows from the fact that A is an open mapping from X into Y. Q.E.D.

Note that the above gives an alternate proof of Proposition 1.7.10 when applied to product measures.

Proposition 1.7.12. *Let* $\mu \in \mathscr{P}(X)$. *Then* $(\operatorname{supp}\mu)^\perp = \{x^* \in X^* : \hat{\mu}(tx^*) = 1 \text{ for all } t \in \mathbb{R}^1\}$.

Proof. If $x^* \in (\operatorname{supp}\mu)^\perp$, then $\langle x^*, tx \rangle = 0$ for all $x \in \operatorname{supp}\mu$ and for all t, so $\hat{\mu}(tx^*) = 1$. If $\hat{\mu}(tx^*) = 1$ for all t, then $\int (1 - \cos\langle tx^*, x \rangle)\mu(dx) = 0$, so $\cos\langle tx^*, x \rangle = 1$ for all $x \in \operatorname{supp}\mu$ and for all t. Hence, $\langle tx^*, x \rangle = 0 \pmod{\pi}$ for all t which implies that $\langle x^*, x \rangle = 0$, so $x^* \in (\operatorname{supp}\mu)^\perp$. Q.E.D.

1.8 INFINITELY DIVISIBLE PROBABILITIES

This section is crucial for the entire subject we will present. Namely, all classes of limit distributions investigated in Chapters 3 and 4 are subclasses of the infinitely divisible class. Let us recall that a probability μ on a Banach space X is said to be *infinitely divisible* if for each integer $n \geq 2$ there exists an element $\mu_n \in \mathscr{P}(X)$ such that $\mu_n^n = \mu$, where the nth power of a probability is taken in the sense of convolution. Note that the characteristic function of μ_n^n is given by the usual power, $(\mu_n^n)\hat{} = (\hat{\mu}_n)^n$. In terms of an X-valued random variable ξ, we have that $\mathscr{L}(\xi)$ is infinitely divisible if and only if for each $n \geq 2$ there are independent copies of ξ, say $\xi_1, \ldots, \xi_n$, such that $\xi \stackrel{d}{=} \xi_1 + \cdots + \xi_n$. Let $\mathrm{ID}(X)$ denote the class of all infinitely divisible probabilities on X. Its algebraic and topological structure are given in part (b) of the following proposition.

Proposition 1.8.1

(a) *For $\mu \in \mathrm{ID}(X)$, we have $\hat{\mu}(x^*) \neq 0$ for all $x^* \in X^*$.*
(b) *$\mathrm{ID}(X)$ is a closed subsemigroup of $\mathcal{P}(X)$.*

Proof

(a) Note that $\hat{\mu}(x^*) \neq 0$ if and only if $(\mu^0)\hat{\,}(x^*) > 0$, where μ^0 is the symmetrization of μ. Also, if $\mu = \mu_n^n$, then $\mu^0 = (\mu_n^0)^n$. Hence, from Proposition 1.7.5,

$$1 - \left((\mu^0)\hat{\,}(x_1^* + x_2^*)\right)^{1/n}$$
$$= 1 - (\mu_n^0)\hat{\,}(x_1^* + x_2^*)$$
$$\leq 2\left[1 - (\mu_n^0)\hat{\,}(x_1^*) + 1 - (\mu_n^0)\hat{\,}(x_2^*)\right]$$
$$= 2\left[1 - \left((\mu^0)\hat{\,}(x_1^*)\right)^{1/n} + 1 - \left((\mu^0)\hat{\,}(x_2^*)\right)^{1/n}\right]$$

for all $n \geq 1$ and for all $x_1^*, x_2^* \in X^*$. Since $(\mu^0)\hat{\,}(x^*) > 0$ is equivalent to $\lim_{n \to \infty}[1 - ((\mu^0)\hat{\,}(x^*))^{1/n}] = 0$, we see that the set

$$\{x^* \in X^*: \hat{\mu}(x^*) \neq 0\} = \{x^* \in X^*: (\mu^0)\hat{\,}(x^*) > 0\}$$

is a subgroup of X^* containing zero. By the continuity of $\hat{\mu}$, this subgroup is an open set. Therefore, it must be X^* itself.

(b) It is obvious from the definition of $\mathrm{ID}(X)$ that it is a subsemigroup. Now, assume that $\{\mu_k\}$ is a sequence in $\mathrm{ID}(X)$ and that $\mu_k \Rightarrow \mu \in \mathcal{P}(X)$. For each $k \geq 1$ and for each $n \geq 2$, there is $\mu_{k,n} \in \mathcal{P}(X)$ such that $\mu_k = \mu_{k,n}^n = \mu_{k,n} * \mu_{k,n}^{n-1}$. Theorem 1.7.1(b) implies that the sequence $\{\mu_{k,n}\}_{k \geq 1}$ is shift conditionally compact. Also, $\lim_{k \to \infty} \hat{\mu}_{k,n}(x^*) = (\hat{\mu}(x^*))^{1/n}$ uniformly over balls in X^* [cf. Proposition 1.7.6(a)]. Finally, from Proposition 1.7.6(c), we obtain some $\mu_n \in \mathcal{P}(X)$ such that $\mu_{k,n} \Rightarrow \mu_n$ as $k \to \infty$. Therefore, $\mu_n^n = \mu$ for each $n \geq 2$, which shows that $\mu \in \mathrm{ID}(X)$, that is, $\mathrm{ID}(X)$ is closed. Q.E.D.

For a finite measure m on X, let

$$e(m) := \exp(-m(X)) \sum_{k=0}^{\infty} \frac{m^k}{k!}, \qquad (1.8.1)$$

with $m^0 := \delta(0)$. Then $e(m) \in \mathcal{P}(X)$ and it is referred to as the *Poisson measure associated with m*. Note that $e(\lambda \delta(a))$ for $\lambda > 0$ and $a \in X$ is a classical Poisson distribution. It is easy to see that

$$(e(m))\widehat{\ }(x^*) = \exp \int_X (e^{i\langle x^*, x\rangle} - 1) m(dx)$$

$$= \exp\left\{i\langle x^*, a_m\rangle + \int_X \left(e^{i\langle x^*, x\rangle} - 1 - i\langle x^*, x\rangle I_{B_1}(x)\right) m(dx)\right\},$$

where $a_m := \int_{B_1} x m(dx)$ and B_1 is the unit ball in X (cf. Proposition 1.3.1). Hence,

$$e(m_1) * e(m_2) = e(m_1 + m_2) \qquad (1.8.2)$$

for finite measures m_1 and m_2. Therefore, we see that all Poisson measures are infinitely divisible. Moreover,

$$m_n \Rightarrow m \quad \text{implies} \quad e(m_n) \Rightarrow e(m) \qquad (1.8.3)$$

for finite measures m_n and m; the weak convergence of finite measures, not necessarily probabilities, was given after (1.6.1). To see (1.8.3), note that from

$$e(m_n)(A) \geq e^{-m_n(X)} m_n(A)$$

for all $A \in \mathcal{B}(X)$ and from the conditional compactness of $\{m_n\}$, we obtain that the sequence $\{e(m_n)\}$ is conditionally compact. Furthermore, $(e(m_n))\widehat{\ }(x^*) \to (e(m))\widehat{\ }(x^*)$ for each $x^* \in X^*$, so Proposition 1.7.6(b) gives $e(m_n) \Rightarrow e(m)$.

A σ-finite measure M on X is said to be a *Lévy spectral measure* if there exists an element $\tilde{e}(M) \in \mathscr{P}(X)$ such that

$$(\tilde{e}(M))^{\wedge}(x^*) = \exp\int_X \left(e^{i\langle x^*, x\rangle} - 1 - i\langle x^*, x\rangle I_{B_1}(x)\right) M(dx). \tag{1.8.4}$$

Note that any finite measure m is a Lévy spectral measure because $\tilde{e}(m) = e(m) * \delta(-a_m)$.

Theorem 1.8.2. *Let M be a σ-finite measure on X. Then the following are equivalent.*

(a) *M is a Lévy spectral measure.*
(b) *For any sequence $\{m_n\}$ of finite measures such that $m_n(A) \le m_{n+1}(A)$ for all $n \ge 1$ and $m_n(A) \to M(A)$ for all $A \in \mathscr{B}(X)$, we have $e(m_n) * \delta(-a_{m_n}) \Rightarrow \tilde{e}(M)$, where $a_m := \int_{B_1} x m(dx)$.*
(c) *There exists a sequence $\{m_n\}$ as in (b) such that the sequence $\{e(m_n) * \delta(-a_{m_n})\}$ is shift conditionally compact.*

Unfortunately, in an arbitrary Banach space there is no complete characterization of a Lévy spectral measure in terms of the integrability of some function. Also, we note that in the case of a Hilbert space, the kernel in (1.8.4) is slightly different. Namely,

$$(\tilde{e}(M))^{\wedge}(y) = \exp\int_H \left(e^{i\langle y, x\rangle} - 1 - \frac{i\langle y, x\rangle}{1 + \|x\|^2}\right) M(dx). \tag{1.8.5}$$

For further reference, we state the following result.

Theorem 1.8.3

(a) *On a real separable Banach space X, each measure M such that $\int_X (1 \wedge \|x\|) M(dx) < \infty$ is a Lévy spectral measure.*
(b) *Let H be a real separable Hilbert space. Then a measure M is a Lévy spectral measure if and only if $\int_H (1 \wedge \|x\|^2) M(dx) < \infty$; equivalently, M is finite outside of every neighborhood of zero and $\int_H \|x\|^2/(1 + \|x\|^2) M(dx) < \infty$.*

Another important class of infinitely divisible measures are the Gaussian ones. First, recall that a probability γ on $\mathbb{R}^1$ is called *normal* or *Gaussian* with mean value a and variance $\sigma^2 > 0$, if, for all Borel sets B,

$$\gamma(B) = \int_B \frac{e^{-(x-a)^2/2\sigma^2}}{\sigma\sqrt{2\pi}}\,dx.$$

We call a probability γ on X a *centered Gaussian* measure if, for every $x^* \in X^*$, $\Pi_{x^*}\gamma$ is a zero-mean Gaussian measure on $\mathbb{R}^1$, where for $x^* \in X^*$, $\Pi_{x^*}: X \to \mathbb{R}^1$ is given by $\Pi_{x^*}(x) := \langle x^*, x \rangle$. Hence, a probability γ on X is called a *Gaussian* measure if there is $a \in X$ such that $\gamma * \delta(a)$ is a centered Gaussian measure.

An alternative definition of Gaussian measures on Banach spaces, even for measurable linear spaces, is provided by the following proposition.

Proposition 1.8.4. *A measure $\gamma \in \mathscr{P}(X)$ is a centered Gaussian measure if and only if whenever ξ_1 and ξ_2 are independent X-valued random variables with $\mathscr{L}(\xi_1) = \mathscr{L}(\xi_2) = \gamma$, and when $s, t \in \mathbb{R}^1$ are such that $s^2 + t^2 = 1$, the X-valued random variables $\eta_1 := s\xi_1 + t\xi_2$ and $\eta_2 := t\xi_1 - s\xi_2$ are independent and $\mathscr{L}(\eta_1) = \mathscr{L}(\eta_2) = \gamma$.*

From this equivalent definition, one obtains the following fact about exponential moments of Gaussian measures.

Proposition 1.8.5. *Let ξ be an X-valued random variable whose distribution is a centered Gaussian measure. Then there is $\lambda_0 > 0$ such that for every $\lambda < \lambda_0$*

$$\mathbf{E}\bigl(\exp(\lambda\|\xi\|^2)\bigr) < \infty.$$

Finally, we recall some known characterizations of Gaussian measures in terms of their characteristic functionals. Once again, as for Lévy spectral measures, a complete description is given only for Hilbert spaces.

Theorem 1.8.6

(a) *Let X be a real separable Banach space. For a centered Gaussian measure γ, we have, for all $x^* \in X^*$,*

$$\hat{\gamma}(x^*) = \exp(-2^{-1}\langle x^*, Rx^* \rangle),$$

where R is a covariance operator, that is, R is a compact operator from X^* into X which is symmetric ($\langle x_1^*, Rx_2^* \rangle = \langle x_2^*, Rx_1^* \rangle$) and nonnegative ($\langle x^*, Rx^* \rangle \geq 0$).

(b) Let H be a real separable Hilbert space. Then a complex-valued function φ on H is the characteristic functional of a centered Gaussian measure if and only if, for all $y \in H$,

$$\varphi(y) = \exp(-2^{-1}\langle y, Dy \rangle),$$

where D is an S-operator, that is, D is symmetric, nonnegative, and $\sum_i \langle e_i, De_i \rangle < \infty$ for an arbitrary base $\{e_i\}$ in H. Furthermore, if $\varphi = \hat{\gamma}$, then D is the covariance operator of γ, that is, $\langle y, Dy \rangle = \int_H \langle y, x \rangle^2 \gamma(dx)$.

Next, we wish to describe the class $\mathrm{ID}(X)$ in terms of characteristic functionals.

Theorem 1.8.7 (Lévy–Khintchine Representation). *A probability μ on X is infinitely divisible if and only if, for every $x^* \in X^*$,*

$$\hat{\mu}(x^*) = \exp\left\{ i\langle x^*, a \rangle - 2^{-1}\langle x^*, Rx^* \rangle + \int_{X \setminus \{0\}} \left(e^{i\langle x^*, x \rangle} - 1 - i\langle x^*, x \rangle I_{B_1}(x) \right) M(dx) \right\},$$

where $a \in X$, R is the covariance operator of a centered Gaussian measure, and M is a Lévy spectral measure. Furthermore, the parameters a, R, and M are uniquely determined by μ and they uniquely determine μ.

Remark 1.8.8

(a) In case of a Hilbert space, the kernel in the above theorem is replaced by the one given in (1.8.5).

(b) Due to the uniqueness of a, R, and M, we write $\mu := [a, R, M]$ if $\hat{\mu}$ is as in Theorem 1.8.7.

As we mentioned, there is no complete description of Gaussian covariance operators and Lévy spectral measures in an arbitrary

Banach space [cf. Theorems 1.8.3(a) and 1.8.6(a)]. In some situations, the following sufficient criteria are useful.

Proposition 1.8.9. *Let M be a Lévy spectral measure and let R be a Gaussian covariance operator on a Banach space X. Then*

(a) *if M_1 is a σ-finite measure such that $M_1(A) \leq M(A)$ for all $A \in \mathscr{B}(X \setminus \{0\})$, then M_1 and $M - M_1$ are also Lévy spectral measures;*

(b) *if R_1 is a symmetric, positive, bounded, linear operator from X^* into X such that $\langle x^*, R_1 x^* \rangle \leq \langle x^*, Rx^* \rangle$ for all $x^* \in X^*$, then R_1 is also a Gaussian covariance operator.*

In Proposition 1.8.1(b) we saw that $\mathrm{ID}(X)$ is a closed subsemigroup of $\mathscr{P}(X)$. We now look at the inner structure of this class.

Proposition 1.8.10

(a) *The class $\mathrm{ID}(\mathbb{R}^1)$ is the smallest closed semigroup in $\mathscr{P}(\mathbb{R}^1)$ which contains all probabilities of the form $[x, 0, \lambda\delta(y)]$ for some $\lambda > 0$ and some $x, y \in \mathbb{R}^1$.*

(b) *For a Banach space X, $\mathrm{ID}(X)$ is the smallest closed semigroup generated by the probabilities of the forms $[x, 0, \lambda\delta(y)]$ and $[x, R, 0]$, where $x, y \in X$, $\lambda > 0$, and R is a Gaussian covariance operator.*

Proof

(a) This is a well-known characterization on $\mathbb{R}^1$.

(b) From Proposition 1.8.1(b), we know that $\mathrm{ID}(X)$ is a closed semigroup. We know that $\tilde{e}(M)$ is the weak limit of a sequence $\{e(m_n) * \delta(x_n)\}$, where $m_n \uparrow M$ (cf. Theorem 1.8.2), and we know that for finite measures m, $e(m)$ is the limit of finite convolutions of $e(\lambda_k \delta(x_k))$ [cf. Proposition 1.6.8 and formula (1.8.3)]. This yields the desired result because each infinitely divisible measure is of the form $\delta(a) * \tilde{e}(M) * \gamma$, where γ is a centered Gaussian measure, $a \in X$, and M is a Lévy spectral measure (cf. Theorem 1.8.7). Q.E.D.

Our next aim is to show that the class ID(X) is equal to a particular class of limit distributions. Recall that a triangular array $\xi_{n,k}$, $1 \le k \le k_n$, $n \ge 1$, of X-valued random variables is called an *infinitesimal system*, or *uniformly asymptotically negligible*, if, for every $\varepsilon > 0$,

$$\lim_{n \to \infty} \max_{1 \le k \le k_n} P[\|\xi_{n,k}\| > \varepsilon] = 0.$$

In terms of the distributions $\mu_{n,k} := \mathcal{L}(\xi_{n,k})$, this becomes

$$\lim_{n \to \infty} \max_{1 \le k \le k_n} \mu_{n,k}(\{\|x\| > \varepsilon\}) = 0 \qquad (1.8.6)$$

for every $\varepsilon > 0$, that is, for any sequence $\{j_n\}$ of integers such that $1 \le j_n \le k_n$ we have $\mu_{n, j_n} \Rightarrow \delta(0)$ as $n \to \infty$.

Theorem 1.8.11. *The class* ID(X) *is equal to the class of all limit distributions of sequences of the form* $\mu_{n,1} * \cdots * \mu_{n,k_n} * \delta(x_n)$, *where the triangular array* $\mu_{n,k}$ *forms an infinitesimal system and* $\{x_n\}$ *is a sequence in* X.

The existence of exponential moments for Gaussian measures was stated in Proposition 1.8.5. For any infinitely divisible measure, we also have an analog of such exponential moments.

Proposition 1.8.12. *If* $\mu = [a, R, M] \in $ ID(X) *is such that* M *is concentrated on some bounded set, then, for every* $\lambda \in \mathbb{R}^1$,

$$\int_X \exp(\lambda \|x\|) \mu(dx) < \infty.$$

This proposition can be generalized to submultiplicative functions. A function $g: X \to [0, \infty)$ is said to be *submultiplicative* if there is $c > 0$ such that, for all $x, y \in X$,

$$g(x + y) \le c g(x) g(y).$$

Similarly, $h: X \to [0, \infty)$ is said to be *subadditive* if there is $a > 0$ such that, for all $x, y \in X$,

$$h(x + y) \leq a(h(x) + h(y)).$$

Note that, if h is subadditive with $a \geq 1$, then $g(x) := a(2 + h(x))$ is submultiplicative with $c = 1$. Also, if g is submultiplicative and continuous at zero, then there are constants $\alpha, \beta \in \mathbb{R}^1$ such that, for all $x \in X$,

$$g(x) \leq \alpha \exp(\beta \|x\|). \tag{1.8.7}$$

Examples of submultiplicative functions on X are

(a) $g_1(x) := \exp(a\|x\|^p)$, $0 < p \leq 1$, $a > 0$;
(b) $g_2(x) := 2^p(2 + \|x\|^p)$, $p > 0$;
(c) $g_3(x) := 2 + \log(1 + \|x\|)$.

In the last two examples, note that $x \mapsto \|x\|^p$ and $x \mapsto \log(1 + \|x\|)$ are subadditive functions on X.

Proposition 1.8.13. *For $\mu = [a, R, M]$ and for the submultiplicative function g which is continuous at zero, we have*

$$\int_X g(x)\mu(dx) < \infty \quad \text{if and only if} \quad \int_{\|x\|>1} g(x)M(dx) < \infty.$$

Proof. This follows from Propositions 1.8.5 and 1.8.12, the inequality (1.8.7), and the following inequalities:

$$\int_X g(x)e(m)(dx) \leq e^{-m(X)}\left[g(0) + \sum_{k=1}^\infty \frac{c^{k-1}}{k!}\left(\int_X g(x)m(dx)\right)^k\right],$$

$$e(m)(A) \geq e^{-m(X)}m(A). \qquad \text{Q.E.D.}$$

Remark 1.8.14. Let

$$\mathrm{ID}_{\log}(X) := \left\{\mu \in \mathrm{ID}(X) : \int_X \log(1 + \|x\|)\mu(dx) < \infty\right\}.$$

Then $\mathrm{ID}_{\log}(X)$ is a semigroup and, for $A \in \mathrm{End}(X)$, $A(\mathrm{ID}_{\log}(X)) \subset \mathrm{ID}_{\log}(X)$. Note that if μ has Lévy spectral measure M, then $A\mu$ has Lévy spectral measure AM.

INFINITELY DIVISIBLE PROBABILITIES 37

Theorem 1.8.15. *Let H be a Hilbert space and let $[a_n, R_n, M_n]$, $[a, R, M] \in \mathrm{ID}(H)$. Then $[a_n, R_n, M_n] \Rightarrow [a, R, M]$ if and only if the following hold.*

(i) $a_n \to a$ in H.
(ii) $M_n \Rightarrow M$ outside of every neighborhood of zero.
(iii) *For the S-operators, T_n, given by*

$$\langle y, T_n y \rangle := \int_{B_1} \langle x, y \rangle^2 M_n(dx) + \langle y, R_n y \rangle,$$

we have

(a) $\lim_{\varepsilon \downarrow 0} \left\{ \begin{array}{c} \liminf \\ \limsup_{n \to \infty} \end{array} \right\} \left(\int_{B_\varepsilon} \langle x, y \rangle^2 M_n(dx) + \langle y, R_n y \rangle \right)$
$= \langle y, Ry \rangle$

for every $y \in H$;

(b) $\sup_n \mathrm{trace}(T_n) < \infty;$

(c) $\lim_{k \to \infty} \sup_n \sum_{i=k+1}^{\infty} \langle e_i, T_n e_i \rangle = 0$

for any orthonormal complete basic $\{e_i\}$ in H.

The next part of this section deals with some specific facts about infinitely divisible measures on $\mathbb{R}^d$. The previous theorem has the following formulation on $\mathbb{R}^d$.

Corollary 1.8.16. *Let $[a_n, R_n, M_n], [a, R, M] \in \mathrm{ID}(\mathbb{R}^d)$. Then $[a_n, R_n, M_n] \Rightarrow [a, R, M]$ if and only if (i) and (ii) in Theorem 1.8.15 hold, and*

(iii') $\lim_{\varepsilon \downarrow 0} \left\{ \begin{array}{c} \liminf \\ \limsup_{n \to \infty} \end{array} \right\} \left(\int_{B_\varepsilon} \langle x, y \rangle^2 M_n(dx) + \langle y, R_n y \rangle \right)$
$= \langle y, Ry \rangle$

for all $y \in \mathbb{R}^d$.

There are also conditions for an infinitesimal system to converge to an infinitely divisible measure.

Proposition 1.8.17. *An infinitesimal system $\{\mu_{n,k}: n \geq 1, 1 \leq k \leq k_n\}$ of probabilities on $\mathbb{R}^d$ is shift weakly convergent to $[a, R, M] \in \text{ID}(\mathbb{R}^d)$ if and only if*

(a) *outside of every neighborhood of zero,*

$$\sum_{k=1}^{k_n} \mu_{n,k} \Rightarrow M;$$

(b) $$\lim_{\varepsilon \downarrow 0} \left\{ \begin{array}{c} \liminf \\ \limsup_{n \to \infty} \end{array} \right\} \sum_{k=1}^{k_n} \int_{B_\varepsilon} \langle y, x \rangle^2 \mu_{n,k}(dx) = \langle y, Ry \rangle$$

for every $y \in \mathbb{R}^d$.

We end this section with some facts about factors and supports of Gaussian measures.

Theorem 1.8.18 (Cramér Theorem). *Let γ be a centered Gaussian measure on a Banach space X. Assume $\gamma = \gamma_1 * \gamma_2$ for some $\gamma_1, \gamma_2 \in \mathscr{P}(X)$. Then both γ_1 and γ_2 are Gaussian measures.*

Theorem 1.8.19 (Skitovich Theorem). *Let ξ be a nondegenerate $\mathbb{R}^d$-valued random variable. Then ξ is Gaussian if and only if there exists a coordinate system for $\mathbb{R}^d$, $\{x_1, \ldots, x_d\}$, such that the projections of ξ onto the x_i-axes are independent, and there exist two more axes, $\{y_1, y_2\}$, which are not contained in any proper hyperplane generated by $\{x_1, \ldots, x_d\}$, such that the projections of ξ onto the y_1-axis and y_2-axis are independent.*

Finally, we wish to specify the support of a Gaussian measure. Let us recall from Section 1.7 that the support of a measure is the smallest closed subset having measure one.

Theorem 1.8.20

(a) *The support of a centered Gaussian measure on a Banach space X is a linear subspace of X.*

(b) *If H is a real separable Hilbert space and γ is a centered Gaussian measure on H with covariance operator D, then*

$$\operatorname{supp} \gamma = (\ker D)^\perp.$$

Proof. We only prove (b) since it is needed later. Let $x \in (\operatorname{supp} \gamma)^\perp$. Then, for all $z \in H$,

$$\langle z, Dx \rangle = \int_H \langle z, y \rangle \cdot \langle x, y \rangle \gamma(dy) = 0.$$

Hence, $(\operatorname{supp} \gamma)^\perp \subset \ker D$. Conversely, for $x \in \ker D$, we have

$$\langle x, Dx \rangle = \int_H \langle x, y \rangle^2 \gamma(dy) = 0,$$

so $\langle x, y \rangle = 0$ for all $y \in H$ γ-a.s. Hence, $\gamma\{y \in H: \langle x, y \rangle = 0\} = 1$ when $x \in \ker D$. Thus,

$$\operatorname{supp} \gamma \subset \bigcap_{x \in \ker D} \{y \in H: \langle x, y \rangle = 0\} = (\ker D)^\perp.$$

Therefore, $\operatorname{supp} \gamma = (\ker D)^\perp$. Q.E.D.

Corollary 1.8.21. *For a centered Gaussian measure γ on $\mathbb{R}^d$ with covariance operator D, we have $\dim(\operatorname{supp} \gamma) = \operatorname{rank}(D)$.*

Note that from the above two facts, the support of a Gaussian measure is always a *linear* space (cf. the second part of Section 3.4).

1.9 THE SKOROHOD SPACE

Let $(\Omega, \mathscr{F}, P)$ be a complete probability space, that is, $\mathscr{F}$ contains all subsets of a set with P-measure zero. Let $(S, \mathscr{S})$ be a measurable space and let T be an abstract set. A mapping $\xi: \Omega \times T \to S$ is said to be a *random function* if, for any $t \in T$, the mapping $\xi(\cdot, t): \Omega \to S$ is an S-valued random variable. Furthermore, if T and S are topological spaces, then a random function ξ is said to be *separable* (relative to open sets in T and closed sets in S) if there exist a dense countable subset T_0 of T and a set N of the P-measure zero such that for any open G in T and any closed F in S we have

$$\{\omega \in \Omega: \xi(t, \omega) \in F \text{ for all } t \in G \cap T_0\} \setminus$$
$$\{\omega \in \Omega: \xi(t, \omega) \in F \text{ for all } t \in G\} \subseteq N.$$

Note that the first set is in $\mathscr{F}$ and, therefore, so is the second one. We will see that in rather general situations we may assume that the random function is separable. More precisely, if we agree to call two random functions ξ_1 and ξ_2 on $\Omega \times T$ *stochastically equivalent* whenever $P\{\xi_1(\cdot, t) = \xi_2(\cdot, t)\} = 1$, for each $t \in T$, then we have the following theorem.

Theorem 1.9.1. *Let S be a separable locally compact metric space and let T be an arbitrary separable metric space. For any random function ξ on $\Omega \times T$ with values in S, there exists a stochastically equivalent separable function $\tilde{\xi}$ taking values in a certain compact extension $\tilde{S}$ of S.*

If T is either the real line or a segment of the real line, the random function is said to be a *random process*, and T may be viewed as time. Furthermore, we say that the random process $\xi(t)$, $t \in [a, b]$, is *without discontinuities of the second kind* on $[a, b]$ if the sample functions $\xi(\omega, \cdot)$ have at each $t \in (a, b)$ with probability one left and right limits and possess at the points a and b a right and left limit, respectively. Many classes of random processes have sample paths (functions) without discontinuities of the second kind. For our needs recall the following theorem.

Theorem 1.9.2. *If $\xi(t)$, $t \in [a, b]$, is a separable stochastically continuous process with independent increments and values in a linear normed space $\mathbb{E}$, then it has no discontinuities of the second kind.*

Theorem 1.9.3. *If $\xi(t)$, $t \in [a, b]$, is a stochastically continuous process without discontinuities of the second kind and with values in a metric space S, then there exists an equivalent process $\tilde{\xi}(t)$, $t \in [a, b]$, whose sample functions are continuous from the right with probability one.*

Let $D(S; [a, b])$ be the set of all functions x defined on $[a, b]$ with values in a complete metric space S that are right-continuous on $[a, b]$, have left limits $x(t-)$ on $(a, b]$, and $x(b) = x(b-)$. From the above discussion [existence of separable version; random functions without discontinuities of the second kind; existence of a version with sample functions in $D(S;[a,b])$], we see that without loss of generality we may usually assume that the random function ξ from $\Omega \times T$ into S is a mapping $\eta \colon \Omega \to D(S;T)$, when T is a closed interval in the real line. Moreover, using the *coordinate mappings* $\Pi^t \colon D(S;T) \to S$ given by $\Pi^t(x) := x(t)$, we have $\omega \mapsto (\Pi^t \eta)(\omega) := \eta(t, \omega)$ is an S-valued random variable, that is, measurable mapping from $(\Omega, \mathscr{F})$ into $(S, \mathscr{S})$.

Remark 1.9.4. Let $\xi(t)$, $t \in [a,b]$, be a random process with values in an infinite-dimensional Banach space X. We cannot directly apply Theorem 1.9.1 to claim that there is a version $\tilde{\xi}$ with sample functions in $D(X;[a,b])$. However, such a version exists, if ξ has independent increments and is stochastically continuous.

The discussion preceding Remark 1.9.4 suggests that it would be helpful to treat a random function as a measurable mapping from $(\Omega, \mathscr{F})$ into $(D(S;[a,b]), \mathscr{D})$ for some σ-field $\mathscr{D}$. A natural requirement is that $\mathscr{D}$ should coincide with the σ-algebra generated by all cylindrical sets

$$\{x \in D(S;[a,b]) \colon x(t_1) \in A_1, \ldots, x(t_n) \in A_n\},$$

where $t_j \in [a,b]$ and $A_j \in \mathscr{B}(S)$ for all $j = 1, \ldots, n$. Note that the coordinate mappings must be measurable. Also, let us note that such a measurable mapping η is a separable random function; as T_0 one can take an arbitrary separable dense subset of $[a,b]$. Moreover, to be able to apply results from Sections 1.6 or 1.7, we should have a metric d on $D(S;[a,b])$ such that $\mathscr{D}$ would be the σ-field generated by the Borel sets with respect to the metric d and $D(S;[a,b])$ would be a separable metric space. Sometimes it also needs to be a complete separable metric space (cf. Theorem 1.6.6). The metric that suits these purposes was defined by A. V. Skorohod and is now referred to as the Skorohod metric J_1. It is defined as follows:

$$J_1(x,y) := \inf\left\{ \sup_{a \leq t \leq b} \rho(x(t), y(\lambda(t))) + \sup_{a \leq t \leq b} |t - \lambda(t)| \colon \lambda \in \Lambda \right\}, \tag{1.9.1}$$

where Λ is the group of all continuous, strictly increasing real functions λ on $[a,b]$ such that $\lambda(a) = a$ and $\lambda(b) = b$, and ρ is the metric on S. From now on, we refer to $D(S;[a,b])$ with the Skorohod metric J_1 as a *Skorohod space*.

Theorem 1.9.5. *The Skorohod space $D(S;[a,b])$ is a separable metric space. The Borel σ-field generated by the Skorohod metric J_1 coincides with the σ-field generated by cylindrical sets, that is, it is generated by the coordinate mappings.*

Remark 1.9.6. The Skorohod space is not complete under J_1. If in (1.9.1) we restrict Λ to Λ_0 defined as follows: $\lambda \in \Lambda_0$ if and only if

$$\sup_{s \neq t} \left| \log \frac{\lambda(t) - \lambda(s)}{t - s} \right| < \infty$$

and if we denote the corresponding metric by d_0, then J_1 and d_0 are equivalent metrics and $(D(S;[a,b]), d_0)$ is a complete metric space.

Remark 1.9.7

(a) Let $\Pi^t: D(S;[a,b]) \to S$ be the coordinate mapping $\Pi^t(x) := x(t)$. Then Π^t is continuous (in the Skorohod metric) at x if and only if x is continuous at t; that is, $x(t-) = x(t)$. So Π^a and Π^b are continuous everywhere.

(b) Let S be a linear normed space. Then the mapping for $x \in D(S;[a,b])$ given by $x \mapsto \int_{(a,s]} f(t)\, dx(t) := f(s)x(s) - f(a)x(a) - \int_{(a,s]} x(t)\, df(t)$, where f is an operator-valued function such that the integral $\int_{(a,s]} x(t)\, df(t)$ is well-defined, is continuous at x if and only if x is continuous at s.

As a matter of fact, we will need to deal with the space $D(S;[0,\infty))$ of all right-continuous functions, with left limits on $(0,\infty)$. To extend the procedure from a finite interval to the half-line, we must be aware of the following small but essential difference. In $D(S;[a,b])$ both points a and b require special treatment [for $\lambda \in \Lambda$, $\lambda(a) = a$ and $\lambda(b) = b$], whereas only 0 has an a priori right to special treatment in $D(S;[0,\infty))$. This suggests that it may be more convenient to study $D(S;[0,\infty))$ directly than to deduce needed properties by taking restrictions to $D(S;[0,b])$ for all b.

THE SKOROHOD SPACE

For each finite $b > 0$ and functions x, y from $D(S; [0, \infty))$, let

$$d_b(x, y) := \inf\{\varepsilon > 0: \text{there exist grids } 0 = t_0 < t_1 < \cdots < t_k,\ 0 = s_0 < s_1 < \cdots < s_k \text{ such that } |t_i - s_i| < \varepsilon \text{ and } \sup\{\rho(x(t), y(s)): t_i \leq t < t_{i+1},\ s_i \leq s < s_{i+1}\} \leq \varepsilon \text{ for all } i = 0, 1, \ldots, k-1 \text{ and } t_k \geq b,\ s_k \geq b\},$$

where ρ is the metric on S. Then each d_b is a metric on $D(S; [0, \infty))$, and note that the assumptions $s_k, t_k \geq b$ prevent discontinuities near b from overinflating the distance $d_b(x, y)$; if, say, y jumps just before b, then it can be matched by a similar jump in x just a little beyond b. By allowing d_b to depend on more than the segment of the path over $[0, \infty)$, we avoid difficulties that were required in J_1, $\lambda(b) = b$, and x, y are left-continuous at b. Finally,

$$d(x, y) := \sum_{n=1}^{\infty} 2^{-n} \min\{1, d_n(x, y)\} \tag{1.9.2}$$

is the *Skorohod metric* on $D(S; [0, \infty))$. Of course, $d(x_n, x) \to 0$ if and only if $d_b(x_n, x) \to 0$ for each finite b. Furthermore, we also have the following that relates J_1 to d [cf. (1.9.1)].

Lemma 1.9.8. *For x_n, x in $D(S; [0, \infty))$, we have $x_n \to x$ in d if and only if there exists a sequence $\{\lambda_n\}$ of strictly increasing continuous functions from $[0, \infty)$ onto itself such that $\lambda_n(t) \to t$ and $\rho(x_n(\lambda_n(t)), x(t)) \to 0$ uniformly over compact sets of t values.*

Finally, we also have an analog of Theorem 1.9.5 for $D(S; [0, \infty))$.

Theorem 1.9.9. *The Skorohod space $(D(S; [0, \infty)), d)$ is a separable metric space. The Borel σ-field, with respect to d, coincides with the σ-field generated by cylindrical sets.*

Let us also mention that, similarly as in the case of $D(S; [a, b])$, there is an equivalent metric d_0 on $D(S; [0, \infty))$ such that it becomes a complete separable metric space.

Theorem 1.9.10. *Let η and η_n be $D(S; [0, \infty))$-valued random variables. Then $\eta_n \to \eta$ in distribution if and only if $r_\alpha \eta_n \to r_\alpha \eta$ in distribution in $D(S; [0, \alpha])$ for any $\alpha \in C_\eta$. Here, $r_\alpha x$ is the restriction of x from $D(S; [0, \infty))$ to $D(S; [0, \alpha])$ and $C_\eta := \{t: P(\eta(t) \neq \eta(t-)) = 0\}$.*

We conclude this section with a remark on convergence a.s. along each subsequence $t_n \to \infty$ and along $t \to \infty$.

Remark 1.9.11. Let ξ be a separable random function from $\Omega \times [0, \infty)$ into a separable Banach space X. If for any sequence $0 < t_1 < t_2 < \cdots$ converging to ∞, the sequence $\xi(t_n)$ converges a.s. (in norm), then $\lim_{t \to \infty} \xi(t)$ exists a.s. as well.

1.10 BIBLIOGRAPHIC COMMENTS

The main result in Section 1.1 is due to Numakuru (1952). Our presentation of this result is from Paalman and de Miranda (1964). Urbanik (1972) was the first to use such algebraic methods in probability theory. Section 1.2 is based on Hille and Phillips (1957) and Hoffman and Kunze (1971). The material in Sections 1.3 and 1.4 is standard and classical. We partially followed Dynkin (1965). However, Lemma 1.4.4 is due to Watanabe (1968). Section 1.6 is a selection of material from Billingsley (1968), whereas Section 1.7 is based on Araujo and Giné (1980) and Parthasarathy (1967). Also, Section 1.8 is a compilation of results from Araujo and Giné (1980). Propositions 1.8.12 and 1.8.13 were proved in de Acosta (1980) and in Jurek and Smalara (1981). Finally, Section 1.9 on Skorohod spaces is based on Gikhman and Skorohod (1969) [cf. also Pollard (1984)].

CHAPTER 2

Convergence of Types Theorems, Symmetry Groups, and Decomposability Semigroups

In any study of limit theorems, a convergence of types theorem is usually one of the basic tools. Before considering the convergence of types problem, the notion of full measures is examined in Section 2.1. The idea of fullness is the natural extension of nondegeneracy on $\mathbb{R}^1$. Several conditions which are equivalent to fullness are obtained and it is shown that the set of all full measures is an open subsemigroup of $\mathscr{P}(\mathbb{R}^d)$.

Section 2.2 presents the necessary convergence of types theorems and introduces an important group of affine transformations, the so-called invariant semigroups. It is shown that for a full measure its invariant semigroup is a compact subgroup of the group of all invertible affine transformations. Next, the decomposability semigroup of a measure due to Urbanik is examined. In Section 2.3, it is established that for a full measure its decomposability semigroup is a compact subsemigroup of the semigroup of all linear transformations. Some characterizations of a measure are obtained from the topological and algebraic properties of these semigroups. Some special results concerning Gaussian measures are also obtained. The decomposability semigroups are the main tools used in Chapter 3 and the symmetry groups are presented in Chapter 4.

In Section 2.4, a characterization of any compact subgroup of the group of all invertible affine transformations is obtained. Several examples are given in Section 2.5 which point out why many of the methods used in $\mathbb{R}^d$ do not work in an infinite-dimensional space. In Section 2.6, the convergence of types theorems and the decomposability semigroup are specialized to the important case of $\mathbb{R}^1$. Finally, in

Section 2.7, the convergence of types theorems are developed in the more usual setting, that is, the affine transformations are restricted to scaling by a constant plus a shift. Here, one need assume neither fullness of the measures nor finite dimensionality of the space.

2.1 FULL MEASURES

We say that a measure μ on $\mathbb{R}^d$ is *full* if its support is not contained in any proper hyperplane of $\mathbb{R}^d$, that is, for any x in $\mathbb{R}^d$ and any subspace W of $\mathbb{R}^d$ with $\dim W < d$, we have $\mu(W + x) < 1$. By $\mathscr{F}(\mathbb{R}^d)$ we denote the set of all full measures on $\mathbb{R}^d$; usually for simplicity, $\mathbb{R}^d$ is omitted. Furthermore, let

$$H(\mu) = \{y \in \mathbb{R}^d : \hat{\mu}(y) = 1\}.$$

It is easy to see that $y \in H(\mu)$ if and only if $\langle y, x \rangle = 0 \pmod{2\pi}$ for μ-a.a. x. Hence, $H(\mu)$ is a subgroup of $\mathbb{R}^d$. The fact that $H(\mu)$ is a subgroup also follows from the inequality $1 - \cos(\theta + \phi) \leq 2(1 - \cos \theta) + 2(1 - \cos \phi)$ for all θ and ϕ. We see that $H(\mu)$ is closed since $\hat{\mu}$ is continuous. The measures $\mu \notin \mathscr{F}$ are referred to as nonfull.

Proposition 2.1.0. *The measure* $\mu = \delta(0)$ *if and only if* $H(\mu)$ *contains an open neighborhood of the origin.*

Note that μ is full if and only if all its translates $\mu * \delta(a)$ are full. Further equivalent conditions are given in the following proposition.

Proposition 2.1.1. *The following statements are equivalent.*

(a) μ *is full.*
(b) μ^0 *is full.*
(c) $H(\mu^0)$ *does not contain any one-dimensional subspace.*
(d) *For each* $y \neq 0$, *the measure* $\Pi_y \mu$ *is nondegenerate on* $\mathbb{R}$ *where* $\Pi_y(x) = \langle x, y \rangle$ *for* $x \in \mathbb{R}^d$.

Proof. (a) $\Rightarrow$ (b) Suppose there is a linear subspace W of $\mathbb{R}^d$ with $\dim W < d$ and $\operatorname{supp} \mu^0 \subseteq W$. Then

$$\int [1 - \mu(W - x)] \mu^-(dx) = 0.$$

Hence, there is $x \in \mathbb{R}^d$ such that $\mu(W - x) = 1$, that is, μ is nonfull.

(b) $\Rightarrow$ (c) Let $H(\mu^0)$ contain a one-dimensional subspace W of $\mathbb{R}^d$. Then for all $y \in W$, $\mu^0(\{x \in \mathbb{R}^d: \langle y, x \rangle = 0 \,(\text{mod}\, 2\pi)\}) = 1$, so that

$$\text{supp}\, \mu^0 \subseteq \{x: \langle y, x \rangle = 0 \,(\text{mod}\, 2\pi) \text{ for all } y \in W\}$$
$$= \{x: \langle y, x \rangle = 0 \text{ for all } y \in W\} = W^\perp,$$

a $(d-1)$-dimensional subspace of $\mathbb{R}^d$, that is, μ^0 is not full.

(c) $\Rightarrow$ (d) Suppose there is $y \neq 0$ such that $\Pi_y \mu = \delta(a)$ for some $a \in \mathbb{R}$. Then

$$\delta(0) = (\Pi_y \mu)^0 = \Pi_y \mu^0$$

and

$$1 = (\Pi_y \mu^0)\hat{}\,(t) = (\mu^0)\hat{}\,(ty) \quad \text{for all } t \in \mathbb{R},$$

that is, $H(\mu^0)$ contains the one-dimensional subspace generated by y.

(d) $\Rightarrow$ (a) Suppose there are an $x \in \mathbb{R}^d$ and a subspace W with $\dim W < d$ and $\text{supp}(\mu * \delta(x)) \subseteq W$. Then for $y \in W$, we have

$$1 = (\Pi_y(\mu * \delta(x)))\hat{}\,(t) = (\Pi_y \mu)\hat{}\,(t) \exp(it\langle y, x \rangle),$$

that is, $\Pi_y \mu = \delta(-x)$. Q.E.D.

Corollary 2.1.2. *The set $\mathscr{F}(\mathbb{R}^d)$ of all full measures on $\mathbb{R}^d$ is an open subsemigroup of $\mathscr{P}(\mathbb{R}^d)$.*

Proof. Let μ_n be nonfull for $n \geq 1$ and $\mu_n \Rightarrow \mu$. Since μ_n is nonfull, for each $n \geq 1$, there is $y_n \in \mathbb{R}^d$ such that $\|y_n\| = 1$ and $\Pi_{y_n} \mu_n$ is degenerate on $\mathbb{R}^1$. Taking a subsequence if necessary, we may assume that $y_n \to$ (some) y. Then $\Pi_{y_n} \mu_n \Rightarrow \Pi_y \mu$ (cf. Corollary 1.7.3), and $\Pi_y \mu$ is degenerate, that is, μ is nonfull. Therefore, $\mathscr{F}$ is an open set.

Suppose there are $\mu, \nu \in \mathscr{F}(\mathbb{R}^d)$ with $\mu * \nu \notin \mathscr{F}(\mathbb{R}^d)$. Proposition 2.1.1 gives that there is $y \neq 0$ such that $|\hat{\mu}(ty)||\hat{\nu}(ty)| = 1$ for all $t \in \mathbb{R}$. Hence, we have $|\hat{\mu}(ty)| = |\hat{\nu}(ty)| = 1$, that is, μ and ν are both nonfull. Thus, $\mu, \nu \in \mathscr{F}$ implies $\mu * \nu \in \mathscr{F}$. Therefore, $\mathscr{F}$ is a subsemigroup of $\mathscr{P}(\mathbb{R}^d)$. Q.E.D.

Corollary 2.1.3. *Let $A \in \text{End}(\mathbb{R}^d)$ and $\mu \in \mathscr{P}(\mathbb{R}^d)$. Then $A\mu$ is full if and only if A is invertible and μ is full.*

Proof. For the sufficiency, suppose $A\mu$ is nonfull. Then for some $y \neq 0$, $\Pi_y(A\mu) = \delta(a)$ for some $a \in \mathbb{R}^d$. Since $A \in \text{Aut}$, $A^*y \neq 0$. But,

$$\delta(a) = \Pi_y(A\mu) = \Pi_{A^*y}(\mu).$$

Therefore, μ is nonfull.

For the necessity, suppose either $A \notin \text{Aut}$ or $\mu \notin \mathscr{F}$. When $A \notin \text{Aut}$, there is $y \neq 0$ such that $A^*y = 0$. Consequently,

$$\delta(0) = \Pi_{A^*y}(\mu) = \Pi_y(A\mu),$$

so $A\mu$ is nonfull. When μ is nonfull, we have $\text{supp}(A\mu) = A(\text{supp }\mu)$ is contained in a proper hyperspace of $\mathbb{R}^d$, so $A\mu$ is nonfull.   Q.E.D.

The preceding corollary may be stated in terms of affine transformations.

Corollary 2.1.4. *Let α be an affine transformation on $\mathbb{R}^d$ and let $\mu \in \mathscr{P}(\mathbb{R}^d)$. Then $\alpha\mu$ is full if and only if α is invertible and μ is full.*

2.2 CONVERGENCE OF TYPES THEOREMS

We denote the affine transformation $\alpha: x \to Ax + a$ by $\langle A; a \rangle$, where $A \in \text{End}(\mathbb{R}^d)$ and $a \in \mathbb{R}^d$. By $\mathscr{A}(\mathbb{R}^d)$ and $\mathscr{A}_I(\mathbb{R}^d)$ we denote the sets of all affine and of all invertible affine transformations on $\mathbb{R}^d$, respectively; briefly, we sometimes write $\mathscr{A}$ or $\mathscr{A}_I$. The transformation $\langle A; a \rangle$ is invertible if and only if A is invertible. With composition as the operation, $\mathscr{A}$ becomes a semigroup and $\mathscr{A}_I$ a group of transformations. Defining for $\alpha = \langle A; a \rangle \in \mathscr{A}$, $\beta = \langle B; b \rangle \in \mathscr{A}$, and $r \in \mathbb{R}$, $\alpha + \beta = \langle A + B; a + b \rangle$, and $r\alpha = \langle rA; ra \rangle$, we obtain that $\mathscr{A}$ is a real linear space. Also, $\mathscr{A}$ becomes a Banach space using the norm $\|\langle A; a \rangle\| = \max\{\|A\|, \|a\|\}$, where $\|A\|$ and $\|a\|$ are some norm on $\text{End}(\mathbb{R}^d)$ and some norm on $\mathbb{R}^d$; all norms on finite-dimensional linear spaces are equivalent. This norm gives $\mathscr{A}$ a topology under which composition is continuous.

If $\mu_n \Rightarrow \mu$ in $\mathscr{P}$ and $\alpha_n \Rightarrow \alpha$ in $\mathscr{A}$, then $\alpha_n\mu_n \Rightarrow \alpha\mu$ in $\mathscr{P}$ (cf. Corollary 1.7.4). Furthermore, since $\mathscr{F}(\mathbb{R}^d)$ is open in $\mathscr{P}(\mathbb{R}^d)$ (cf. Corollary 2.1.2), we obtain from Corollary 2.1.4 the following corollary.

Corollary 2.2.1. If $\alpha_n \mu_n \Rightarrow \mu$ with $\mu_n \in \mathscr{P}$, $\mu \in \mathscr{F}$, and $\alpha_n \in \mathscr{A}$, then $\alpha_n \in \mathscr{A}_I$ and μ_n is full for all sufficiently large n.

In this section, we investigate the following problem. Let $\mu_n \Rightarrow \mu$ and $\alpha_n \mu_n \Rightarrow \nu$. What can one say about $\{\alpha_n\}$, μ, and ν? We begin with the following auxiliary lemmas.

Lemma 2.2.2. Let $\mu_n, \mu \in \mathscr{P}(\mathbb{R}^d)$, $y_n \in \mathbb{R}^d$, and $r_n \in \mathbb{R}$. If $\mu_n \Rightarrow \mu$ with μ full, and if $\{\Pi_{y_n}\mu_n * \delta(r_n)\}$ is tight on $\mathbb{R}$, then

$$\sup\|y_n\| < \infty \quad \text{and} \quad \sup|r_n| < \infty.$$

Proof. Let $s_n = \|y_n\| + |r_n|$. Suppose $\{s_n\}$ is unbounded. Choose a subsequence $\{s_{n'}\}$ tending to infinity. Since the balls are compact in a finite-dimensional linear space, we may choose a further subsequence $\{s_{n''}\}$ such that

$$\frac{y_n}{s_n} \to y_0 \quad \text{and} \quad \frac{r_n}{s_n} \to r_0 \quad \text{with } \|y_0\| + |r_0| = 1, \qquad (2.2.1)$$

suppressing the primes on n for notational convenience. This together with tightness gives

$$\Pi_{y_n/s_n}\mu_n * \delta\!\left(\frac{r_n}{s_n}\right) \Rightarrow \Pi_{y_0}\mu * \delta(r_0). \qquad (2.2.2)$$

On the other hand, since $\{\Pi_{y_n}\mu_n * \delta(r_n)\}$ is tight and $s_n \to \infty$, we infer that

$$\Pi_{y_n/s_n}\mu_n * \delta\!\left(\frac{r_n}{s_n}\right) = T_{1/s_n}\!\big(\Pi_{y_n}\mu_n * \delta(r_n)\big) \Rightarrow \delta(0) \qquad (2.2.3)$$

(cf. Corollary 1.7.3), where T_a is scalar multiplication by a. But, (2.2.2) and (2.2.3) together imply that $\Pi_{y_0}\mu = \delta(-r_0)$. From (2.2.1) we know that y_0 and r_0 cannot both be equal to zero. Thus, $y_0 \neq 0$ and, by Proposition 2.1.1, μ is nonfull. This contradiction shows that $\{s_n\}$ is bounded. Q.E.D.

Lemma 2.2.3. If $\mu_n \Rightarrow \mu$ with μ full and if $\{\alpha_n \mu_n\}$ is tight, where $\alpha_n \in \mathscr{A}$, then $\sup\|\alpha_n\| < \infty$, that is, $\{\alpha_n\}$ is conditionally compact in $\mathscr{A}$.

Proof. Let $\alpha_n = \langle A_n; a_n \rangle$. Then, for $x, y \in \mathbb{R}^d$,

$$\Pi_y(\alpha_n x) = \Pi_{A_n^* y}(x) + \Pi_y(a_n),$$

so

$$\Pi_y(\alpha_n \mu_n) = \Pi_{A_n^* y} \mu_n * \delta(\Pi_y a_n).$$

Since $\{\alpha_n \mu_n\}$ is tight, for each $\varepsilon > 0$, there is a compact set K_ε such that for all n, $\alpha_n \mu_n(K_\varepsilon) > 1 - \varepsilon$. Since Π_y is continuous, $\Pi_y(K_\varepsilon)$ is compact. Note that $\Pi_y^{-1}(\Pi_y(K_\varepsilon)) \supseteq K_\varepsilon$. Thus,

$$(\Pi_y \alpha_n \mu_n)(\Pi_y K_\varepsilon) \geq \alpha_n \mu_n(K_\varepsilon) > 1 - \varepsilon \quad \text{for all } n.$$

Therefore, $\{\Pi_y(\alpha_n \mu_n)\}$ is tight. By Lemma 2.2.2, $\sup \|A_n^* y\| < \infty$ and $\sup \|\Pi_y a_n\| < \infty$ for all $y \in \mathbb{R}^d$. Hence, $\sup \|A_n\| < \infty$ and $\sup \|a_n\| < \infty$, that is, $\sup \|\alpha_n\| < \infty$. Q.E.D.

Corollary 2.2.4. *Let μ be full and $\alpha_n \in \mathcal{A}$. Then $\{\alpha_n \mu\}$ is conditionally compact in $\mathcal{P}$ if and only if $\{\alpha_n\}$ is conditionally compact in $\mathcal{A}$.*

In the convergence of types theorems, a fundamental role is played by the set of operators having the property that the limit measure μ is unchanged by the action of one of these operators. More formally, we define the *invariant* semigroup of μ, $\text{Inv}(\mu)$, to be

$$\text{Inv}(\mu) = \{\alpha \in \mathcal{A} : \mu = \alpha \mu\}.$$

Note that the identity transformation is always in $\text{Inv}(\mu)$, so $\text{Inv}(\mu)$ is never empty. Also, $\text{Inv}(\mu)$ is clearly closed under composition. The next result gives a more precise description of the properties of $\text{Inv}(\mu)$.

Theorem 2.2.5. *If μ is full, then $\text{Inv}(\mu)$ is a compact subgroup of $\mathcal{A}_1$. Conversely, if μ is nonfull, then $\text{Inv}(\mu)$ is neither a group nor compact.*

Proof. Assume $\alpha \in \text{Inv}(\mu)$ and α is not invertible. Since $\alpha = \langle A; a \rangle$ is not invertible, neither is A. Thus, there is $y \neq 0$ such that $A^* y = 0$. Hence,

$$\Pi_y \mu = \Pi_{A^* y} \mu * \delta(\langle a, y \rangle) = \delta(\langle a, y \rangle),$$

so by Proposition 2.1.1, μ is nonfull. Therefore, μ is full implies

$\text{Inv}(\mu) \subseteq \mathscr{A}_I$. Clearly, $\alpha^{-1} \in \text{Inv}(\mu)$ for each $\alpha \in \text{Inv}(\mu)$. This establishes that $\text{Inv}(\mu)$ is a subgroup of $\mathscr{A}_I$. To show that $\text{Inv}(\mu)$ is compact, it suffices to show that $\text{Inv}(\mu)$ is closed and bounded, since we are in a finite-dimensional space. The closedness follows from the fact that $\alpha_n \in \text{Inv}(\mu)$ with $\alpha_n \to \alpha$ implies $\mu = \alpha_n \mu \Rightarrow \alpha \mu$, so $\alpha \in \text{Inv}(\mu)$. To show $\text{Inv}(\mu)$ is bounded, let $\{\alpha_n\}$ be an arbitrary sequence in $\text{Inv}(\mu)$. Then $\alpha_n \mu = \mu$ for all n, so by Lemma 2.2.3, we have $\sup\|\alpha_n\| < \infty$. Thus, $\text{Inv}(\mu)$ is bounded. This establishes the first part of the theorem.

Now, assume μ is nonfull. Without loss of generality, we may assume, and do, that μ is concentrated on a proper subspace W of $\mathbb{R}^d$. Let P be the orthogonal projection onto $W^\perp$. For each $c \in \mathbb{R}$ set $A_c = I - P + cP$. Then $A_c W = W$, so $A_c \mu = \mu$, that is, if $\alpha_c = \langle A_c; 0 \rangle$, then $\alpha_c \in \text{Inv}(\mu)$. Since $\|\alpha_c\| = \|I - P\| + |c|\|P\| \geq |c|$, we see that $\text{Inv}(\mu)$ is unbounded. Hence, $\text{Inv}(\mu)$ cannot be compact. Also, since $I - P \in \text{Inv}(\mu)$ and is not invertible, $\text{Inv}(\mu)$ is not a group.
Q.E.D.

Lemma 2.2.6. *Let $\mu \in \mathscr{P}$ and $\alpha \in \mathscr{A}_I$. Then*

$$\text{Inv}(\alpha \mu) = \alpha(\text{Inv}(\mu))\alpha^{-1}.$$

Proof. The following sequence of statements are equivalent and establish the lemma. $\beta \in \text{Inv}(\alpha \mu)$ if and only if $\alpha \mu = \beta(\alpha \mu)$ if and only if $\mu = (\alpha^{-1} \beta \alpha)\mu$ if and only if $\alpha^{-1} \beta \alpha \in \text{Inv}(\mu)$ if and only if $\beta \in \alpha(\text{Inv}(\mu))\alpha^{-1}$.
Q.E.D.

We say that two measures μ and ν are of the *same operator type* provided there is $\alpha \in \mathscr{A}$ such that $\mu = \alpha \nu$. For full measures, "being of the same operator type" is an equivalence relation. Of importance is the relationship between two measures μ and ν when $\beta_n \mu_n \Rightarrow \mu$ and $\alpha_n \mu_n \Rightarrow \nu$; note the same sequence $\{\mu_n\}$. The convergence of types theorem states that for full measures μ and ν, it is necessary and sufficient that they be of the same operator type. Furthermore, it gives a relationship between the norming sequence $\{\beta_n\}$ and $\{\alpha_n\}$ which involves the invariant group of μ. Preliminary to the convergence of types theorem, we deal with the following special case.

Theorem 2.2.7. *Assume that $\mu_n \Rightarrow \mu$ with μ full. Then in order that $\alpha_n \mu_n \Rightarrow \mu$, it is necessary and sufficient that α_n have the form $\alpha_n = \eta_n \gamma_n$ for sufficiently large n, where $\eta_n \Rightarrow \eta_0 = \langle I; 0 \rangle$ and $\gamma_n \in \text{Inv}(\mu)$.*

Proof. For the sufficiency, by Theorem 2.2.5, $\{\gamma_n\}$ is precompact in $\mathscr{A}$ and all of its limit points are in $\mathbf{Inv}(\mu)$. Since $\eta_n \to \eta_0$, $\{\alpha_n\}$ is also precompact in $\mathscr{A}$ and all of its limit points are in $\mathbf{Inv}(\mu)$. Since $\mu_n \Rightarrow \mu$ and any subsequence of $\{\alpha_n\}$ contains a further subsequence converging to some $\alpha \in \mathbf{Inv}(\mu)$, we see that $\alpha_n \mu_n \Rightarrow \mu$.

For the necessity, by Lemma 2.2.3, we have that $\{\alpha_n\}$ is conditionally compact in $\mathscr{A}$. Let F denote the set of all limit points of the sequence $\{\alpha_n\}$. We now show that $F \subseteq \mathbf{Inv}(\mu)$. Let $\alpha \in F$. Then $\alpha_{n'} \to \alpha$ for some subsequence $\{\alpha_{n'}\}$ of $\{\alpha_n\}$. Thus, $\alpha_{n'}\mu_{n'} \Rightarrow \alpha\mu$ and $\alpha_{n'}\mu_{n'} \Rightarrow \mu$, so $\mu = \alpha\mu$, that is, $\alpha \in \mathbf{Inv}(\mu)$. Since μ is full, each $\alpha \in F$ is invertible (cf. Corollary 2.1.4). Since F is a closed subset of a compact set, F is compact. Therefore, for each $n \geq 1$, there is a $\gamma_n \in F$ such that $\min\{\|\alpha_n - \gamma\|: \gamma \in F\} = \|\alpha_n - \gamma_n\|$. Note that $\|\alpha_n - \gamma_n\| \to 0$. Setting $\eta_n = \alpha_n \gamma_n^{-1}$, we have $\alpha_n = \eta_n \gamma_n$ and

$$0 \leq \|\eta_n - \eta_0\| \leq \|\alpha_n - \gamma_n\| \|\gamma_n^{-1}\|$$
$$\leq \|\alpha_n - \gamma_n\| \sup\{\|\gamma\|: \gamma \in \mathbf{Inv}(\mu)\} \to 0.$$

Therefore, $\eta_n \to \eta_0$. Q.E.D.

Remark 2.2.8. In the preceding theorem, it is possible to interchange the order of η_n and γ_n, perhaps with new η_n and γ_n.

Lemma 2.2.9. *If $\mu_n \Rightarrow \mu$ and $\alpha_n \mu_n \Rightarrow \nu$, with $\alpha_n \in \mathscr{A}$ and μ and ν full, then there is $\alpha \in \mathscr{A}_I$, a limit point of $\{\alpha_n\}$, such that $\alpha\mu = \nu$.*

Proof. Since $\alpha_n \mu_n \Rightarrow \nu$, $\{\alpha_n \mu_n\}$ is tight. Thus, $\{\alpha_n\}$ is conditionally compact (cf. Lemma 2.2.3). Let $\{\alpha_{n'}\}$ be a subsequence of $\{\alpha_n\}$ such that $\alpha_{n'} \to$ (some) $\alpha \in \mathscr{A}$. Then $\alpha_{n'}\mu_{n'} \Rightarrow \alpha\mu$ and $\alpha_{n'}\mu_{n'} \Rightarrow \nu$ so $\alpha\mu = \nu$. Since μ and ν are full, by Corollary 2.1.4 we have $\alpha \in \mathscr{A}_I$. Q.E.D.

We are now ready to state and prove the main convergence of operator types theorem.

Theorem 2.2.10. *Assume that $\beta_n \mu_n \Rightarrow \mu$, where $\beta_n \in \mathscr{A}$, $\mu_n \in \mathscr{P}$, and μ full. In order that $\alpha_n \mu_n \Rightarrow \nu$, with $\alpha_n \in \mathscr{A}$ and ν full, it is necessary and sufficient that $\nu = \alpha\mu$ for some $\alpha \in \mathscr{A}_I$, that is, μ and ν are of the same operator type, and, for all sufficiently large n,*

$$\alpha_n = \alpha\eta_n\gamma_n\beta_n,$$

where $\eta_n \to \eta_0 = \langle I; 0 \rangle$ and $\gamma_n \in \mathbf{Inv}(\mu)$.

Proof. The sufficiency follows by an argument similar to the proof of the sufficiency of Theorem 2.2.7. To prove the necessity, note that β_n is invertible for all sufficiently large n, by Corollary 2.1.4. Let $\rho_n = \beta_n \mu_n$, so $\rho_n \Rightarrow \mu$ and $\alpha_n \beta_n^{-1} \rho_n \Rightarrow \nu$. By Lemma 2.2.9, there is $\alpha \in \mathscr{A}_I$ such that $\alpha \mu = \nu$. Since $\alpha^{-1} \alpha_n \beta_n^{-1} \rho_n \Rightarrow \mu$ and $\rho_n \Rightarrow \mu$, by Theorem 2.2.7, we have that $\alpha^{-1} \alpha_n \beta_n^{-1} = \eta_n \gamma_n$, with $\eta_n \to \eta_0$ and $\gamma_n \in \text{Inv}(\mu)$. Q.E.D.

2.3 URBANIK DECOMPOSABILITY SEMIGROUP

Let $\mu \in \mathscr{P}$ and $A \in \text{End}$. We say that μ is *A-decomposable*, or A *divides* μ, provided there is $\mu_A \in \mathscr{P}$ such that $\mu = A\mu * \mu_A$. In terms of random elements, this means that if ξ_1 and ξ_2 are independent with common distribution μ, then there is a random element ζ_A, independent of ξ_1 and ξ_2, such that $\xi_1 = A\xi_2 + \zeta_A$ in distribution. Clearly, every μ is I-decomposable and 0-decomposable, where I and 0 are the identity and zero operators, respectively. Let $\mathbf{D}(\mu)$ denote the set of all $A \in \text{End}$ such that A divides μ, that is,

$$\mathbf{D}(\mu) := \{A \in \text{End}: \mu = A\mu * \mu_A \text{ for some } \mu_A \in \mathscr{P}\}.$$

We call $\mathbf{D}(\mu)$ the *Urbanik decomposability semigroup of* μ. It is easy to see that $\mathbf{D}(\mu)$ is a semigroup. Simply note that if $\mu = A\mu * \mu_A$ and $\mu = B\mu * \mu_B$, then $\mu = (AB)\mu * (A\mu_B * \mu_A)$.

An important subset of $\mathbf{D}(\mu)$ is formed by those operators A which divide μ and the measure μ_A is a point-mass measure. This subset is called the *symmetry semigroup of* μ and is denoted by $\mathbf{A}(\mu)$, that is,

$$\mathbf{A}(\mu) := \{A \in \text{End}: \mu = A\mu * \delta(a) \text{ for some vector } a\}.$$

Theorem 2.3.1. *For μ full, the Urbanik decomposability semigroup $\mathbf{D}(\mu)$ is a compact subsemigroup of* $\text{End}(\mathbb{R}^d)$. *For μ nonfull, $\mathbf{D}(\mu)$ is not compact.*

Proof. Assume μ is full and let $A_n \in \mathbf{D}(\mu)$ for each $n \geq 1$. Then there are $\mu_{A_n} \in \mathscr{P}$ such that $\mu = A_n \mu * \mu_{A_n}$. By Theorem 1.7.1, there are $a_n \in \mathbb{R}^d$ such that $\{A_n \mu * \delta(a_n)\}$ and $\{\mu_{A_n} * \delta(-a_n)\}$ are conditionally compact. By Lemma 2.2.3, $\{A_n\}$ is conditionally compact in End. Let A be a limit point of the sequence $\{A_n\}$, so $A_{n'} \to A$ for some subsequence $\{A_{n'}\}$ of $\{A_n\}$. Let ν be a limit point of $\{\mu_{A_{n'}} * \delta(-a_{n'})\}$, so $\mu_{A_{n''}} * \delta(-a_{n''}) \Rightarrow \nu$ for some subsequence $\{A_{n''}\}$ of

$\{A_{n'}\}$. Then $\mu = A_{n''}\mu * \mu_{A_{n''}} \Rightarrow A\mu * \nu$, that is, $A \in \mathbf{D}(\mu)$. This shows that every sequence in $\mathbf{D}(\mu)$ contains a subsequence which converges in $\mathbf{D}(\mu)$, so $\mathbf{D}(\mu)$ is compact.

Now assume μ is nonfull. We may assume μ is concentrated in a proper subspace W. Let J be the orthogonal projection onto $W^{\perp}$. Then for each $c \in \mathbb{R}$, we have $\mathrm{supp}(cJ(\mu)) = cJ(\mathrm{supp}\,\mu) = \{0\}$, so $(cJ)\mu = \delta(0)$. Hence, $\mu = (cJ)\mu * \mu$, that is, $cJ \in \mathbf{D}(\mu)$. Thus, $\mathbf{D}(\mu)$ is unbounded. Q.E.D.

Corollary 2.3.2. *If μ is full, then $\mathbf{A}(\mu)$ is a compact subgroup of $\mathrm{Aut}(\mathbb{R}^d)$. Conversely, if μ is nonfull, then $\mathbf{A}(\mu)$ is neither a group nor compact.*

This follows from Theorem 2.2.5.

Lemma 2.3.3. *Let $\mu \in \mathscr{P}$ and $\alpha := \langle A, a \rangle \in \mathscr{A}_I$. Then*

$$\mathbf{D}(\alpha\mu) = A(\mathbf{D}(\mu))A^{-1} \quad \text{and} \quad \mathbf{A}(\alpha\mu) = A(\mathbf{A}(\mu))A^{-1}.$$

Proof. The second statement is immediate from Lemma 2.2.6. The first statement is obtained by an easy modification of the proof of Lemma 2.2.6. Q.E.D.

By our definition, each $A \in \mathbf{A}(\mu)$ belongs to the Urbanik semigroup $\mathbf{D}(\mu)$. Two conditions on $A \in \mathbf{D}(\mu)$ which imply that $A \in \mathbf{A}(\mu)$ are given in the following two propositions.

Proposition 2.3.4

(a) *If $A, A^{-1} \in \mathbf{D}(\mu)$, then $A \in \mathbf{A}(\mu)$.*
(b) *For any measure μ, $\mathbf{A}_0(\mu) := \mathbf{A}ut \cap \mathbf{A}(\mu)$ is the largest subgroup of $\mathbf{D}(\mu)$.*

Proof

(a) Since $A, A^{-1} \in \mathbf{D}(\mu)$, $\mu = A^{-1}\mu * \nu_1$ and $\mu = A\mu * \nu_2$ for some $\nu_1, \nu_2 \in \mathscr{P}$. Hence $\mu = \mu * A\nu_1 * \nu_2$, so

$$|\hat{\mu}(y)| = |\hat{\mu}(y)| \cdot |(A\nu_1)\hat{\,}(y)| \cdot |\hat{\nu}_2(y)| \le |\hat{\mu}(y)| \cdot |\hat{\nu}_2(y)|$$

for all $y \in \mathbb{R}^d$. Since $|\hat{\mu}(y)| > 0$ in some neighborhood of the origin, $|\hat{\nu}_2(y)| = 1$ on an open set containing the origin. Therefore, by Proposition 2.1.0, ν_2 is a point-mass measure. Thus, $A \in \mathbf{A}(\mu)$.

(b) If $A \in \mathbf{A}_0(\mu)$, then A^{-1} exists and is clearly in $\mathbf{A}_0(\mu)$. Thus, $\mathbf{A}_0(\mu)$ is a subgroup in $\mathbf{D}(\mu)$. Let $\mathbb{G}$ be any subgroup of $\mathbf{D}(\mu)$. Each $A \in \mathbb{G}$ is in $\mathbf{A}_0(\mu)$ by (a). Hence, $\mathbb{G} \subset \mathbf{A}_0(\mu)$. Q.E.D.

Proposition 2.3.5. *Let μ be full. If $A \in \mathbf{D}(\mu)$ with determinant equal to either 1 or -1, then $A \in \mathbf{A}(\mu)$.*

Proof. Let sem$\{A\}$ be the monothetic subsemigroup of $\mathbf{D}(\mu)$ generated by A, that is, the closure of the set $\{A^k : k \geq 1\}$. By Theorem 2.3.1, we see that sem$\{A\}$ is compact and from the Numakura theorem (cf. Corollary 1.1.3), we obtain an idempotent J_0 and a B, both in sem$\{A\}$, such that $J_0 = AB = BA$. Since $|\det A| = 1$, $|\det A^k| = 1$ for all $k \geq 1$. Hence, $|\det J_0| = 1$; actually, $\det J_0 = 1$ since $J_0^2 = J_0$. Thus, J_0 is an invertible idempotent. Hence,

$$I = J_0^{-1} J_0 = J_0^{-1} J_0^2 = (J_0^{-1} J_0) J_0 = J_0,$$

that is, $J_0 = I$. This implies that $A^{-1} = B \in \text{sem}\{A\} \subseteq \mathbf{D}(\mu)$. By Proposition 2.3.4, $A \in \mathbf{A}(\mu)$. Q.E.D.

Later, a special role is played by idempotents in $\mathbf{D}(\mu)$, that is, operators $J \in \mathbf{D}(\mu)$ such that $J^2 = J$. Recall that the idempotents I and 0 are always in $\mathbf{D}(\mu)$.

Theorem 2.3.6. *Let $\mu \in \mathscr{P}(\mathbb{R}^d)$.*

(a) *For every idempotent $J \in \mathbf{D}(\mu)$, we have $\mu = J\mu * (I - J)\mu$. Consequently, $I - J \in \mathbf{D}(\mu)$.*

(b) *Let $J_1, \ldots, J_n \in \mathbf{D}(\mu)$ be idempotents with the property that $J_r J_s = 0$ for $r \neq s$. Then*

$$\mu = J_1 \mu * \cdots * J_n \mu * \left(I - \sum_{k=1}^n J_k\right)\mu.$$

(c) *Let $J_1, \ldots, J_n$ be as in (b). Then for every $A_1, \ldots, A_n \in \mathbf{D}(\mu)$ satisfying the conditions $A_k J_k = J_k A_k$ for $k = 1, \ldots, n$, we have*

$$\sum_{k=1}^n J_k A_k \in \mathbf{D}(\mu).$$

(d) *Again, let $J_1, \ldots, J_n$ be as in (b). Then for every $B_1, \ldots, B_n \in \mathbf{D}(\mu)$, we have*

$$\sum_{k=1}^{n} J_k B_k J_k \in \mathbf{D}(\mu).$$

Proof

(a) Since $J \in \mathbf{D}(\mu)$, $\mu = J\mu * \mu_J$ for some $\mu_J \in \mathcal{P}$. Hence, $J\mu = J^2\mu * J\mu_J = J\mu * J\mu_J$. Consequently, $H(J\mu_J)$ contains an open neighborhood of the origin, so by Proposition 2.1.0, $J\mu_J = \delta(0)$. Therefore, $\mathrm{supp}(\mu_J) \subseteq (I - J)\mathbb{R}^d$, so $(I - J)\mu_J = \mu_J$. Since $(I - J)\mu = (I - J)J\mu * (I - J)\mu_J = \mu_J$, we obtain the desired formula that $\mu = J\mu * (I - J)\mu$.

(b) For $n = 1$, this reduces to (a). Assume (b) holds for some $m \geq 1$, that is,

$$\mu = J_1\mu * \cdots * J_m\mu * \left(I - \sum_{k=1}^{m} J_k\right)\mu,$$

with $J_r J_s = 0$ for $r \neq s$ and each J_k is an idempotent in $\mathbf{D}(\mu)$. Let J_{m+1} be an idempotent in $\mathbf{D}(\mu)$ such that $J_r J_{m+1} = 0$ for $r \leq m$. By (a),

$$\mu = J_{m+1}\mu * (I - J_{m+1})\mu.$$

Apply $I - \sum_{k=1}^{m} J_k$ to this equation to obtain

$$\left(I - \sum_{k=1}^{m} J_k\right)\mu = J_{m+1}\mu * \left(I - \sum_{k=1}^{m+1} J_k\right)\mu.$$

Substituting this result into the induction hypothesis, we obtain (b) for $m + 1$.

(c) For each k, A_k divides μ, so $\mu = A_k\mu * \mu_k$ for some $\mu_k \in \mathcal{P}$. Hence, $J_k\mu = J_k A_k\mu * J_k\mu_k$. From (b),

$$\mu = J_1\mu * \cdots * J_n\mu * \left(I - \sum_{k=1}^{n} J_k\right)\mu.$$

Therefore,
$$\mu = J_1 A_1 \mu * \cdots * J_n A_n \mu * \nu,$$
where
$$\nu = J_1 \mu_1 * \cdots * J_n \mu_n * \left(I - \sum_{k=1}^{n} J_k\right)\mu.$$

Set $B := \sum_{k=1}^{n} J_k A_k$. Then $BJ_k = J_k A_k$ for all k, and $B(I - \sum_{k=1}^{n} J_k) = 0$. Thus,

$$B\mu = BJ_1\mu * \cdots * BJ_n\mu * B\left(I - \sum_{k=1}^{n} J_k\right)\mu = J_1 A_1 \mu * \cdots * J_n A_n \mu.$$

Therefore, $\mu = B\mu * \nu$, so $B \in \mathbf{D}(\mu)$.

(d) For $1 \le k \le n$, set $A_k := J_k B_k J_k$. Since $J_k, B_k \in \mathbf{D}(\mu)$ and $\mathbf{D}(\mu)$ is a semigroup, we have $A_k \in \mathbf{D}(\mu)$ for each k. Since $J_k A_k = A_k = A_k J_k$, by (c), $\sum_{k=1}^{n} J_k B_k J_k = \sum_{k=1}^{n} J_k A_k \in \mathbf{D}(\mu)$.
Q.E.D.

In many situations it is very useful to know whether the characteristic function $\hat{\mu}$ never vanishes. The following proposition gives a sufficient condition in terms of $\mathbf{D}(\mu)$.

Proposition 2.3.7. *If $\mathbf{D}(\mu)$ contains a sequence $\{B_n\}$ such that $B_n \to I$ as $n \to \infty$, for each n, $B_n^k \to 0$ as $k \to \infty$ and the set $\{B_n^{*k}: k \ge 0, n \ge 1\}$ is conditionally compact in $\mathrm{End}(\mathbb{R}^d)$, then $\hat{\mu}$ never vanishes.*

Proof. Let $\mu_n \in \mathscr{P}$ be such that $\mu = B_n \mu * \mu_n$. Since
$$\hat{\mu}(y) = \hat{\mu}(B_n^* y)\hat{\mu}_n(y)$$
and $B_n \to I$, we have $\hat{\mu}_n(y) \to 1$ in some neighborhood of the origin. Hence, $\mu_n \Rightarrow \delta(0)$ since $\{\mu_n\}$ is conditionally compact. Therefore, $\hat{\mu}_n(y) \to 1$ uniformly on compact sets.

Now, suppose to the contrary of the proposition that there is $a \in \mathbb{R}^d$ such that $\hat{\mu}(a) = 0$ and $\hat{\mu}(y) \ne 0$ for all y with $\|y\| < \|a\|$. Let
$$C := \mathrm{closure}\{B_n^{*k} a : k \ge 0, n \ge 1\}.$$

Since C is compact,

$$\limsup_{n \to \infty} \sup_{y \in C} |1 - \hat{\mu}_n(y)| = 0.$$

Hence, we may assume that $\hat{\mu}_n(y) \neq 0$ for all $y \in C$ and for all $n \geq 1$. Thus, if $y \in C$ and $\hat{\mu}(y) = 0$, then $\hat{\mu}(B_n^* y) = 0$, because $\hat{\mu}(y) = \hat{\mu}(B_n^* y)\hat{\mu}_n(y)$. Since $a \in C$ and $\hat{\mu}(a) = 0$, $\hat{\mu}(B_n^* a) = 0$ for all $n \geq 1$. Since $B_n^* a \in C$, $\hat{\mu}(B_n^{*2} a) = 0$ for all $n \geq 1$. Continuing this argument, we see that $\hat{\mu}$ vanishes on $\{B_n^{*k} a : k \geq 0, n \geq 1\}$. By continuity, $\hat{\mu}$ vanishes on all of C. But, $0 \in C$. This contradiction establishes the proposition. Q.E.D.

The Gaussian measures are always of particular interest. We characterize the Urbanik semigroups and give sufficient conditions in terms of $\mathbf{D}(\nu)$ for ν to be Gaussian. Before stating them, we recall that ν is Gaussian, if and only if, whenever $\nu = \nu_1 * \nu_2$, both ν_1 and ν_2 are Gaussian, by the Cramér theorem (cf. Theorem 1.8.18). Moreover, the covariance operator of ν is the sum of the covariance operators of ν_1 and ν_2. Furthermore, if $A \in \text{End}(\mathbb{R}^d)$ and if R is the covariance operator of ν, then $A\nu$ is again Gaussian with covariance operator ARA^*.

Proposition 2.3.8. *Let ν be Gaussian with covariance operator R. Then*

$$\mathbf{D}(\nu) = \{A \in \text{End}(\mathbb{R}^d): R - ARA^* \text{ is positive-semidefinite}\}.$$

Proof. Let $A \in \mathbf{D}(\nu)$. Then there is $\nu_A \in \mathscr{P}$ such that $\nu = A\nu * \nu_A$. Hence, ν_A is Gaussian by the Cramér theorem, and its covariance operator, $R - ARA^*$, is positive-semidefinite.

Conversely, if $A \in \text{End}(\mathbb{R}^d)$ is such that $R - ARA^*$ is positive-semidefinite, then there is a Gaussian measure ν_A with covariance operator $R - ARA^*$ (cf. Theorem 1.8.6). Note that $R - ARA^*$ is symmetric. Consequently, $A \in \mathbf{D}(\nu)$. Q.E.D.

Proposition 2.3.9. *Let $\nu \in \mathscr{P}(\mathbb{R}^d)$. Assume $\mathbf{D}(\nu)$ contains orthogonal projections $J_1, \ldots, J_d, Q_1, Q_2$ satisfying the conditions $J_i J_j = 0$ for all $i \neq j$, $Q_1 Q_2 = 0$, and $Q_k J_i \neq 0$ for all k and i. Then ν is Gaussian.*

Proof. This is a direct application of the Skitovich theorem (cf. Theorem 1.8.19). Q.E.D.

Proposition 2.3.10. *Let μ be a full Gaussian measure on $\mathbb{R}^d$. Then, for some $W \in \mathbf{A}ut(\mathbb{R}^d)$,*

$$\mathbf{A}(\mu) = W^{-1}\mathscr{O}W,$$

where $\mathscr{O}$ is the group of all orthogonal transformations.

Proof. Let $W \in \mathrm{Aut}(\mathbb{R}^d)$ be such that, for some $w \in \mathbb{R}^d$, $W\mu * \delta(w)$ has the property that any orthogonal projection of $W\mu * \delta(w)$ onto any one-dimensional subspace of $\mathbb{R}^d$ is a standard Gaussian measure. Then $\mathbf{A}(W\mu) = \mathscr{O}$. Hence, by Lemma 2.3.3, $W\mathbf{A}(\mu)W^{-1} = \mathscr{O}$, that is, $\mathbf{A}(\mu) = W^{-1}\mathscr{O}W$. Q.E.D.

We conclude this section with some elementary properties of $\mathbf{D}(\mu)$.

Proposition 2.3.11

(a) $\mathbf{D}(\mu) = \mathbf{D}(\mu * \delta(x))$ *for all vectors x.*
(b) $\mathbf{D}(\mu_1) \cap \mathbf{D}(\mu_2) \subseteq \mathbf{D}(\mu_1 * \mu_2)$.
(c) *If $B \in \mathbf{A}(\mu)$, then $\mathbf{D}(B\mu) = \mathbf{D}(\mu)$.*
(d) *For $\alpha > 0$ and $\mu \in \mathrm{ID}$, $\mathbf{A}(\mu^\alpha) = \mathbf{A}(\mu)$ and $\mathbf{D}(\mu^\alpha) = \mathbf{D}(\mu)$.*

Proof

(a) Since $\mu = A\mu * \nu$ if and only if $\mu * \delta(x) = A(\mu * \delta(x)) * \nu * \delta(x - Ax)$ for all vectors x, we see that $\mathbf{D}(\mu) = \mathbf{D}(\mu * \delta(x))$.
(b) If $\mu_1 = A\mu_1 * \nu_1$ and $\mu_2 = A\mu_2 * \nu_2$, then $\mu_1 * \mu_2 = A(\mu_1 * \mu_2) * \nu_1 * \nu_2$, so $\mathbf{D}(\mu_1) \cap \mathbf{D}(\mu_2) \subseteq \mathbf{D}(\mu_1 * \mu_2)$.
(c) If $B \in \mathbf{A}(\mu)$, then $B\mu = \mu * \delta(b)$ for some b. Hence, $\mathbf{D}(B\mu) = \mathbf{D}(\mu * \delta(b)) = \mathbf{D}(\mu)$, by (a).
(d) If $B \in \mathbf{A}(\mu)$, then $\mu^\alpha = B\mu^\alpha * \delta(\alpha b)$ for some b, so $B \in \mathbf{A}(\mu^\alpha)$. The reverse inclusion is done in the same manner. Thus, $\mathbf{A}(\mu^\alpha) = \mathbf{A}(\mu)$. Similarly, $\mathbf{D}(\mu^\alpha) = \mathbf{D}(\mu)$. Q.E.D.

2.4 COMPACT SUBGROUPS OF $\mathscr{A}_I(\mathbb{R}^d)$ AND $\mathrm{Aut}(\mathbb{R}^d)$

With a probability measure μ on $\mathbb{R}^d$ we associated the following sets of mappings: the invariant semigroup $\mathbf{Inv}(\mu)$ of affine transformations; the Urbanik semigroup $\mathbf{D}(\mu)$ of linear operators; and the symmetry semigroup $\mathbf{A}(\mu)$ of linear operators. In case μ is full, we know that

Inv(μ) is a compact subgroup of $\mathscr{A}_I(\mathbb{R}^d)$, $\mathbf{D}(\mu)$ is a compact subsemigroup of $\text{End}(\mathbb{R}^d)$, and $\mathbf{A}(\mu)$ is a compact subgroup of $\text{Aut}(\mathbb{R}^d)$. This suggests the problem of how to characterize those compact subgroups of $\mathscr{A}_I$ or compact subsemigroup of End which are the invariant groups or decomposability semigroups of some probability measures. We are not going to investigate this important problem but only recall a classical theorem on compact subgroups of $\mathscr{A}_I(\mathbb{R}^d)$ which we need later on.

First, we note that not every compact subgroup of $\mathscr{A}_I$ can be the invariant group for some μ. For instance, the compact group of rotations on $\mathbb{R}^d$ cannot be the invariant group for any measure μ. To see this, assume μ has the group of rotations contained in its invariant group. For any $y \in \mathbb{R}^d$ and any one-dimensional subspace W, there is a rotation taking y into W. Hence, $\hat{\mu}$ is a function only of the norm of y. Thus, its invariant group also contains all reflections about any subspace.

Let $\mathscr{O}$ denote the group of real orthogonal operators on $\mathbb{R}^d$. For $A \in \mathscr{O}$, we have $\|A\| = 1$ and $A^{-1} = A^*$, so $\mathscr{O}$ is a compact subgroup of $\text{Aut}(\mathbb{R}^d)$.

Theorem 2.4.1. *Any compact subgroup $\mathscr{G}$ of $\mathscr{A}_I$ has the following form: there exist a closed (hence compact) subgroup $\mathscr{O}_0$ of $\mathscr{O}$, a point $x_0 \in \mathbb{R}^d$, and a symmetric positive-definite operator V such that $\mathscr{G}$ consists of exactly those affine transformations α such that*

$$\alpha x = VWV^{-1}(x - x_0) + x_0, \qquad W \in \mathscr{O}_0.$$

Proof. For $\alpha = \langle A; a \rangle$, set $I(\alpha) = \det A$. From the formulas

$$\alpha^n x = A^n x + A^{n-1}a + \cdots + Aa + a \quad \text{for } n \geq 1$$

and $\alpha^{-1}x = A^{-1}x - A^{-1}a$, we obtain $I(\alpha^n) = (I(\alpha))^n$ and $I(\alpha^{-1}) = (I(\alpha))^{-1}$. Since $I(\ )$ is continuous in the topology of $\mathscr{A}$, it is bounded on the compact group $\mathscr{G}$. It follows that $I(\alpha)$ is either 1 or -1 for any $\alpha \in \mathscr{G}$.

Let L be any bounded open set in $\mathbb{R}^d$ and set

$$M := \bigcup_{\alpha \in \mathscr{G}} \alpha(L).$$

Then M is bounded and open, and $\alpha M = M$ for all α in $\mathscr{G}$. Let

$$x_0 := |M|^{-1} \int_M x\, dx \in \mathbb{R}^d,$$

where $|M|$ denotes the Lebesgue measure of M. Since for $\alpha \in \mathscr{G}$ the Jacobian is $|I(\alpha)| = 1$ and since $\alpha M = M$, we have $\alpha x_0 = x_0$ for all $\alpha \in \mathscr{G}$. Therefore, $\mathscr{G}$ consists of all affine transformations α of the form

$$\alpha(x) = A(x - x_0) + x_0 \quad \text{for } A \in \mathscr{G}_1,$$

where $\mathscr{G}_1$ is a compact subgroup of $\mathrm{Aut}(\mathbb{R}^d)$.

Let L be bounded and open, as before. Then the set

$$N := \bigcup_{A \in \mathscr{G}_1} A^*(L)$$

is bounded and open, and $A^*N = N$ for all $A \in \mathscr{G}_1$. Let U be the bounded linear operator defined by

$$\langle x, Uy \rangle := \int_N \langle x, z \rangle \langle y, z \rangle \, dz.$$

Hence, U is symmetric and $\langle x, Ux \rangle \geq 0$. Moreover, $\langle x, Ux \rangle = 0$ implies $\langle x, z \rangle = 0$ for all z in N, and since N is open, $x = 0$. Thus, U is a symmetric positive-definite operator, so U has a square root V, that is, $V^2 = U$, which is also a symmetric positive-definite operator. Since for $A \in \mathscr{G}_1$ the determinant of A is either plus or minus one and since N is A^*-invariant, we have, for all $A \in \mathscr{G}_1$,

$$\langle Ax, UAy \rangle = \langle VAx, VAy \rangle = \int_N \langle x, u \rangle \langle y, u \rangle \, d(A^{*-1}u) = \langle Vx, Vy \rangle.$$

Setting $x = V^{-1}u$ and $y = V^{-1}v$ for $u, v \in \mathbb{R}^d$, we see that VAV^{-1} is orthogonal for each $A \in \mathscr{G}_1$. Finally, the family $\{VAV^{-1} : A \in \mathscr{G}_1\}$ forms a compact subgroup $\mathscr{O}_0$ of the orthogonal group. Q.E.D.

Corollary 2.4.2. *Any compact subgroup $\mathscr{G}$ of $\mathrm{Aut}(\mathbb{R}^d)$ has the form $\mathscr{G} = V\mathscr{O}_0 V^{-1}$, where $\mathscr{O}_0$ is a closed subgroup of the group of all orthogonal operators on $\mathbb{R}^d$ and V is a symmetric positive-definite operator.*

2.5 EXAMPLES IN INFINITE-DIMENSIONAL SPACES

In this section, H denotes an infinite-dimensional real separable Hilbert space with complete orthonormal basis $\{e_n\}$. Thus, each vector $x \in H$ has a unique representation $x = \Sigma x_n e_n$, where $x_n \in \mathbb{R}^1$ and $\|x\|^2 = \Sigma x_n^2$. More precisely, we have $x_n = \langle x, e_n \rangle$ for $n \geq 1$. Example 2.5.1 shows that the convergence of the characteristic functions does not imply the convergence of the corresponding measures. This is Example 1.7.7, again. Example 2.5.2 shows that Lemma 2.2.3 and Theorem 2.2.7 do not hold in H. Example 2.5.3 shows that the set of all full measures $\mathscr{F}(H)$ is not open in $\mathscr{P}(H)$, that is, Corollary 2.1.2 does not extend to H. Example 2.5.4 shows that $\text{Inv}(\mu)$ need neither be compact nor a group, even when μ is a full Gaussian measure. Example 2.5.5 shows that the convergence of types theorem 2.2.10 does not hold in H.

Example 2.5.1. Let $\mu_n := \delta(e_n)$ and $\mu := \delta(0)$. Then for each $y \in H$, $\hat{\mu}_n(y) = \exp i \langle y, e_n \rangle \to 1 = \hat{\mu}(y)$ as $n \to \infty$. It is easy to see that for point measures $\delta(x_n)$ to converge to $\delta(x)$ it is necessary and sufficient that $x_n \to x$. Here, $\|e_n - e_m\| = \sqrt{2}$ for all $n \neq m$. Hence, $\delta(e_n)$ cannot converge to $\delta(0)$. Therefore, the continuity theorem does not hold in infinite-dimensional spaces.

Example 2.5.2. Let $a_n > 0$ with $\Sigma a_n = 1$, and let $\mu := \Sigma a_n \delta(e_n)$. Then μ is full on H. Let $A_n: H \to H$ be given by

$$A_n x := \sum_{i \neq n} x_i e_i + n x_n e_n,$$

when $x = \Sigma x_i e_i$. Then each A_n is a bounded linear invertible operator with $\|A_n\| = n$. For arbitrary $f: H \to \mathbb{R}^1$ bounded and continuous, we have

$$\int f(x)(A_n \mu)(dx) = \sum_{i \neq n} a_i f(e_i) + a_n f(n e_n) \to \Sigma a_i f(e_i)$$

$$= \int f(x) \mu(dx),$$

so $A_n \mu \Rightarrow \mu$. Since $\|A_n\| = n$, we see that Theorem 2.2.7 and Lemma 2.2.3 do not hold in infinite-dimensional Hilbert spaces.

Example 2.5.3. Let

$$P_n x := \sum_{i=1}^{n} x_i e_i \quad \text{for } x = \sum x_i e_i.$$

Clearly, $\|P_n - I\| \to 0$ as $n \to \infty$, so $P_n \mu \Rightarrow \mu$ for all $\mu \in \mathcal{P}(H)$. Since the $P_n \mu$ are concentrated in finite-dimensional subspaces of H, we see that $\mathcal{F}(H)$ is not open in $\mathcal{P}(H)$ (cf. Corollary 2.1.2).

Example 2.5.4. Define D on H by

$$Dx := \sum k^{-2} x_k e_k.$$

Then D is a positive-definite Hermitian operator with finite trace. Let

$$A_n x := \sum (n+k) k^{-1} x_{n+k} e_k \quad \text{for } n \geq 1.$$

Hence,

$$A_n^* x = \sum_{k=n+1}^{\infty} k(k-n)^{-1} x_{k-n} e_k = \sum_{k=1}^{\infty} \tilde{x}_k e_k,$$

where $\tilde{x}_k$ is 0 for $1 \leq k \leq n$ and is $k(k-n)^{-1} x_{k-n}$ for $k > n$. Consequently, for $x, y \in H$,

$$\langle A_n D A_n^* x, y \rangle = \langle D A_n^* x, A_n^* y \rangle$$
$$= \sum_{k=n+1}^{\infty} (k-n)^{-2} x_{k-n} y_{k-n} = \langle Dx, y \rangle,$$

so $A_n D A_n^* = D$. Let μ be the symmetric Gaussian measure with the operator D for its covariance operator (cf. Theorem 1.8.6). Then μ is full (cf. Theorem 1.8.20), and $\langle A_n; 0 \rangle \in \text{Inv}(\mu)$. But, each A_n is noninvertible and $\|A_n\| = n + 1 \to \infty$. Therefore, $\text{Inv}(\mu)$ is neither compact nor a group although μ is full (cf. Theorem 2.2.5).

Example 2.5.5. Let

$$\nu := \sum a_n \delta(e_n),$$

where $a_n > 0$ and $\Sigma a_n = 1$. Also, let

$$\tilde{e}_n := \frac{e_n}{n} \quad \text{and} \quad \mu := \Sigma a_n \delta(\tilde{e}_n).$$

Let

$$A_n x := \sum_{k=1}^{n} kx_k e_k + \sum_{k=n+1}^{\infty} x_k e_k.$$

Then $\|A_n\| = n$, A_n are invertible, and for bounded continuous f we have

$$\int f(x)(A_n\mu)(dx) = \sum_{k=1}^{\infty} a_k f(k^{-1} A_n e_k)$$

$$= \sum_{k=1}^{n} a_k f(e_k) + \sum_{k=n+1}^{\infty} a_k f(k^{-1} e_k) \to \Sigma a_k f(e_k)$$

$$= \int f(x)\nu(dx).$$

Therefore, $A_n\mu \Rightarrow \nu$, μ and ν are full measures, and $\|A_n\| = n$ (cf. Lemma 2.2.3). Now, we show that there does not exist a sequence $\{B_n\}$ of bounded linear operators on H such that $B_n\mu \Rightarrow \nu$ and $\sup\|B_n\| < \infty$. Thus, in particular, there is no bounded linear operator B such that $B\mu = \nu$. To the contrary, assume $B_n\mu \Rightarrow \nu$ and $\sup\|B_n\| < r$, with r an integer. To obtain a contradiction, we find a bounded continuous f such that $\int f(x)(B_n\mu)(dx)$ does not converge to $\int f(x)\nu(dx)$. Define $f(x) := \min\{1, \|x\|\}$. Then

$$\int f(x)\nu(dx) = \Sigma a_n f(e_n) = 1.$$

On the other hand, for $k > 2r$, we have $f(B_n \tilde{e}_k) = \|B_n \tilde{e}_k\| < 1/2$ for all $n \geq 1$. Hence,

$$\int f(x)(B_n\mu)(dx) \leq \sum_{k=1}^{2r} a_k + \frac{1}{2} \sum_{k=2r+1}^{\infty} a_k < 1,$$

that is, $\int f(x)(B_n\mu)(dx)$ is bounded away from one.

This concludes the examples. All the above examples show that in the case of operator-limit theorems in infinite-dimensional spaces we must add some special assumptions on the norming sequences.

2.6 THE CASE WHEN $d = 1$; COMMENTS

In this section, we consider the case opposite to the one in the previous section. Namely, we assume that $d = 1$ and we work with probabilities on the real line. The idea of fullness reduces to nondegeneracy, that is, μ is degenerate if and only if $\mu = \delta(a)$ for some $a \in \mathbb{R}^1$.

If $\alpha = \langle a; b \rangle$ is an affine transformation on $\mathbb{R}^1$, that is, $\alpha x = ax + b$, and if $\alpha \in \text{Inv}(\mu)$, then

$$|\hat{\mu}(t)| = |\hat{\mu}(ta^n)| \quad \text{for all } t \in \mathbb{R}^1 \text{ and for all } n \geq 1.$$

Hence, when μ is full, $|a| = 1$. Furthermore, letting $T_a x = ax$, we have $\mu = T_{-1}\mu * \delta(b)$ for some $b \in \mathbb{R}^1$ if and only if μ is a translate by $\delta(b/2)$ of a symmetric measure ν, that is, $\mu = \nu * \delta(b/2)$. Hence, either

$$\text{Inv}(\mu) = \{\langle 1; 0 \rangle\} \quad \text{or} \quad \text{Inv}(\mu) = \{\langle 1; 0 \rangle, \langle -1; b \rangle\};$$

note that $\langle -1; b \rangle^2 = \langle 1; 0 \rangle$. Consequently, we have that the symmetry group $\mathbf{A}(\mu)$ of a nondegenerate μ is either $\{1\}$ when μ is not a translate of a symmetric measure or $\{1, -1\}$ when μ is. From this, we now obtain the classical convergence of types theorem on $\mathbb{R}^1$.

Corollary 2.6.1. *If $\mu_n \Rightarrow \mu$ and $T_{a_n}\mu_n * \delta(b_n) \Rightarrow \mu$, where μ is nondegenerate, $a_n > 0$, and $b_n \in \mathbb{R}^1$, then $a_n \to 1$ and $b_n \to 0$.*

Proof. Let $\alpha_n = \langle a_n; b_n \rangle$. By Theorem 2.2.7, we have $\alpha_n = \eta_n \gamma_n$ for some $\eta_n \to \langle 1; 0 \rangle$ and $\gamma_n \in \text{Inv}(\mu)$. Since $a_n > 0$, we see that $\gamma_n = \langle 1; 0 \rangle$ for sufficiently large n. Consequently, $a_n \to 1$ and $b_n \to 0$.
Q.E.D.

Corollary 2.6.2. *If $\mu_n \Rightarrow \mu$ and*

$$T_{a_n}\mu_n * \delta(b_n) \Rightarrow \nu,$$

where μ and ν are nondegenerate, $a_n > 0$ and $b_n \in \mathbb{R}^1$, then there are

$a > 0$ and $b \in \mathbb{R}^1$ such that $a_n \to a$, $b_n \to b$, and

$$\nu = T_a\mu * \delta(b).$$

Proof. From Theorem 2.2.7 and Lemma 2.2.9, we have that $\alpha_n = \langle a_n; b_n \rangle$ is conditionally compact in $\mathscr{A}_1$ and there is $\alpha = \langle a; b \rangle$ with $a > 0$ such that $\nu = \alpha\mu$. Consequently, Theorem 2.2.10 implies $\alpha_n = \alpha\eta_n\gamma_n$, where $\gamma_n \in \mathbf{Inv}(\mu)$ and $\eta_n \to \langle 1; 0 \rangle$. Thus, $\gamma_n = \langle 1; 0 \rangle$ and $\alpha_n \to \alpha$. Q.E.D.

Let us now consider the Urbanik semigroups on $\mathbb{R}$. If for some $a \in \mathbb{R}^1$ and some $\mu_a \in \mathscr{P}(\mathbb{R}^1)$, we have

$$\mu = T_a\mu * \mu_a,$$

then

$$\mu = T_{a_n}\mu * \nu_{a,n},$$

where

$$\nu_{a,n} = T_a\mu_a * T_{(a^2)}\mu_a * \cdots * T_{(a^n)}\mu_a.$$

Hence,

$$\left|\hat{\mu}\left(\frac{t}{a^n}\right)\right| = |\hat{\mu}(t)|\left|\hat{\nu}_{a,n}\left(\frac{t}{a^n}\right)\right|,$$

so if $a > 1$, we see that

$$1 = |\hat{\mu}(t)||\hat{\nu}(t)| \quad \text{for all } t,$$

where ν is the limit of $\nu_{a,n}$ as $n \to \infty$, that is, μ is degenerate. Similarly, if $a < -1$, taking a subsequence we again see that μ must be degenerate. Therefore, for μ nondegenerate, we have

$$\mathbf{D}(\mu) \subseteq [-1, 1].$$

Also, $\mathbf{D}(\mu)$ is a compact semigroup by Theorem 2.3.1.

From Proposition 2.3.8, we see that when ν is Gaussian, we have

$$\mathbf{D}(\nu) = [-1, 1].$$

Further examples of the decomposability semigroups on $\mathbb{R}^1$ are presented in Urbanik (1976). Ilinskii (1978) proved that each compact subsemigroup **D** of $[0, 1]$ with $0, 1 \in$ **D** is a decomposability semigroup of some measure on $\mathbb{R}^1$. Some further results in this area are also given in Niedbalsha-Rajba (1981).

2.7 STANDARD CONVERGENCE OF TYPES

The examples discussed in Section 2.5 show how much operator type convergence theorems depend on the dimension of the underlying space and on fullness of the measures in question. Roughly speaking, when we allow normalization to be by arbitrary affine transformations, then we have to consider only full measures on finite-dimensional linear spaces. In this section, we will impose restrictions on the form of affine mappings (arbitrary lines); on the other hand, we relax conditions on the measures (nondegenerate, i.e., not concentrated at a single vector) and the underlying space (arbitrary Banach space).

Let X be a Banach space. Affine mappings α of the form

$$\alpha(x) := ax + v, \quad a \in \mathbb{R}^1, v, x \in X, \tag{2.7.1}$$

are called *lines* in X. Lines with $a \neq 0$ form a subgroup in **Aff**(X) with composition as the operator. Mappings α given by (2.7.1) will be written as

$$\alpha(x) = T_a x + v. \tag{2.7.2}$$

Measures $\mu, \nu \in \mathscr{P}(X)$ are said to be of the *same type* if there exist $a \in \mathbb{R}^1$ and $v \in X$ such that

$$\mu = T_a \nu * \delta(v), \tag{2.7.3}$$

where $\delta(v)$ is the degenerate measure concentrated at $v \in X$. Let us note that degenerate measures form a closed (in weak convergence topology) subsemigroup of $\mathscr{P}(X)$ (cf. Remark 1.6.7). Also, one usually assumes that $a > 0$ in (2.7.3) in order to obtain the facts below. At the end of this section, we consider arbitrary $a_n \in \mathbb{R}$.

Proposition 2.7.1. *Let X be a Banach space and assume $\mu_n \Rightarrow \mu$ in $\mathscr{P}(X)$, where μ is nondegenerate. If $a_n > 0$, $v_n \in X$, then*

$$T_{a_n}\mu_n * \delta(v_n) \Rightarrow \mu \quad \textit{if and only if } a_n \to 1 \textit{ and } v_n \to 0 \textit{ in } X.$$

Proof. The direct part is a particular case of Corollary 1.7.4. For the converse part, taking symmetrizations we have $\mu^0 \neq \delta(0)$, $\mu_n^0 \Rightarrow \mu^0$, and $\rho_n := T_{a_n}\mu_n^0 \Rightarrow \mu^0$ as $n \to \infty$.

If zero is a limit point of $\{a_n\}$, then Corollary 1.7.4 implies $\mu^0 = \delta(0)$. Since $T_{1/a_n}\rho_n \Rightarrow \mu^0$, infinity cannot be a limit point of $\{a_n\}$. Finally, if c_1 and c_2 are two limit points of $\{a_n\}$, then Corollary 1.7.4 gives $\mu^0 = T_{c_1}\mu^0 = T_{c_2}\mu^0$ and hence

$$\mu^0 = T_{(c_1/c_2)^n}\mu^0, \qquad n = 1, 2, \ldots .$$

Since $\mu^0 \neq \delta(0)$, we have $c_1/c_2 = 1$. Consequently, $a_n \to 1$.

To show that $v_n \to 0$, first observe that $\{v_n\}$ is conditionally compact in X, by Theorem 1.7.1(b). Furthermore, if ω is a limit point of $\{v_n\}$, then we have $\mu * \delta(\omega) = \mu$.

Taking Fourier transforms, we arrive at $\exp i\langle x^*, \omega \rangle = 1$ in some open neighborhood U of zero in X^*. Thus, for $x^* \in U$, there exists an integer k such that $\langle x^*, \omega \rangle = 2\pi k(x^*)$. Since $k(0) = 0$ and $k(\cdot)$ is continuous, we conclude that $k(x^*) = 0$ for $x^* \in U$. Consequently, $\langle x^*, \omega \rangle = 0$ for all $x^* \in X^*$ and therefore $\omega = 0$. This proves that $v_n \to 0$ in X. Q.E.D.

Remark. The above proposition is an analog of Theorem 2.2.7 where one deals with operator types and full measures on $\mathbb{R}^d$. Also, see Example 2.5.1 for infinite-dimensional linear spaces.

Corollary 2.7.2. *Let $\mu_n \Rightarrow \mu$ in $\mathscr{P}(X)$, with μ nondegenerate. Then $T_{c_n}\mu_n * \delta(u_n) \Rightarrow v$ for some $c_n > 0$, $u_n \in X$, with nondegenerate v if and only if $c_n \to c_0 > 0$, $u_n \to u_0$ in X, and $v = T_{c_0}\mu * \delta(u_0)$.*

Proof. First, taking symmetrizations and reasoning as in the above proof of Proposition 2.7.1, we conclude that $c_n \to c_0 > 0$. This implies that $\{u_n\}$ is conditionally compact in X and once again repeating the arguments from the above, we obtain $u_n \to u_0$ in X. Finally, we appeal to Corollary 1.7.4. Q.E.D.

Corollary 2.7.3 (Convergence of Types). *Assume $T_{a_n}\mu_n * \delta(v_n) \Rightarrow \mu$, μ is nondegenerate, and $a_n > 0$, $v_n \in X$. In order for $T_{b_n}\mu_n * \delta(w_n) \Rightarrow v$, with v nondegenerate, it is necessary and sufficient that $b_n a_n^{-1} \to c_0 > 0$, $w_n - b_n a_n^{-1} v_n \to u_0$ in X, and $v = T_{c_0}\mu * \delta(u_0)$.*

Proof. Note that

$$T_{b_n}\mu_n * \delta(u_n) = T_{b_n a_n^{-1}}\left(T_{a_n}\mu_n * \delta(v_n)\right) * \delta(w_n - b_n a_n^{-1} v_n)$$

and apply Corollary 2.7.2. Q.E.D.

Now, let us consider normalization T_a with $a \in \mathbb{R}$, not necessarily positive.

Lemma 2.7.4. *If $\mu_n \Rightarrow \mu$ and $T_{a_n}\mu_n * \delta(u_n) \Rightarrow \mu$ with μ nondegenerate and $a_n \in \mathbb{R}^1$, $u_n \in X$, then $|a_n| \to 1$, $u_{n'} \to 0$ when $n' \in \{n: a_n > 0\}$, and $u_{n''} \to u_0$ when $n'' \in \{n: a_n < 0\}$. Moreover, if -1 is a limit point of $\{a_n\}$, then μ is the translation of a symmetric measure.*

Proof. Taking symmetrizations we have

$$\mu_n^0 \Rightarrow \mu^0 \quad \text{and} \quad T_{a_n}\mu_n^0 \Rightarrow \mu^0 \neq \delta(0). \tag{2.7.4}$$

As in the proof of Proposition 2.7.1, we see that neither 0 nor $\pm\infty$ can be limit points of $\{a_n\}$. Furthermore, if c is a limit point of $\{a_n\}$ and $0 < |c| < \infty$, then (2.7.4) and Corollary 1.7.4 give

$$T_c\mu^0 = \mu^0 \quad \text{and} \quad \mu^0 = T_{c^n}\mu^0 = T_{|c|^n}\mu^0, \quad n \geq 1.$$

Thus, $|c| = 1$ and $|a_n| \to 1$. Hence, $\{T_{a_n}\mu_n\}$ is conditionally compact in $\mathscr{P}(X)$ and, by Theorem 1.7.1, so is $\{u_n\}$ in X.

Let $\{n'\}$ and $\{n''\}$ be subsequences corresponding to the positive and the negative terms of $\{a_n\}$, respectively. Note $a_n \neq 0$ for all sufficiently large n. If $\{n'\}$ is infinite, then, by Proposition 2.7.1, $u_{n'} \to 0$. In the opposite case, when $\{n''\}$ is infinite, suppose w_1 and w_2 are limit points of $\{u_{n''}\}$, then we obtain

$$T_{-1}\mu * \delta(\omega_1) = \mu = T_{-1}\mu * \delta(w_2).$$

Hence, $w_1 = w_2$, that is, $u_{n''} \to u_0$ for some $u_0 \in X$. From the above, we also conclude the last part of the lemma. Q.E.D.

Remark 2.7.5. *If -1 is a limit point of $\{a_n\}$, then $\nu = \mu * \delta(x)$ is symmetric for some $x \in X$. Thus, our setting can be changed to the following one.*

If $\mu_n \Rightarrow \nu$ and $T_{a_n}\mu_n * \delta(w_n) \Rightarrow \nu$, with ν nondegenerate and symmetric, then we have $|a_n| \to 1$ and $w_n \to 0$. Also, the converse is true.

So, we are in the same position as in Proposition 2.7.1 when symmetric nondegenerate measures are limits.

2.8 BIBLIOGRAPHIC COMMENTS

Proposition 2.1.1 is from Sharpe (1969). The material on the convergence of operator types in Section 2.2 is based on Billingsley (1966). Some of these results were also proved in Sharpe (1969) and in Weissman (1976). In fact, the Khintchine theorem was proved by Fisz as early as 1954. All the results in Section 2.3 are due to Urbanik. We collect them from his papers: (1972), (1975), (1976), (1978), and (1979).

Characterization of the compact affine groups on $\mathbb{R}^d$, Section 2.4, is well known. Originally, it is due to Fenchal (1936), but we quote its proof after Billingsley (1966). Examples in Section 2.5 are from Jurek (1981). They show the importance of fullness of the measures and of the finite dimensionality of the linear spaces. Results in Section 2.7 are well known and standard, except Lemma 2.7.4 which seems not to be stated explicitly elsewhere.

CHAPTER 3

Operator-Selfdecomposable Measures

The Urbanik decomposability semigroup $\mathbf{D}(\mu)$ of a measure μ is the set of all linear operators A such that $\mu = A\mu * \nu$ for some measure ν. In Section 3.3, it is shown that for a full operator-selfdecomposable measure μ, $\mathbf{D}(\mu)$ always contains a one-parameter semigroup of the form $\{\exp(-tQ): t \geq 0\}$ for some linear operator Q such that $\exp(-tQ) \to 0$ as $t \to \infty$. In fact, this condition is sufficient for μ to be operator-selfdecomposable. Preliminary to this, some basic properties of norming sequences are obtained in Section 3.2. Section 3.4 deals with the representation of the characteristic function of a measure whose decomposability semigroup contains the one-parameter semigroup $\{\exp(-tQ): t \geq 0\}$. A useful new norm associated with this semigroup is introduced. Section 3.5 is concerned with the question of the generators for the class $L_0(Q)$ of all μ whose semigroup $\mathbf{D}(\mu)$ contains $\{\exp(-tQ): t \geq 0\}$ for a fixed Q. The result obtained is analogous to the characterization of the class of all infinitely divisible laws as the smallest class closed with respect to weak limits which contains the compound Poisson laws. In Section 3.6, a random integral representation is obtained for these $\exp(-tQ)$-decomposable laws, that is, laws in $L_0(Q)$. This random integral representation shows that any measure in $L_0(Q)$ is the limit in distribution of a particular Markov process as time goes to infinity. Section 3.7 deals with the corresponding Markov semigroups. A characterization of their infinitesimal generators is obtained. In Section 3.8, it is shown that all full measures in $L_0(Q)$ are absolutely continuous.

3.1 STATEMENT OF THE PROBLEM

Let $\{\xi_n\}$ be a sequence of independent $\mathbb{R}^d$-valued random variables and let $\{A_n\}$ be a sequence of bounded linear operators on $\mathbb{R}^d$.

Assume that the triangular array $\{A_n \xi_k : 1 \le k \le n, 1 \le n\}$ is infinitesimal. The problem is to characterize the class of all limit distributions of sequences

$$A_n(\xi_1 + \cdots + \xi_n) + a_n, \qquad (3.1.1)$$

where each $a_n \in \mathbb{R}^d$. In terms of affine transformations, (3.1.1) becomes $\langle A_n; a_n \rangle$ applied to the partial sums of $\{\xi_n\}$. Equivalently, (3.1.1) expressed as measures is

$$A_n(\mu_1 * \cdots * \mu_n) * \delta(a_n). \qquad (3.1.2)$$

The limit measures of (3.1.2) are called *operator-selfdecomposable* and $\mathrm{OL}(\mathbb{R}^d)$, briefly OL, denotes the class of all possible limit measures of (3.1.2). The infinitesimal assumption of $\{A_n \mu_k\}$ implies that $\mathrm{OL} \subseteq \mathrm{ID}$, the class of all infinitely divisible measures. Therefore, to characterize OL, it suffices to characterize the Gaussian covariance operators and the Lévy spectral measures corresponding to operator-selfdecomposable measures.

In order to do this, we need certain properties of the sequence $\{A_n\}$. Such a sequence $\{A_n\}$ is called a *norming sequence* for μ if (3.1.2) converges to μ. When the limit measure is full, we have that A_n is invertible for all sufficiently large n (cf. Corollary 2.2.1). Thus, under the assumption of fullness, without loss of generality we assume that each A_n is invertible. These basic properties of norming sequences are used to obtain a one-parameter semigroup of operators contained in the Urbanik decomposability semigroup of an operator-selfdecomposable measure. These in turn lead to the solution of the above-mentioned problem. Further characterizations and results are also derived from these facts.

3.2 NORMING SEQUENCES

The sequence of linear operators $\{A_n\}$ is called a *norming* sequence for the operator-selfdecomposable measure μ provided there exist a sequence $\{\mu_n\}$ in $\mathscr{P}$ and a sequence of vectors $\{a_n\}$ such that the triangular array $\{A_n \mu_k : 1 \le k \le n, n \ge 1\}$ is infinitesimal and $A_n(\mu_1 * \cdots * \mu_n) * \delta(a_n) \Rightarrow \mu$. For μ full, we may assume that every norming sequence is in **Aut**.

Proposition 3.2.1. *For every norming sequence $\{A_n\}$ of a full operator-selfdecomposable measure μ, we have*

$$\lim_{n \to \infty} A_n = 0.$$

Proof. Suppose to the contrary that for some sequence $\{A_n\}$ the proposition fails to hold. Without loss of generality, we assume $A_n(\mu_1 * \cdots * \mu_n) * \delta(a_n) \Rightarrow \mu$ and

$$\lim_{n \to \infty} \|A_n\| > 0.$$

Select vectors z_n so that $\|z_n\| = 1$ and $\|A_n^*\| = \|A_n^* z_n\|$ for all $n \geq 1$. Set

$$u_n := \|A_n^*\|^{-1} A_n^* z_n \quad \text{for } n \geq 1.$$

Then $\|u_n\| = 1$, so without loss of generality we assume

$$u_n \to (\text{some}) \, u \quad \text{with } \|u\| = 1.$$

By the infinitesimal assumption, $A_n \mu_j \Rightarrow \delta(0)$ for each $j \geq 1$. This together with the fact that the sequence $\{\|A_n^*\|^{-1} z_n\}$ is bounded implies that, for all $c \in \mathbb{R}^1$ and all $j \geq 1$,

$$\lim_{n \to \infty} (A_n \mu_j)\hat{\,}\left(c\|A_n^*\|^{-1} z_n\right) = 1.$$

On the other hand, we have

$$(A_n \mu_j)\hat{\,}\left(c\|A_n^*\|^{-1} z_n\right) = \hat{\mu}_j(c u_n) \to \hat{\mu}_j(cu).$$

Thus, for all $c \in \mathbb{R}^1$ and all $j \geq 1$,

$$\hat{\mu}_j(cu) = 1 \quad \text{with } u \neq 0.$$

Hence,

$$(\mu_1 * \cdots * \mu_n)\hat{\,}(cu) = \prod_{j=1}^{n} \hat{\mu}_j(cu) = 1$$

for all $c \in \mathbb{R}^1$ and all $n \geq 1$, with $u \neq 0$. This implies that μ is nonfull (cf. Corollaries 2.1.2 and 2.1.3). This contradiction establishes the proposition. Q.E.D.

Proposition 3.2.2. *For every full operator-selfdecomposable measure μ, there is a norming sequence $\{A_n\}$ such that*

$$\lim_{n \to \infty} A_{n+1} A_n^{-1} = I.$$

Proof. Let $\rho_n := B_n \nu_n * \delta(a_n)$ with $\nu_n := \mu_1 * \cdots * \mu_n$ and assume $\rho_n \Rightarrow \mu$ and $\{B_n \mu_k : 1 \le k \le n, n \ge 1\}$ is infinitesimal. Since

$$\rho_{n+1} = B_{n+1} B_n^{-1}(B_n \nu_n * \delta(a_n)) * B_{n+1} \mu_{n+1} * \delta(c_n)$$

with

$$c_n := a_{n+1} - B_{n+1} B_n^{-1} a_n$$

and since $B_{n+1} \mu_{n+1} \Rightarrow \delta(0)$, we have

$$B_{n+1} B_n^{-1} \rho_n * \delta(c_n) \Rightarrow \mu. \qquad (3.2.1)$$

Consequently, by Theorem 2.2.6, $\{B_{n+1} B_n^{-1}\}$ is conditionally compact in **Aut** and the set **T** of all its limit points is contained in the compact symmetry group $\mathbf{A}(\mu)$. For each $n \ge 1$, choose $F_n \in \mathbf{T}$ such that

$$\varepsilon_n = \|F_n - B_{n+1} B_n^{-1}\| = \min\{\|F - B_{n+1} B_n^{-1}\| : F \in \mathbf{T}\}.$$

Obviously, $\varepsilon_n \to 0$. Set $H_1 := I$ and for $n \ge 2$, set

$$H_n := F_1^{-1} \cdots F_{n-1}^{-1}.$$

Then each $H_n \in \mathbf{A}(\mu)$. Now, set

$$A_n := H_n B_n.$$

Using the compactness of $\mathbf{A}(\mu)$ and the usual subsequence argument (cf. Corollary 1.6.3), we see that the sequence $A_n \nu_n * \delta(H_n a_n)$ is conditionally compact and all of its limit points are of the form $\mu * \delta(b)$ for some b. Hence, there is a sequence b_n such that

$$A_n \nu_n * \delta(b_n) \Rightarrow \mu.$$

Therefore, $\{A_n\}$ is a norming sequence for μ. Furthermore,

$$\|A_{n+1}A_n^{-1} - I\| \le \|H_n\|\|H_{n+1}^{-1}\|\varepsilon_n \le \varepsilon_n \sup{}^2\{\|A\|: A \in \mathbf{A}(\mu)\}.$$

Since $\varepsilon_n \to 0$ and the supremum is finite, we have $A_{n+1}A_n^{-1} \to I$.
Q.E.D.

Proposition 3.2.3. *Let $\{m_k\}$ and $\{n_k\}$ be sequences of positive integers with $m_k \le n_k$ and $m_k \to \infty$. For every norming sequence $\{A_n\}$ for a full operator-selfdecomposable μ, the sequence $\{A_{n_k}A_{m_k}^{-1}\}$ is conditionally compact in* **End**. *Moreover, all of its limit points belong to* $\mathbf{D}(\mu)$.

Proof. Let $\rho_n := A_n \nu_n * \delta(a_n) \Rightarrow \mu$, where $\nu_n := \mu_1 * \cdots * \mu_n$ and the triangular array $\{A_n\mu_k : 1 \le k \le n, n \ge 1\}$ is infinitesimal. Setting

$$\omega_k := A_{n_k}(\mu_{m_k+1} * \cdots * \mu_{n_k}) * \delta\left(a_{n_k} - A_{n_k}A_{m_k}^{-1}a_{m_k}\right),$$

we have

$$\rho_{n_k} = A_{n_k}A_{m_k}^{-1}\rho_{m_k} * \omega_k. \tag{3.2.2}$$

By Theorem 1.7.1, $\{A_{n_k}A_{m_k}^{-1}\rho_{m_k}^0\}$ is conditionally compact in **End**. Since $\rho_{m_k}^0 \Rightarrow \mu^0$, by Lemma 2.2.3, we have that $\{A_{n_k}A_{m_k}^{-1}\}$ is conditionally compact in **End**.

Let $A \in $ **End** be a limit point of $\{A_{n_k}A_{m_k}^{-1}\}$. Without loss of generality, we assume $A_{n_k}A_{m_k}^{-1} \to A$. Then

$$\rho_{n_k} \Rightarrow \mu \quad \text{and} \quad A_{n_k}A_{m_k}^{-1}\rho_{m_k} \Rightarrow A\mu. \tag{3.2.3}$$

By Theorem 1.7.1, (3.2.2) and (3.2.3) imply that $\{\omega_k\}$ is conditionally compact. Taking a subsequence if necessary, we may assume $\omega_k \Rightarrow$ (some)ω. Therefore, by this convergence and (3.2.2) and (3.2.3), we have $\mu = A\mu * \omega$, that is, $A \in \mathbf{D}(\mu)$.
Q.E.D.

Earlier, we used sem$\{\mathbf{F}\}$ to denote the smallest closed multiplicative subsemigroup of operators containing **F**, where **F** is a subset of **End**. From Proposition 3.2.3 we get the following corollary.

Corollary 3.2.4. *If $\{A_n\}$ is a norming sequence for a full operator-selfdecomposable measure μ, then* sem$\{A_nA_m^{-1}: 1 \le m \le n, n \ge 1\}$ *is compact in* **End**.

Proposition 3.2.5. *For every norming sequence $\{A_n\}$ of a full operator-selfdecomposable measure μ, we have that*

$$\mathbf{A}(\mu) \cap \mathrm{sem}\{A_n A_m^{-1}: 1 \le m \le n, n \ge 1\}$$

is a compact group containing all the limit points of the sequence $\{A_{n+1}A_n^{-1}\}$.

Proof. Let $\mathbb{B}$ denote the set of interest. The compactness of $\mathbb{B}$ is an immediate consequence of Corollaries 3.2.4 and 2.3.2. From the proof of Proposition 3.2.2 and Theorem 2.2.6, we have that $A_{n+1}A_n^{-1} = D_n C_n$, where $D_n \to I$ and each $C_n \in \mathbf{A}(\mu)$. Hence, the limit points of $\{A_{n+1}A_n^{-1}\}$ are contained in $\mathbb{B}$.

Next, we show that $\mathbb{B}$ is a group. It suffices to show that each $B \in \mathbb{B}$ has an inverse in $\mathbb{B}$. Consider a $B \in \mathbb{B}$ and its compact semigroup sem$\{B\}$. By the Numakura theorem (cf. Corollary 1.1.3), the set G of all limit points of the sequence $\{B^n\}$ forms a group and it is the minimal ideal of sem$\{B\}$. Moreover, sem$\{B\}$ contains exactly one idempotent J, namely the unit of G. Hence, there is $C \in G \subseteq \mathbb{B}$ such that

$$CB = BC = J.$$

Since $J \in \mathbb{B} \subseteq \mathbf{A}(\mu)$, $\mu = J\mu * \delta(x)$ for some vector x. But, μ is full. Therefore, J is invertible. Since J is idempotent, it must be the identity. Therefore, $C = B^{-1}$, so $B^{-1} \in \mathbb{B}$. Q.E.D.

3.3 THE URBANIK SEMIGROUP OF AN OPERATOR-SELFDECOMPOSABLE MEASURE AND ITS EXPONENTS

The main result in this section is that the Urbanik decomposability semigroup of a full operator-selfdecomposable measure always contains a one-parameter semigroup of the form $\exp(-tQ)$, $t \ge 0$, with Q invertible and $\exp(-tQ) \to 0$ as $t \to \infty$. By Theorem 2.3.1, $\mathbf{D}(\mu)$ is a compact subsemigroup of **End** whenever μ is full. The nonzero idempotents in $\mathbf{D}(\mu)$ play an important role in the theory. The two idempotents, I and 0, are always in $\mathbf{D}(\mu)$. For a given idempotent

$J \in \mathbf{D}(\mu)$, we define $\mathbf{D}_J(\mu)$ and $\mathbf{A}_J(\mu)$ as follows:

$$\mathbf{D}_J(\mu) := \{A \in \mathbf{D}(\mu): AJ = JA = A\},$$
$$\mathbf{A}_J(\mu) := \{A \in \mathbf{D}_J(\mu): J\mu = A\mu * \delta(x) \text{ for some vector } x\}.$$

Clearly, $\mathbf{D}_J(\mu)$ and $\mathbf{A}_J(\mu)$ are closed, hence compact, subsemigroups of $\mathbf{D}(\mu)$ and $\mathbf{D}_J(\mu)$, respectively. For $\mathbf{A}_J(\mu)$ we have even more.

Lemma 3.3.1. *Let μ be full and let J be an idempotent in $\mathbf{D}(\mu)$. Then $\mathbf{A}_J(\mu)$ is a compact group with the unit J.*

Proof. By the definition of $\mathbf{D}_J(\mu)$, J is a unit in $\mathbf{A}_J(\mu)$. Let $A \in \mathbf{A}_J(\mu)$. Then sem$\{A\}$ is a compact semigroup and, from the Numakura theorem, we see that there are an idempotent J_1 and an operator B, both in sem$\{A\}$, such that

$$AB = BA = J_1.$$

Since J_1 is in $\mathbf{A}_J(\mu)$, we have $J_1 J = JJ_1 = J_1$ and $J\mu = J_1\mu * \delta(x)$ for some vector x. Since μ is full, the images of J and J_1 are the same. Consequently, $J = J_1$ and we have $AB = BA = J$, that is, B is an inverse of A in $\mathbf{A}_J(\epsilon)$. Thus, $\mathbf{A}_J(\epsilon)$ is a group with unit J.   Q.E.D.

A result which is analogous to Proposition 2.3.4 is the following lemma.

Lemma 3.3.2. *Let μ be full and let J be an idempotent in $\mathbf{D}(\mu)$. If $A \in \mathbf{D}_J(\mu)$ and $J \in$ sem$\{A\}$, then $A \in \mathbf{A}_J(\mu)$.*

Proof. Let $k_1 \le k_2 \le \cdots$ be positive integers such that $A^{k_n} \to J$. We may assume that $k_n \ge 2$ and that the sequence $\{A^{k_n - 1}\}$ converges to some operator B in sem$\{A\}$. Then we have

$$J = AB, \quad \mu = A\mu * \mu_A, \quad \text{and} \quad \mu = B\mu * \mu_B$$

for some measures μ_A and μ_B. Consequently, $A\mu = J\mu * A\mu_B$, so $\mu = J\mu * A\mu_B * \mu_A$. Since $JA = A$, the last equality implies

$$J\mu = J\mu * A\mu_B * J\mu_A.$$

Hence, $|(J\mu_A)\hat{\ }(y)| = 1$ for all y in some open neighborhood of the origin, so $J\mu_A = \delta(x)$ for some vector x; note that $H((J\mu_A)^0)$ contains

an open set containing the origin (cf. Proposition 2.1.1). Finally, the equation $\mu = A\mu * \mu_A$ gives $J\mu = A\mu * \delta(x)$, that is, $A \in \mathbf{A}_J(\mu)$.
Q.E.D.

Now we are ready for the main steps in the construction of the uniformly continuous one-parameter semigroup of the linear operators belonging to $\mathbf{D}(\mu)$. Since the construction is rather long, we divide it into three steps in the following lemma.

Lemma 3.3.3. *Let μ be full and operator-selfdecomposable and let J be a nonzero idempotent in $\mathbf{D}(\mu)$.*

   (i) *For any c in $(0, 1)$, there exists an operator B_c in $\mathbf{D}_J(\mu) \setminus \mathbf{A}_J(\mu)$ such that*

   $$c = \|J - B_c\| = \min\{\|J - GB_c\|: G \in \mathbf{A}_J(\mu)\}.$$

   (ii) *There are a constant $a > 0$ and a subset $\{E_d: 0 < d < a\}$ of $\mathbf{D}_J(\mu) \setminus \mathbf{A}_J(\mu)$ such that*

   $$d = \min\{\|J - GE_d\|: G \in \mathbf{A}_J(\mu)\}$$

   and each E_d is the limit of a sequence of the form $\{B_{c_n}^{m_n}\}$, where $0 < c_n \to 0$, the positive integers $m_n \to \infty$, and B_{c_n} comes from part (i) of this lemma.

   (iii) *The semigroup $\mathbf{D}_J(\mu)$ contains a one-parameter semigroup $\{J\exp(tV): t \geq 0\}$ with $V \in \mathbf{End}$ such that $JV = VJ = V$. Moreover, $\mathbf{D}_J(\mu)$ contains an idempotent J_1, possibly 0, $J_1 \neq J$, $J_1V = VJ_1$, and $\lim_{t \to \infty}(J - J_1)\exp(tV) = 0$.*

Proof

   (i) Let $\{A_n\}$ be a fixed norming sequence for μ such that $A_{n+1}A_n^{-1} \to I$ as $n \to \infty$ (cf. Proposition 3.2.2). For integers $m \geq n \geq 1$, let

   $$a_{m,n} := \min\{\|J - GA_m A_n^{-1} J\|: G \in \mathbf{A}_J(\mu)\}.$$

   Since $\mathbf{A}_J(\mu)$ is compact by Lemma 3.3.1, $a_{m,n}$ are well defined. Since $J \in \mathbf{A}_J(\mu)$, we have

   $$a_{n,n} = 0 \quad \text{for all } n \geq 1. \tag{3.3.1}$$

The compactness of $\mathbf{A}_J(\mu)$ coupled with the fact that $A_m \to 0$ as $m \to \infty$, by Proposition 3.2.1, gives

$$\lim_{m \to \infty} a_{m,n} = \|J\| \geq 1 \quad \text{for all } n \geq 1. \tag{3.3.2}$$

Furthermore, the compactness of $\mathbf{A}_J(\mu)$ and of $\text{sem}\{A_m A_n^{-1} : 1 \leq n \leq m\}$, by Corollary 3.2.4, implies that, for $n \leq m$,

$$a_{m+1,n} \leq \min\{\|J - GA_m A_n^{-1}J\|$$
$$+ \|G(A_{m+1}A_m^{-1} - I)A_m A_n^{-1}J\| : G \in \mathbf{A}_J(\mu)\}$$
$$\leq a_{m,n} + b^3 \|A_{m+1}A_m^{-1} - I\|,$$

where b is an upper bound for the norms of the operators in these compact sets. Similarly, for $n \leq m$,

$$a_{m,n} \leq a_{m+1,n} + b^3 \|A_{m+1}A_m^{-1} - I\|.$$

Consequently, for positive integers $1 \leq n_m \leq m$, we have

$$\lim_{m \to \infty} |a_{m+1, n_m} - a_{m, n_m}| = 0. \tag{3.3.3}$$

From (3.3.1) and (3.3.2), we see that given any c in $(0, 1)$, we can find integers $1 \leq n \leq m_n$ such that

$$a_{m_n, n} < c \leq a_{m_n + 1, n}.$$

By (3.3.3), this implies that

$$\lim_{n \to \infty} a_{m_n, n} = c. \tag{3.3.4}$$

Let A_c be a limit point of $\{A_{m_n} A_n^{-1}\}$. By Proposition 3.2.3, A_c is in $\mathbf{D}(\mu)$. By (3.3.4),

$$c = \min\{\|J - GA_c J\| : G \in \mathbf{A}_J(\mu)\}.$$

From the compactness of $\mathbf{A}_J(\mu)$, we know that, for some D_c in $\mathbf{A}_J(\mu)$,

$$c = \|J - D_c A_c J\|.$$

Set $B_c := D_c A_c J$. Then B_c is in $\mathbf{D}(\mu)$ and $JB_c = B_c J = B_c$ since $D_c \in \mathbf{D}_J(\mu)$. Thus, $B_c \in \mathbf{D}_J(\mu)$. Since $\mathbf{A}_J(\mu)$ is a group and $D_c \in \mathbf{A}_J(\mu)$, we see that

$$c = \min\{\|J - GD_c A_c J\|: G \in \mathbf{A}_J(\mu)\}$$
$$= \min\{\|J - GB_c\|: G \in \mathbf{A}_J(\mu)\}.$$

Since $c > 0$ we have that B_c is not in $\mathbf{A}_J(\mu)$. This establishes part (i) of the lemma.

(ii) Let B_c be as in part (i) and for $n \geq 1$ set

$$b_{n,c} := \min\{\|J - GB_c^n\|: G \in \mathbf{A}_J(\mu)\}.$$

Then $b_{1,c} = c$. Consider the compact semigroup sem$\{B_c\}$. From the Numakura theorem, there is an idempotent J_c in sem$\{B_c\}$. Furthermore,

$$\limsup_{n \to \infty} b_{n,c} \geq \min\{\|J - GJ_c\|: G \in \mathbf{A}_J(\mu)\}$$

and J_c is in $\mathbf{D}_J(\mu)$. Since $JJ_c = J_c$ and J and J_c commute, $J - J_c$ is also idempotent. Moreover, $J \neq J_c$; otherwise, Lemma 3.3.2 implies B_c is in $\mathbf{A}_J(\mu)$, contradicting part (i). Consequently,

$$\|J - J_c\| \geq 1.$$

Set

$$a := \inf\{\|J - GJ_c\|: G \in \mathbf{A}_J(\mu) \text{ and } 0 < c < 1\}.$$

Suppose $a = 0$. Then, by the compactness of $\mathbf{A}_J(\mu)$ and $\mathbf{D}_J(\mu)$, we could find D in $\mathbf{A}_J(\mu)$ and a limit point R of $\{J_c: 0 < c < 1\}$ such that $J = DR$. Since R is idempotent and in $\mathbf{D}_J(\mu)$, we get $R = JR = DR = J$, that is, J is a limit point of $\{J_c: 0 < c < 1\}$. But, $\|J - J_c\| \geq 1$ for all c in $(0, 1)$. Thus, $a > 0$ and

$$\limsup_{n \to \infty} b_{n,c} \geq a > 0 \quad \text{for all } c \text{ in } (0, 1).$$

Let $b := \max\{\|A\|: A \in \mathbf{D}(\mu)\}$. For $n \geq 1$,

$$\|J - GB_c^{n+1}\| \leq \|J - GB_c^n\| + \|G(J - B_c)B_c^n\|,$$

so

$$b_{n+1,c} \leq b_{n,c} + b^2 c.$$

Similarly, we obtain the same equality with $n+1$ and n interchanged. Thus, for all $n \geq 1$,

$$|b_{n+1,c} - b_{n,c}| \leq b^2 c.$$

Hence, for every sequence $\{m_n\}$ of positive integers and every sequence $\{c_n\}$ in $(0, 1)$ with $c_n \to 0$, we have

$$\lim_{n \to \infty} |b_{m_n+1, c_n} - b_{m_n, c_n}| = 0. \tag{3.3.5}$$

Thus, given d in $(0, a)$, we can find positive integers m_n and numbers c_n in $(0, 1)$ such that $c_n \to 0$ and $b_{m_n, c_n} < d \leq b_{m_n+1, c_n}$. By (3.3.5),

$$\lim_{n \to \infty} b_{m_n, c_n} = d.$$

Since $\{B_{c_n}^{m_n}\}$ is contained in the compact set $\mathbf{D}_J(\mu)$, it is conditionally compact, so we may assume

$$B_{c_n}^{m_n} \to \text{(some) } E_d \text{ in } \mathbf{D}_J(\mu).$$

The compactness of $\mathbf{A}_J(\mu)$ implies

$$d = \min\{\|J - GE_d\|: G \in \mathbf{A}_J(\mu)\}. \tag{3.3.6}$$

Since $0 < d$, E_d is not in $\mathbf{A}_J(\mu)$, which completes the proof of part (ii).

(iii) Let $\{E_d: 0 < d < a\}$ be as in part (ii) and let E_0 be a limit point of this set for some sequence $d_n \to 0$. Compactness of $\mathbf{A}_J(\mu)$ and (3.3.6) give that $J = G_0 E_0$ for some $G_0 \in \mathbf{A}_J(\mu)$. From Lemma 3.3.1, $\mathbf{A}_J(\mu)$ is a group, so E_0 is in $\mathbf{A}_J(\mu)$. Furthermore, the Numakura theorem implies that J is a limit point of the sequence $\{E_0^n\}$. Select a positive integer q such that $\|J - E_0^q\| < 1/4$ and select d_0 in $(0, a)$ such that $\|E_0^q - E_{d_0}^q\| < 1/4$. Let $W := E_{d_0}^q$. Then $\|J - W\| < 1/2$ and

$$B_{c_n}^{r_n} \to W, \qquad (3.3.7)$$

where the positive integers $r_n \to \infty$ and $0 < c_n \to 0$, from part (ii).

In a Banach algebra with unit J, $\log A$ is well defined whenever $\|J - A\| < 1$ and $\exp(\log A) = A$ [cf. Hille and Phillips (1957), Sections 9.5 and 5.4]. Thus, we have the following representations of W and B_{c_n}:

$$B_{c_n} = J\exp(U_n) \quad \text{and} \quad W = J\exp(V), \qquad (3.3.8)$$

where $U_n, V \in \text{End}$, $JU_n = U_n J = U_n$, $JV = VJ = V$, and $WV = VW$. From (3.3.7) and (3.3.8), we have $r_n U_n \to V$. This and (3.3.7) imply that, for any $t \geq 0$,

$$B_{c_n}^{[tr_n]} \to J\exp(tV),$$

where $[x]$ denotes the greatest integer less than or equal to x. From part (i), B_{c_n} are in $D_J(\mu)$, so the one-parameter semigroup $\{J\exp(tV): t \geq 0\}$ is contained in $\mathbf{D}_J(\mu)$.

Now consider the compact semigroup, $\text{sem}\{W\}$. By the Numakura theorem, it contains an idempotent J_1. Since V and W commute, $J_1 V = VJ_1$, and since $W = E_{d_0}^q$, we have J_1 is in $\text{sem}\{E_{d_0}\}$. Consequently, $J \neq J_1$ because otherwise by Lemma 3.3.2 we would have that E_{d_0} is in $\mathbf{A}_J(\mu)$ which is false. Since $J - J_1 = (J - J_1)J$, we have that the semigroup $\{(J - J_1)\exp(tV): t \geq 0\}$ is conditionally compact. Let $0 < t_n \to \infty$ so that

$$(J - J_1)\exp(t_n V) \to H, \qquad (3.3.9)$$

where H is some element of $\mathbf{D}_J(\mu)$. Without loss of generality, we may assume that

$$W^{[t_n]} = J\exp([t_n]V) \to H_1,$$
$$J\exp((t_n - [t_n])V) \to H_2,$$

where both H_1 and H_2 belong to $\mathbf{D}_J(\mu)$. But, H_1 is a limit point of $\{W^n\}$. Hence, $J_1 H_1 = H_1 J_1 = H_1$, so $(J - J_1)H_1 = 0$. Finally, (3.3.9) implies that $(J - J_1)H_1 H_2 = H$, so $H = 0$. Therefore, all limit points of $\{(J - J_1)\exp(tV): t \geq 0\}$ as $t \to \infty$ are 0. Q.E.D.

Lemma 3.3.4. *If J_1 and J_2 are idempotents in $\mathbf{D}(\mu)$ and $J_2 \in \mathbf{D}_{J_1}(\mu)$, then $\mathbf{D}_{J_2}(\mu) \subseteq \mathbf{D}_{J_1}(\mu)$.*

Proof. Let $A \in \mathbf{D}_{J_2}(\mu)$. Then $AJ_2 = J_2 A = A$. Since $J_2 \in \mathbf{D}_{J_1}(\mu)$, $J_1 J_2 = J_2 J_1 = J_2$. Hence, $A = AJ_2 = A(J_2 J_1) = (AJ_2)J_1 = AJ_1$. Similarly, $A = J_1 A$. Therefore, $A \in \mathbf{D}_{J_1}(\mu)$. Q.E.D.

We are now ready to establish the main characterization of an operator-selfdecomposable measure in terms of its decomposability semigroup.

Theorem 3.3.5. *A full measure μ on $\mathbb{R}^d$ is operator-decomposable if and only if its Urbanik decomposability semigroup $\mathbf{D}(\mu)$ contains a one-parameter semigroup $\{\exp(-tQ): t \geq 0\}$ with Q invertible and $\exp(-tQ) \to 0$ as $t \to \infty$. Furthermore, if $\mu = \exp(-tQ)\mu * \mu_t$ for $t \geq 0$, then each μ_t is infinitely divisible.*

Proof. Assume μ is full and operator-selfdecomposable. Since I is idempotent and belongs to $\mathbf{D}(\mu)$, by consecutive applications of Lemma 3.3.3(iii), we obtain a sequence $\{J_n\}_0^\infty$ of idempotents with $J_0 = I$, a sequence $\{V_j\}_1^\infty$ in End such that $\mathbf{D}_{J_i}(\mu)$ contains the one-parameter semigroup $\{J_i \exp(tV_{i+1}): t \geq 0\}$, and

$$J_i V_{i+1} = V_{i+1} J_i = V_{i+1}, \qquad J_{i+1} V_{i+1} = V_{i+1} J_{i+1},$$
$$J_{i+1} \in \mathbf{D}_{J_i}(\mu), \qquad \lim_{t \to \infty}(J_i - J_{i+1})\exp(tV_{i+1}) = 0,$$

and

$$J_i \neq J_{i+1} \quad \text{unless} \quad J_i = 0.$$

From Lemma 3.3.4, $\mathbf{D}_{J_{i+1}}(\mu) \subseteq D_{J_i}(\mu)$. Hence, $J_n \in \mathbf{D}_{J_i}(\mu)$ for $n > i$, and so

$$J_n J_i = J_i J_n = J_{\max(n,i)}. \tag{3.3.10}$$

Consequently, $J_i - J_n$ are idempotents and either $J_n = 0$ eventually or $\|J_i - J_n\| \geq 1$ for $n \neq i$. However, $\{J_n\} \subseteq \mathbf{D}(\mu)$ and $\mathbf{D}(\mu)$ is compact (cf. Theorem 2.3.1). Thus, $J_n = 0$ eventually. Let r be such that

$J_n = 0$ for all $n \geq r$. Set $K_i := J_{i-1} - J_i$ for $1 \leq i \leq r$. Since $K_i = J_{i-1}(I - J_i)$, by Theorem 2.3.6(a), $K_i \in \mathbf{D}(\mu)$. From (3.3.10), we have $K_k K_s = 0$ for $k \neq s$. Finally, note that $K_i \exp(tV_i) \in \mathbf{D}(\mu)$ and $K_i V_i = V_i K_i$ for $1 \leq i \leq r$. From Theorem 2.3.6(c), we obtain

$$\sum_{i=1}^{r} K_i \exp(tV_i) \in \mathbf{D}(\mu).$$

Setting $Q := -\sum_{i=1}^{r} K_i V_i$, we have $(-Q)^m = \sum_{i=1}^{r} K_i V_i^m$ for $m \geq 1$. Consequently, $\exp(-tQ) = \sum_{i=1}^{r} K_i \exp(tV_i) \in \mathbf{D}(\mu)$ and $\exp(-tQ) \to 0$ as $t \to \infty$. This last property shows that $\int_0^\infty \exp(-tQ)\, dt$ is well defined, and a simple calculation shows that $Q^{-1} = \int_0^\infty \exp(-tQ)\, dt$.

Now, for the sufficiency, assume μ is full and $\mathbf{D}(\mu)$ contains $\{\exp(-tQ) : t \geq 0\}$ for some Q with $\exp(-tQ) \to 0$ as $t \to \infty$. Set $B_n := \exp(-(1/n)Q)$. Then for each $n \geq 1$ there is a measure ν_n such that

$$\mu = B_n \mu * \nu_n.$$

From Theorem 1.7.1, $\{\nu_n\}$ is conditionally compact in $\mathscr{P}$. By Proposition 2.3.7, $\hat{\mu}$ never vanishes. Hence,

$$\hat{\nu}_n(y) = \frac{\hat{\mu}(y)}{\hat{\mu}(B_n^* y)} \qquad (3.3.11)$$

and $\hat{\nu}_n(y) \to 1$ as $n \to \infty$ for all y. Consequently, $\nu_n \Rightarrow \delta(0)$. Set $A_n := \exp(-Q \sum_{j=1}^{n} (1/j))$, $n \geq 1$. Also set,

$$\mu_1 := A_1^{-1} \mu, \qquad \mu_n := A_n^{-1} \nu_n \quad \text{for } n \geq 2. \qquad (3.3.12)$$

Since $A_n \to 0$, we have $A_n \mu_{j_n} \Rightarrow \delta(0)$ whenever the sequence $\{j_n\}$ is bounded. When $j_n \to \infty$ with $j_n \leq n$, we have $A_n \mu_{j_n} = A_n A_{j_n}^{-1} \nu_{j_n}$ for $j_n \geq 2$. Since $\{A_n A_{j_n}^{-1}\}$ is conditionally compact in Aut and $\nu_n \Rightarrow \delta(0)$, we have $A_n \mu_{j_n} \Rightarrow \delta(0)$. Therefore, the triangular array $\{A_n \mu_j : 1 \leq j \leq n, n \geq 1\}$ is infinitesimal. Finally, from (3.3.11) and (3.3.12),

$$\hat{\mu}_1(y) = \hat{\mu}\big((A_1^{-1})^* y\big)$$

and, for $k \geq 2$,

$$\hat{\mu}_k(y) = \hat{\nu}_k\big((A_k^{-1})^* y\big) = \frac{\hat{\mu}\big((A_k^{-1})^* y\big)}{\hat{\mu}\big((A_{k-1}^{-1})^* y\big)},$$

and, consequently, $A_n(\mu_1 * \cdots * \mu_n) = \mu$, which shows that μ is operator-selfdecomposable.

Finally, we need to show that if $\mu = \exp(-tQ)\mu * \mu_t$ for $t \geq 0$, with Q such that $\exp(-tQ) \to 0$ as $t \to \infty$, then each μ_t is infinitely divisible. Given $t > 0$, select a sequence of positive integers $\{k_n\}$ such that

$$k_n \geq n+1 \quad \text{and} \quad \sum_{j=n+1}^{k_n} \frac{1}{j} \to t \quad \text{as } n \to \infty.$$

Let A_n be the operators defined in the proof of the sufficiency. Then

$$A_{k_n} A_n^{-1} \to \exp(-tQ) \quad \text{as } n \to \infty.$$

Since

$$\mu = A_{k_n} A_n^{-1} \mu * A_{k_n}(\mu_{n+1} * \cdots * \mu_{k_n}),$$

we have $\{A_{k_n}(\mu_{n+1} * \cdots * \mu_{k_n})\}$ is conditionally compact (cf. Theorem 1.7.1). Moreover, since $\hat{\mu}$ never vanishes (cf. Proposition 2.3.7),

$$\left(A_{k_n}(\mu_{n+1} * \cdots * \mu_{k_n})\right)^{\hat{}}(y) = \frac{\hat{\mu}(y)}{\hat{\mu}\left(\left(A_{k_n} A_n^{-1}\right)^* y\right)}$$

has limit $\hat{\mu}(y)/\hat{\mu}(\exp(-tQ^*)y)$ as $n \to \infty$. Consequently, $A_{k_n}(\mu_{n+1} * \cdots * \mu_{k_n}) \Rightarrow \mu_t$, so μ_t is infinitely divisible [cf. Proposition 1.7.6(b) and Theorem 1.8.11]. Q.E.D.

We conclude this section by introducing the notion of a U-exponent of a probability measure. Namely, we say that $Q \in \mathbf{End}$ is a *U-exponent of a measure* μ provided its Urbanik decomposability semigroup $\mathbf{D}(\mu)$ contains the one-parameter semigroup $\{e^{-tQ}: t \geq 0\}$. By Theorem 3.3.5, we see that full operator-selfdecomposable measures have U-exponents with the property that $\lim_{t \to \infty} e^{-tQ} = 0$. However, we do not require this condition in the definition of U-exponents. Also, fullness of μ is not necessary for some properties discussed below.

Let $\mathcal{E}_u(\mu)$ denote the set of all U-exponents of a measure μ, and let $\mathcal{T}(\mathbf{H})$ denote the tangent space at the identity of the semigroup $\mathbf{H}$ of operators (cf. Section 1.5).

Proposition 3.3.6. *The set $\mathcal{E}_u(\mu)$ has the following properties.*

(a) $0 \in \mathcal{E}_u(\mu)$ and $\alpha\mathcal{E}_u(\mu) = \mathcal{E}_u(\mu)$ for all $\alpha > 0$.
(b) $\mathcal{E}_u(\mu)$ is closed in the operator norm topology.
(c) $\mathcal{E}_u(\mu) + \mathcal{E}_u(\mu) = \mathcal{E}_u(\mu)$.
(d) For $A \in \mathbf{A}_0(\mu) := \mathbf{A}(\mu) \cap \text{Aut}$, $A\mathcal{E}_u(\mu)A^{-1} = \mathcal{E}_u(\mu)$.
(e) $\mathcal{E}_u(\mu) \cap (-\mathcal{E}_u(\mu)) = \mathcal{T}(\mathbf{A}(\mu)) \cap (-\mathcal{T}(\mathbf{A}(\mu)))$, and this is the largest linear subspace contained in $\mathcal{E}_u(\mu)$, while $\mathcal{E}_u(\mu) - \mathcal{E}_u(\mu)$ is the smallest linear space containing $\mathcal{E}_u(\mu)$.

Proof

(a) Since the identity is in $\mathbf{D}(\mu)$, 0 is always a U-exponent of any μ. From $e^{-tQ}\mu * \mu_t = e^{-t'(\alpha Q)}\mu * \nu_{t'}$ where $t' := t/\alpha$ and $\nu_{t'} := \mu_t$, we obtain $\mathcal{E}_u(\mu) = \alpha\mathcal{E}_u(\mu)$ for $\alpha > 0$.

(b) Let $Q_n \in \mathcal{E}_u(\mu)$ for all $n \geq 1$ and assume $Q_n \to Q$ in End. Let $\nu_{t,n}$ be such that $\mu = e^{-tQ_n}\mu * \nu_{t,n}$. By the continuous mapping theorem (cf. Theorem 1.6.4), $e^{-tQ_n}\mu \Rightarrow e^{-tQ}\mu$ for all $t \geq 0$. This convergence and Theorem 1.7.1(b) imply that $\{\nu_{t,n}: n \geq 1\}$ is conditionally compact in $\mathcal{P}$. [In fact, $\nu_{t,n} \Rightarrow \nu_t$ (some) whenever the Fourier transform $\hat{\mu}$ never vanishes as when μ is infinitely divisible; cf. Proposition 1.7.6(b).] Hence, $\mu = e^{-tQ}\mu * \nu_t$ for any limit point ν_t of the set $\{\nu_{t,n}: n \geq 1\}$, that is, $Q \in \mathcal{E}_u(\mu)$.

(c) Let $Q_1, Q_2 \in \mathcal{E}_u(\mu)$. Then $e^{-tQ_1}e^{-tQ_2} \in \mathbf{D}(\mu)$ for all $t \geq 0$ and

$$t^{-1}[e^{-tQ_1}e^{-tQ_2} - I] \to -(Q_1 + Q_2)$$

as $t \to 0$. By Lemma 1.5.1, we see that $\mathcal{E}_u(\mu) = -\mathcal{T}(\mathbf{D}(\mu))$, so $Q_1 + Q_2 \in \mathcal{E}_u(\mu)$, that is, $\mathcal{E}_u(\mu) + \mathcal{E}_u(\mu) \subset \mathcal{E}_u(\mu)$. The converse inclusion follows from the fact that $0 \in \mathcal{E}_u(\mu)$.

(d) For $A \in \mathbf{A}(\mu) \cap \text{Aut}$, $\mu = A\mu * \delta(a)$ for some a, and for $Q \in \mathcal{E}_u(\mu)$, $\mu = e^{tQ}\mu * \mu_t$ for some μ_t. Hence, $\mu = e^{-tQ}A\mu * \mu_t * \delta(e^{-tQ}a)$. Together with $\mu = A^{-1}\mu * \delta(-A^{-1}a)$ we have

$$\mu = A^{-1}e^{-tQ}A\mu * A^{-1}\mu_t * \delta(A^{-1}e^{-tQ}a - A^{-1}a).$$

Therefore, $A^{-1}QA \in \mathcal{E}_u(\mu)$, that is, $A^{-1}\mathcal{E}_u(\mu)A \subset \mathcal{E}_u(\mu)$. The converse inclusion follows from $\mathbf{A}(\mu) \cap \text{Aut}$ being a group (cf. Proposition 2.3.4).

(e) From (a) and (c), we have that $\mathscr{E}_u^\circ(\mu) \cap (-\mathscr{E}_u^\circ(\mu))$ is a linear subspace of $\mathscr{E}_u^\circ(\mu)$. Clearly, it is the largest subspace of $\mathscr{E}_u^\circ(\mu)$. Similarly, $\mathscr{E}_u^\circ(\mu) - \mathscr{E}_u^\circ(\mu)$ is the smallest linear space containing $\mathscr{E}_u^\circ(\mu)$. The inclusion $\mathscr{T}(\mathbf{A}(\mu)) \cap (-\mathscr{T}(\mathbf{A}(\mu))) \subset \mathscr{E}_u^\circ(\mu) \cap (-\mathscr{E}_u^\circ(\mu))$ follows from Lemma 1.5.1, because $\mathscr{E}_u^\circ(\mu) = -\mathscr{T}(\mathbf{D}(\mu))$ and, of course, $\mathbf{A}(\mu) \subset \mathbf{D}(\mu)$. Conversely, if Q and $-Q$ are U-exponents, then Lemma 1.5.1 gives that, for $t \geq 0$, $\exp(-tQ)$ and $\exp(tQ)$ are in $\mathbf{D}(\mu)$. From Proposition 2.3.4, $\exp(\pm tQ)$ are in $\mathbf{A}(\mu)$, so $Q \in \mathscr{T}(\mathbf{A}(\mu)) \cap (-\mathscr{T}(\mathbf{A}(\mu)))$.

Q.E.D.

Corollary 3.3.7. *If μ is full on $\mathbb{R}^d$, then*

(a) $\mathscr{E}_u^\circ(\mu) \cap (-\mathscr{E}_u^\circ(\mu)) = \mathscr{T}(\mathbf{A}(\mu))$ *and is a Lie algebra;*
(b) $A\mathscr{E}_u^\circ(\mu)A^{-1} = \mathscr{E}_u^\circ(\mu)$ *for all $A \in \mathbf{A}(\mu)$;*
(c) $e^{[A]}\mathscr{E}_u^\circ(\mu) = \mathscr{E}_u^\circ(\mu)$ *for all $A \in \mathscr{T}(\mathbf{A}(\mu))$, where $[A](B) := [A,B] := AB - BA$ for $B \in$ End.*

Proof. By Corollary 2.3.2, $\mathbf{A}(\mu)$ is a compact subgroup of $\mathrm{Aut}(\mathbb{R}^d)$, so $\mathscr{T}(\mathbf{A}(\mu))$ is a Lie algebra (cf. Section 1.5). Properties (a) and (b) follow from (d) and (e) of Proposition 3.3.6. To see (c), note that $e^{[A]}B = e^A B e^{-A}$. Since the exponential map takes the tangent space back into the group, (c) is a consequence of (d) in Proposition 3.3.6.

Q.E.D.

Proposition 3.3.8. *Let μ be full on $\mathbb{R}^d$. Then*

(a) $\mathscr{E}_u^\circ(\mu) = \mathscr{T}(\mathbf{D}(\mu) \cap \{A: 0 < \det(A) \leq 1\})$;
(b) *for a given $Q \in \mathscr{E}_u^\circ(\mu)$,*

$$Q_c := \int_{\mathbf{A}(\mu)} gQg^{-1} H(dg) \in \mathscr{E}_u^\circ(\mu)$$

and commutes with symmetry group $\mathbf{A}(\mu)$. [Here, H is the normalized Haar measure on $\mathbf{A}(\mu)$.]

Proof

(a) From μ being full on a finite-dimensional space, we have that $\mathbf{D}(\mu)$ is a compact subsemigroup of $\mathrm{End}(\mathbb{R}^d)$ (cf. Theorem

2.3.1). Hence, for $Q \in \mathscr{E}_u(\mu)$,

$$\sup_{s \geq 0}(e^{-s \, \text{trace} \, Q}) = \sup_{s \geq 0} \det(e^{-sQ}) < \infty,$$

so trace $Q \geq 0$ and $0 < \det(e^{-sQ}) \leq 1$. Therefore, $Q \in \mathscr{T}(\mathbf{D}(\mu) \cap \{A: 0 < \det(A) \leq 1\})$, which establishes (a).

(b) Since $\mathbf{A}(\mu)$ is a compact group in Aut (cf. Corollary 2.3.2), H exists. By Proposition 1.6.8, we see that Q_c is the limit of a sequence of the form $\sum \alpha_k g_k Q g_k^{-1}$, with $\alpha_k \geq 0$, $\sum \alpha_k = 1$, and $g_k \in \mathbf{A}(\mu)$. [Note that the functions from $\mathbf{A}(\mu)$ to $\mathbb{R}^1$ given by $g \mapsto \langle gQg^{-1}x, y \rangle$ for x, y fixed are continuous and bounded.] Consequently, (a), (b), and (c) of Proposition 3.3.6 imply that $Q_c \in \mathscr{E}_u(\mu)$. Commutativity of Q_c with $\mathbf{A}(\mu)$ is a simple consequence of the invariance of the Haar measure. Q.E.D.

Let $\mathscr{E}_{cu}(\mu)$ denote the set of all U-exponents which commute with the symmetry semigroup $\mathbf{A}(\mu)$, and let

$$\mathbf{D}_c(\mu) := \{A \in \mathbf{D}(\mu): AB = BA \text{ for all } B \in \mathbf{A}(\mu)\}.$$

Clearly, $\mathbf{D}_c(\mu)$ is a closed subsemigroup of $\mathbf{D}(\mu)$ and $I \in \mathbf{D}_c(\mu)$.

Corollary 3.3.9. *Let μ be full on $\mathbb{R}^d$. Then*

$$\mathscr{E}_{cu}(\mu) = \mathscr{T}(\mathbf{D}_c(\mu) \cap \{A: 0 < \det A \leq 1\}).$$

Proof. From Section 1.5, we have that $\mathscr{E}_{cu}(\mu) = \mathscr{T}(\mathbf{D}_c(\mu))$. Note that $AQ = QA$ if and only if $Ae^{tQ} = e^{tQ}A$ for all $t \geq 0$. Hence, analogous to the proof of (a) in Proposition 3.3.8, we obtain the corollary. Q.E.D.

Proposition 3.3.10. *Let γ be a zero-mean Gaussian measure on $\mathbb{R}^d$ with covariance S. Then*

(a) $\mathscr{E}_u(\gamma) = \{Q \in \text{End}(\mathbb{R}^d): QS + SQ^* \text{ is nonnegative definite}\};$
(b) *the set $\{Q \in \text{End}: QS + SQ^* = 0\}$ is the largest Lie algebra in $\mathscr{E}_u(\gamma)$;*
(c) *when S is the identity, the algebra of all skew-symmetric matrices is the largest Lie subalgebra of $\mathscr{E}_u(\gamma)$.*

Proof

(a) Note that $A\gamma$ is also Gaussian with covariance ASA^*. Furthermore, if $\gamma = \nu_1 * \nu_2$, then both ν_1 and ν_2 are Gaussian measures by the Cramér theorem (cf. Theorem 1.8.18). Hence,

$$\mathbf{D}(\mu) = \{A \in \mathrm{End}(\mathbb{R}^d): ASA^* \leq S\}.$$

Thus, $Q \in \mathscr{E}_u^c(\gamma)$ if and only if $S - e^{-tQ}Se^{-tQ^*} \geq 0$. Differentiation with respect to t gives

$$e^{-tQ}(QS + SQ^*)e^{-tQ^*} \geq 0$$

for all $t \geq 0$. In particular, $QS + SQ^* \geq 0$. Conversely, if $QS + SQ^* \geq 0$, then

$$\langle e^{-tQ}(QS + SQ^*)e^{-tQ^*}x, x \rangle \geq 0$$

for all $t \geq 0$ and for all $x \in \mathbb{R}^d$. Equivalently,

$$\frac{d}{dt}[\langle Sx, x \rangle - \langle e^{-tQ}Se^{-tQ^*}x, x \rangle] \geq 0,$$

and so $S - e^{-tQ}Se^{-tQ^*} \geq 0$ for all $t \geq 0$. Therefore, $e^{-tQ} \in \mathbf{D}(\gamma)$ and $Q \in \mathscr{E}_u^c(\gamma)$.

Since (b) is a consequence of (e) in Proposition 3.3.6, and (c) follows from (b), the proof is complete. Q.E.D.

3.4 CHARACTERISTIC FUNCTIONALS OF exp(−tQ)-DECOMPOSABLE MEASURES

In this section, we specialize the concept of operator-selfdecomposable to a particular Q with $\exp(-tQ) \to 0$ as $t \to \infty$. The assumption of fullness in Theorem 3.3.5 was used only in the proof of the necessity, so we may drop that assumption in this section. For Q with $\exp(-tQ) \to 0$ as $t \to \infty$, we say that a measure μ is $\exp(-tQ)$-*decomposable* if for each $t \geq 0$, there is a measure ν_t such that

$$\mu = \exp(-tQ)\mu * \nu_t. \tag{3.4.1}$$

From the last part of Theorem 3.3.5, each ν_t is infinitely divisible. The assumption on Q implies $\nu_t \Rightarrow \mu$ as $t \to \infty$, so μ is also infinitely divisible [cf. Proposition 1.8.1(b)].

For an infinitely divisible $\mu = [a, R, M]$ to be $\exp(-tQ)$-decomposable, there must be some compatibility conditions between Q and R, and between Q and M. If $\mu = [a, R, M]$ and $A \in \text{End}$, then

$$(A\mu)\hat{\ }(y) = \exp\left\{i\langle y, Aa\rangle - \frac{1}{2}\langle ARA^*y, y\rangle \right.$$

$$\left. + \int \left[\exp i\langle y, Ax\rangle - 1 - \frac{i\langle y, Ax\rangle}{1 + \|x\|^2}\right]M(dx)\right\} = [\tilde{a}, ARA^*, AM],$$

where

$$\tilde{a} := Aa + \int \left[\frac{\|x\|^2 - \|Ax\|^2}{(1 + \|Ax\|^2)(1 + \|x\|^2)}\right]AxM(dx)$$

and the integral exists because the Lévy measure M has the property that $\int \|x\|^2/(1 + \|x\|^2)M(dx) < \infty$. To obtain the parameters of the convolution of two infinitely divisible measures, one simply adds the individual parameters. Consequently, by Proposition 1.8.9, (3.4.1) is equivalent to the following proposition.

Proposition 3.4.1. *A measure $\mu = [a, R, M]$ is $\exp(-tQ)$-decomposable if and only if*

(a) *for each $t \geq 0$ and each $y \in \mathbb{R}^d$,*

$$\langle Ry, y\rangle \geq \langle \exp(-tQ)R\exp(-tQ^*)y, y\rangle;$$

(b) *for each $t \geq 0$ and each Borel subset B of $\mathbb{R}^d \setminus \{0\}$,*

$$M(B) \geq M(\exp(tQ)B).$$

In order to characterize $\exp(-tQ)$-decomposable measures, in terms of characteristic functions, we shall solve these inequalities.

Theorem 3.4.2. *Let Q be an operator with $\exp(-tQ) \to 0$ as $t \to \infty$. A Gaussian covariance operator R satisfies the inequality*

$$R \geq \exp(-tQ)R\exp(-tQ^*)$$

for all $t \geq 0$ if and only if

$$R = \int_0^\infty \exp(-tQ) T \exp(-tQ^*) \, dt, \tag{3.4.2}$$

where T is a unique Gaussian covariance operator; in which case,

$$T = QR + RQ^*.$$

Proof. Assume R is of the form (3.4.2) and fix y in $\mathbb{R}^d$. Then, for $t \geq 0$,

$$\langle \exp(-tQ) R \exp(-tQ^*) y, y \rangle$$
$$= \int_0^\infty \langle \exp(-sQ) T \exp(-sQ^*) \exp(-tQ^*) y, \exp(-tQ^*) y \rangle \, ds$$
$$= \int_0^\infty \langle \exp(-(t+s)Q) T \exp(-(t+s)Q^*) y, y \rangle \, ds$$
$$\leq \langle Ry, y \rangle.$$

Thus, $\exp(-tQ) R \exp(-tQ^*) \leq R$.

Now, assume that R satisfies the inequality. Define T to be $QR + RQ^*$. We will show that R and T satisfy (3.4.2) and that T is unique. For fixed y, set

$$g(t) := \langle (R - \exp(-tQ) R \exp(-tQ^*)) y, y \rangle$$

for $t \geq 0$. Then $g(0) = 0$, $g(t) \geq 0$ for all $t \geq 0$, and $g(t) \to \langle Ry, y \rangle$ as $t \to \infty$. Also, g is differentiable and

$$g'(t) = \langle \exp(-tQ)(QR + RQ^*) \exp(-tQ^*) y, y \rangle.$$

Hence,

$$g(t) = \int_0^t \langle \exp(-sQ) T \exp(-sQ^*) y, y \rangle \, ds + \text{constant}.$$

Since $g(0) = 0$, the constant must be zero. Letting $t \to \infty$, we obtain

$$\int_0^\infty \langle \exp(-sQ) T \exp(-sQ^*) y, y \rangle \, ds = \langle Ry, y \rangle,$$

establishing (3.4.2). Clearly, $T = QR + RQ^*$ is a symmetric operator.

To see that T is a Gaussian covariance operator, it suffices to show that $\langle Ty, y \rangle \geq 0$ for all y (cf. Theorem 1.8.6). Suppose to the contrary that $\langle Ty_0, y_0 \rangle < 0$ for some vector y_0. By continuity, there is $t_0 > 0$ such that, for all s in $[0, t_0]$,

$$\langle \exp(-sQ)T \exp(-sQ^*)y_0, y_0 \rangle < 0.$$

Using this y_0 to define the function g above, we have that $g(t_0) < 0$. However, $g(t) \geq 0$ for all $t \geq 0$. This contradiction establishes that T is a Gaussian covariance operator.

For R given by (3.4.2), we obtain

$$QR + RQ^* = \int_0^\infty (-\exp(-tQ))' T \exp(-tQ^*)\, dt$$

$$+ \int_0^\infty \exp(-tQ) T \exp(-tQ^*) Q^* \, dt$$

$$= -\exp(-tQ) T \exp(-tQ^*)\Big|_{t=0}^{t=\infty} = T$$

and hence we also infer that T is unique in (3.4.2). Q.E.D.

Before solving the second inequality in Proposition 3.4.1, we introduce a new norm $\|\cdot\|_Q$ associated with the one-parameter semigroup $\{\exp(-tQ): t \geq 0\}$. We always assume $\exp(-tQ) \to 0$ as $t \to \infty$. This new norm leads to the natural polar coordinates for $\mathbb{R}^d$. In these polar coordinates, orbits of the form $\|t^Q y\|_Q$, $t \geq 0$, are strictly increasing, so they hit the new unit sphere exactly once. For x in $\mathbb{R}^d$, define $\|x\|_Q$ by

$$\|x\|_Q := \int_0^\infty \|\exp(-tQ)x\|\, dt = \int_0^1 \|s^Q x\| s^{-1}\, dx. \qquad (3.4.3)$$

Since $\|\exp(-tQ)\| \leq a \exp(-bt)$ for some positive constants a and b, the integral is finite. Clearly, $\|\cdot\|_Q$ is a norm. Also,

$$\|x\|_Q \leq \int_0^\infty a e^{-bt} \|x\|\, dt = \|x\| \frac{a}{b}.$$

Since $\|x\| \leq \|\exp(tQ)\|\, \|\exp(-tQ)x\|$, we have

$$\|x\|_Q \geq \|x\| \int_0^\infty \|\exp(tQ)\|^{-1}\, dt.$$

Thus, for all x,

$$\left(\int_0^\infty \|\exp(tQ)\|^{-1} dt\right)\|x\| \leq \|x\|_Q \leq \frac{a}{b}\|x\|.$$

Let S_Q be the unit sphere in the norm $\|\ \|_Q$, that is,

$$S_Q := \{z \in \mathbb{R}^d : \|z\|_Q = 1\}.$$

Define $\Phi_Q : S_Q \times \mathbb{R}^+ \to \mathbb{R}^d \setminus \{0\}$ by

$$\Phi_Q(z, t) := t^Q z$$

(the index Q will be omitted if Q is fixed).

Proposition 3.4.3. *The function Φ is a homeomorphism and, for fixed x in $\mathbb{R}^d \setminus \{0\}$, the function*

$$t \to \|t^Q x\|_Q$$

for $t > 0$, is strictly increasing.

Proof. For $t > 0$ and x in $\mathbb{R}^d \setminus \{0\}$, the equality

$$\|t^Q x\|_Q = \int_0^t \|s^Q x\| s^{-1} ds$$

shows that $(d/dt)\|t^Q x\|_Q > 0$, so the functions $t \to \|t^Q x\|_Q$, $x \neq 0$, are strictly increasing. Furthermore, since $\|t^Q\| \to 0$ as $t \to 0$, from the inequality $\|x\| \leq \|t^Q\| \|t^{-Q} x\|$, we get

$$\|t^Q x\|_Q \to \infty \quad \text{as } t \to \infty.$$

Thus, the orbits $\{t^Q x, t \geq 0\}$, for $x \neq 0$, intersect S_Q exactly once. Also, Φ is one-to-one and onto. Clearly, it is continuous. Now, we show that Φ^{-1} is also continuous. Let $z_n \to z_0$ and let $t_n > 0$ and x_n in S_Q be such that $z_n = t_n^Q x_n$ for $n \geq 0$. We show that $(x_n, t_n) \to (x_0, t_0)$. Suppose $\{t_n\}$ contains a subsequence which converges to zero. Then z_0 would be zero. Hence, $\{t_n\}$ is bounded away from zero. Similarly, $\{t_n\}$ is bounded above. Let t be any limit point of the set

$\{t_n: n \geq 1\}$, say $t_{n'} \to t \in \mathbb{R}^+$ as $n' \to \infty$. Then

$$x_{n'} = t_{n'}^{-Q} z_{n'} \to t^{-Q}(t_0^Q x_0).$$

Hence, $\|(t/t_0)^{-Q} x_0\|_Q = 1$ and $x_0 \in S_Q$. Thus, $t/t_0 = 1$, that is, $t = t_0$. Consequently, $t_n \to t_0 > 0$. This implies $x_n \to x_0$ in S_Q. Therefore, Φ^{-1} is continuous and Φ is homeomorphism. Q.E.D.

Later, we use the following notation for B and C, Borel subsets of S_Q and $\mathbb{R}^+$, respectively,

$$[B, C] := \Phi(B \times C)$$
$$= \{t^Q x : x \in B \text{ and } t \in C\}.$$

By the previous proposition, for $s > 0$,

$$s^Q[B, C] = [B, sC].$$

Let $\mu = [a, R, M]$. Define its Q-Lévy spectral function L_M (for simplicity we suppress the index Q) as follows. For $r > 0$ and B a Borel subset of S_Q,

$$L_M(B; r) := -M(\Phi(B \times [r, \infty))) = -M([B, [r, \infty)]).$$

Since M is a finite measure outside every neighborhood of the origin, Proposition 3.4.3 implies that $-L_M(\cdot; r)$ is a finite measure on S_Q for each $r > 0$. Furthermore, L_M uniquely describes M. This Q-Lévy spectral function determines whether $[a, 0, M]$ is $\exp(-tQ)$-decomposable.

Theorem 3.4.4. *A Lévy measure M satisfies the inequality $M \geq \exp(-tQ)M$ for all $t \geq 0$ if and only if, for each Borel subset B of S_Q, its Q-Lévy spectral function $L_M(B; \cdot)$ has both right and left derivatives everywhere on $(0, \infty)$ and the function*

$$r \to r \frac{\partial}{\partial r} L_M(B; r)$$

is nonincreasing. Here, the derivative denotes either the right or left derivative, possibly changing with r or changing from left to right at the same r.

Proof. Assume that $r(\partial/\partial r)L_M(B;r)$ is nonincreasing. Then $(\partial/\partial r)L_M(B;r)$ is also nonincreasing. Hence, $L_M(B; \;)$ is continuous and concave. Furthermore, for $h \in (0,1)$ and $r > 0$,

$$\frac{\partial}{\partial r}L_M(B;r) \geq \frac{1}{h}\frac{\partial}{\partial r}L_M\left(B;\frac{r}{h}\right).$$

Hence, for $0 < a < b$,

$$M([B;[a,b]]) = \int_{(a,b]}\frac{\partial}{\partial r}L_M(B;r)\,dr$$

$$\geq \int_{[a,b)}\frac{1}{h}\frac{\partial}{\partial r}L_M\left(B;\frac{r}{h}\right)dr$$

$$= \int_{[a/h,b/h)}\frac{\partial}{\partial r}L_M(B;r)\,dr$$

$$= h^Q M([B,[a,b]]).$$

Since Φ is a homeomorphism, $M(A) \geq h^Q M(A)$ for all Borel subsets A of $\mathbb{R}^d \setminus \{0\}$. This establishes the sufficiency.

Now, assume that $M \geq \exp(-tQ)M$ for all $t \geq 0$. Suppose there is $t_0 > 0$ such that $M([B;\{t_0\}]) > 0$. Then, for h in $(0,1)$,

$$M([b,\{ht_0\}]) \geq h^Q M([B,\{ht_0\}]) = M([B,\{t_0\}]) > 0,$$

which is impossible. Hence, $L_M(B; \;)$ is continuous. For $t < s$ and $h > 0$,

$$L_M(B;e^t) - L_M(B;e^s) \geq e^{-hQ}M([B,[e^t,e^s)])$$

$$= L_M(B;e^{t+h}) - L_M(B;e^{s+h}).$$

Hence, $t \to L_M(B;e^t)$ is continuous and concave on $\mathbb{R}^1$. The well-known properties of such functions yield the condition. Q.E.D.

Theorem 3.4.5. *A Lévy measure M satisfies the inequality*

$$M \geq \exp(-tQ)M$$

for all $t \geq 0$ if and only if there exists a unique Lévy measure G with

$$\int_{\|x\|>1} \log(1 + \|x\|) G(dx) < \infty$$

and

$$M(B) = \int_0^\infty G(e^{sQ}B)\, ds$$

for any Borel subset B of $\mathbb{R}^d \setminus \{0\}$.

Proof. Assume M and G are related as in the theorem. Then

$$\exp(-tQ)M(B) = \int_0^\infty G(\exp((s+t)Q)B)\, ds \leq M(B).$$

Now, assume $M \geq \exp(-tQ)M$ for all $t \geq 0$. Let L_M be the Q-Lévy spectral function of M given by (3.4.3). Define g and $\tilde{g}$ on $S_Q \times \mathbb{R}^+$ as follows:

$$g(A, t) := \frac{\partial}{\partial t} L_M(A; t),$$

$$\tilde{g}(A, t) := tg(A, t),$$

where the partial derivative denotes the left derivative which is left-continuous. By Theorem 3.4.4, the functions $g(A, \ )$ and $\tilde{g}(A, \ )$ are nonincreasing and left-continuous. Also,

$$M([A, [t, \infty)]) = -L_M(A; t)$$
$$= \int_t^\infty g(A, s)\, ds = \int_t^\infty s^{-1}\tilde{g}(A, s)\, ds. \quad (3.4.4)$$

Since M is finite outside every neighborhood of the origin, these integrals are finite, so

$$\lim_{t \to \infty} g(A, t) = 0 = \lim_{t \to \infty} \tilde{g}(A, t).$$

Therefore, for each A in $\mathscr{B}(S_Q)$, there exists a unique Borel measure on $\mathbb{R}^+$ such that, for $0 < a < b$,

$$\mu_A([s,b)) := \tilde{g}(A,a) - \tilde{g}(A,b).$$

In particular, $\mu_A([a,\infty)) < \infty$ for all $a > 0$.

We now show that for fixed $0 < a < b$, the function

$$A \to \mu_A([a,b))$$

is a finite Borel measure on $\mathcal{B}(S_Q)$. By the definitions of L_M and $\tilde{g}$, we see that $\mu_A([a,b))$ is finitely additive in A. Hence, it suffices to show that it is continuous at the empty set, that is, for every decreasing sequence $\{A_n\}$ in $\mathcal{B}(S_Q)$ which converges to the empty set, we have $\mu_{A_n}([a,b)) \to 0$. This would follow from $g(A_n, t) \to 0$ for each $t > 0$. Suppose to the contrary that there exist $t > 0$, $\varepsilon > 0$, and sequence $\{A_n\}$ in $\mathcal{B}(S_Q)$ decreasing to the empty set such that $g(A_n; t) \geq \varepsilon$ for all $n \geq 1$. Then, for all $n \geq 1$,

$$M([A_n, [t/2, t)]) = \int_{t/2}^{t} g(A_n, s)\, ds \geq \varepsilon t/2 > 0,$$

since $g(A_n, \cdot)$ is nonincreasing. But, $M([A_n, [t/2, t)]) \to 0$. Therefore, for fixed $0 < a < b$, $\mu_A([a,b))$ is a finite Borel measure on $\mathcal{B}(S_Q)$.

We now extend $\mu_A(I)$, $A \in \mathcal{B}(S_Q)$, $I = [a,b) \subseteq \mathbb{R}^+$, $0 < a < b < \infty$, to finite Borel measure on $\mathcal{B}(S_Q \times [\eta, \infty))$, where $\eta > 0$ is arbitrary. Let

$$\mathcal{R}_0 := \{A \times I : A \in \mathcal{B}(S_Q), I = [a,b) \subseteq \mathbb{R}^+ \text{ for } 0 < a < b < \infty\}.$$

Define $\tilde{\rho}$ on $\mathcal{R}_0$ by

$$\tilde{\rho}(A \times I) := \mu_A(I).$$

Note that $\tilde{\rho}$ is finitely additive on $\mathcal{R}_0$, that is, if

$$A_0 \times I_0 = \sum_{j=1}^{m} A_j \times I_j,$$

where Σ denotes that the summands are disjoint, then $\tilde{\rho}(A_0 \times I_0) = \sum_{j=1}^{m} \tilde{\rho}(A_j \times I_j)$. This is true immediately, if $A_0 := \sum_{k=1}^{n} A_k$ and $I_0 := \sum_{l=1}^{s} I_l$ and consequently $A_0 \times I_0 = \Sigma_k \Sigma_l A_k \times I_l$, because of the properties of $\mu_A(I)$. The general case can be reduced to this simple case by taking the components of A_0 and I_0. Consequently, we get that $A_j \times I_j = (\sum_r A'_{r(j)}) \times (\sum_s I'_{s(j)})$ and the indices $\{(r(j), s(j))\}$ are different for different $j \in \{1, \ldots, m\}$.

Now, let $\mathscr{R}$ be the collection of finite disjoint unions of sets in $\mathscr{R}_0$. Define $\tilde{\rho}$ on $\mathscr{R}$ by

$$\tilde{\rho}\left(\sum_{j=1}^m (A_j \times I_j)\right) := \sum_{j=1}^m \tilde{\rho}(A_j \times I_j).$$

Clearly, $\tilde{\rho}$ is finitely additive on $\mathscr{R}$. We now show that $\tilde{\rho}$ is countably additive on $\mathscr{R}$ by showing that $\tilde{\rho}$ is continuous at the empty set, that is, if $\{C_n\}$ is a sequence in $\mathscr{R}$ with $C_n \supseteq C_{n+1}$ and $\tilde{\rho}(C_n) \geq \varepsilon > 0$ for all $n \in \mathbb{N}$ and for some $\varepsilon > 0$, then $\bigcap_{n=1}^\infty C_n \neq \emptyset$.

Let $\mathscr{F}_1$ be the collection of all sets of the form $\sum_{j=1}^m A_j \times \tilde{I}_j$, where $\tilde{I}_j$ are closed intervals in $[\eta, \infty)$. Since $\mu_A(I) \leq \mu_{S_Q}(I)$ for all $A \in \mathscr{B}(S_Q)$ and μ_{S_Q} is a finite and tight measure on $[\eta, \infty)$, for given $\alpha > 0$ and $C \in \mathscr{R}$, there is $D \in \mathscr{F}_1$ such that $D \subseteq C$ and $\tilde{\rho}(C \setminus D) < \alpha$.

Using this property for $C_1 \in \mathscr{R}$, we select $D_1 \in \mathscr{F}_1$ such that $D_1 \subseteq C_1$ and $\tilde{\rho}(C_1 \setminus D_1) < \varepsilon/2^2$. Then

$$\tilde{\rho}(D_1) = \tilde{\rho}(C_1) - \tilde{\rho}(C_1 \setminus D_1) \geq \varepsilon - \frac{\varepsilon}{2^2} > \frac{\varepsilon}{2}.$$

Also, $\tilde{\rho}(C_2 \setminus D_1) < \varepsilon/2^2$. Let $C_2^* := C_2 \cap D_1$. Then $C_2^* \in \mathscr{R}$ and

$$\tilde{\rho}(C_2^*) = \tilde{\rho}(C_2) - \tilde{\rho}(C_2 \setminus D_1) \geq \varepsilon - \frac{\varepsilon}{2^2} > \frac{\varepsilon}{2}.$$

Let $D_2 \in \mathscr{F}_1$ be such that $D_2 \subseteq C_2^*$ and $\tilde{\rho}(C_2^* \setminus D_2) < \varepsilon/2^3$. Then $D_2 \subseteq C_2$, $D_2 \subseteq D_1$, and

$$\tilde{\rho}(D_2) = \tilde{\rho}(C_2^*) - \tilde{\rho}(C_2^* \setminus D_2) \geq \varepsilon - \frac{\varepsilon}{2^2} - \frac{\varepsilon}{2^3} > \frac{\varepsilon}{2}.$$

By induction, we obtain a sequence $\{D_n\}$ in $\mathscr{F}_1$ such that

$$D_{n+1} \subseteq C_n, \qquad D_{n+1} \subseteq D_n, \qquad \tilde{\rho}(D_n) \geq \frac{\varepsilon}{2} \quad \text{for } n \in \mathbb{N}.$$

Let $\mathscr{F}_2$ be the collection of all sets of the form $\sum_{j=1}^m K_j \times \tilde{I}_j$, where K_j are closed subsets of S_Q and $\tilde{I}_j$ are closed finite subintervals of $[\eta, \infty)$. Since we also have that $A \to \mu_A(I)$ is a tight finite measure on

S_Q, for a fixed interval on $[\eta, \infty)$, repeating the previous construction for $\{D_n\}$, we can find a sequence $\{E_n\}_{n \in \mathbb{N}}$ in $\mathscr{F}_2$ such that

$$E_n \subseteq D_n, \quad E_{n+1} \subseteq E_n, \quad \tilde{\rho}(E_n) \geq \frac{\varepsilon}{4} \quad \text{for } n \in \mathbb{N}.$$

The compactness of the sets $\{E_n\}$ implies that $\bigcap_{n=1}^{\infty} E_n \neq \varnothing$ and consequently $\bigcap_{n=1}^{\infty} C_n \neq \varnothing$, which proves that $\tilde{\rho}$ is countably additive on $\mathscr{R}$.

Let ρ be the unique extension of $\tilde{\rho}$ to a finite Borel measure on $\mathscr{B}(S_Q \times [\eta, \infty))$. By allowing $\eta \to 0$, we obtain ρ as a σ-finite measure on $\mathscr{B}(S_Q \times \mathbb{R}^+)$.

Set

$$G := \Phi \rho.$$

For $A \in \mathscr{B}(S_Q)$ and $t > 0$,

$$M(\Phi(A \times [t, \infty))) = \int_{\log t}^{\infty} \tilde{g}(A, e^s)\, ds = \int_0^{\infty} \tilde{g}(A, te^s)\, ds.$$

Consequently, for $A \in \mathscr{B}(S_Q)$ and $B = [t, u]$ with $0 < t$,

$$M([A, B]) = \int_0^{\infty} \mu_A(e^r[t, u])\, dr$$
$$= \int_0^{\infty} \rho(A \times e^r[t, u])\, dr = \int_0^{\infty} G(\Phi(A \times e^r[t, u]))\, dr$$
$$= \int_0^{\infty} G(e^{rQ}[A, B])\, dr.$$

By Proposition 3.4.3, this last equality holds for arbitrary $F \in \mathscr{B}(\mathbb{R}^d \setminus \{0\})$.

Next, we show that $\int_{\|x\| > 1} \log(1 + \|x\|) G(dx) < \infty$. Let $F_0 := [S_Q, (1, \infty)]$, $f(t) := G([S_Q, (t, \infty)])$, and $\tilde{f}(t) := G(\{x: \|x\|_Q > t\})$ for $t > 0$. Then $M(F_0) < \infty$. For $t > 1$, we have $\|t^Q\| \leq t^{\|Q\|}$, so for $t > 1$, letting $\gamma := 1/\|Q\|_Q$,

$$\{\|x\|_Q > t\} = \{s^Q z: \|s^Q z\|_Q > t,\, s > 1,\, z \in S_Q\},$$
$$\{s^Q z: s > t^{\gamma},\, z \in S_Q\} = [S_Q, (t^{\gamma}, \infty)].$$

Consequently, for $t > 1$,
$$\tilde{f}(t) \le f(t^\gamma).$$

Furthermore,
$$\infty > M(F_0) = \int_0^\infty G(e^{rQ}F_0)\,dr = \int_1^\infty G(s^Q F_0)s^{-1}\,ds$$
$$= \int_1^\infty f(s)s^{-1}\,ds = \gamma \int_1^\infty f(u^\gamma)u^{-1}\,du.$$

Thus,
$$\int_{\|x\|_Q > 1} \log\|x\|_Q G(dx) = \int_1^\infty \tilde{f}(s)s^{-1}\,ds \le \int_1^\infty f(s^\gamma)s^{-1}\,ds < \infty.$$

Hence, $\int_{\|x\|_Q > 1} \log(1 + \|x\|_Q)G(dx) < \infty$. Since the norms $\|\cdot\|$ and $\|\cdot\|_Q$ are equivalent, we have $\int_{\|x\| > 1} \log(1 + \|x\|)G(dx) < \infty$.

To prove that G is a Lévy measure, we see
$$\infty > \int_{0 < \|x\|_Q \le 1} \|x\|^2 M(dx) = \int_0^\infty \int_{[S_Q, (0, t]]} \|e^{-tQ}x\|^2 G(dx)\,dt$$
$$\ge \int_1^2 \|e^{sQ}\|^{-2}\,ds \int_{0 < \|x\|_Q \le 1} \|x\|^2 G(dx),$$

since $\|x\|\|e^{tQ}\|^{-1} \le \|e^{-tQ}x\|$.

Finally, we show that G is unique. Suppose there are G_1 and G_2 which both satisfy the condition of the theorem. Since
$$L_M(A, r) = -M([A, [r, \infty)]) = -\int_0^\infty G_i(e^{tQ}[A, [r, \infty)])\,dt$$
$$= \int_0^\infty L_{G_i}(A, e^t r)\,dt = \int_r^\infty s^{-1}L_{G_i}(A, s)\,ds,$$

we have
$$-r \frac{\partial L_M(A; r)}{\partial r} = L_{G_1}(A; r) = L_{G_2}(A; r) \quad \text{for } A \in \mathcal{B}(S_Q) \text{ and } r \in \mathbb{R}^+,$$

where $\partial L_M/\partial r$ denotes the left derivative. Therefore, G_1 and G_2 are the same. Q.E.D.

From Proposition 3.4.1 and Theorems 3.4.2 and 3.4.5, we obtain the basic characterization of $\exp(-tQ)$-decomposable measures in terms of their characteristic functions.

Theorem 3.4.6. *A measure μ on $\mathbb{R}^d$ is $\exp(-tQ)$-decomposable if and only if its characteristic function $\hat{\mu}$ is of the form*

$$\hat{\mu}(y) = \exp\left\{ i\langle y, b\rangle - \frac{1}{2}\int_0^\infty \langle e^{-tQ}Te^{-tQ^*}y, y\rangle\, dt \right.$$

$$\left. + \int_{\mathbb{R}^d\setminus\{0\}}\int_0^\infty \left(\exp i\langle y, e^{-tQ}x\rangle - 1 - \frac{i\langle y, e^{-tQ}x\rangle}{1+\|e^{-tQ}x\|^2} \right) dt \frac{m(dx)}{\log(1+\|x\|^2)} \right\},$$
(3.4.5)

where b is a vector, T is a Gaussian covariance operator, and m is a finite Borel measure on $\mathbb{R}^d\setminus\{0\}$. Moreover, b, T, and m are unique.

Proof. For $F \in \mathscr{B}(\mathbb{R}^d\setminus\{0\})$, using G from Theorem 3.4.5, define

$$m(F) := \int_F \log(1+\|x\|^2)G(dx).$$

Clearly, m is a finite measure because G has a logarithmic moment outside every neighborhood of the origin and G integrates $\|x\|^2$ in every neighborhood of the origin. Equation (3.4.5) easily follows from the basic Lévy–Khintchine formula and Theorems 3.4.2 and 3.4.5. Q.E.D.

Before obtaining another characterization of $\exp(-tQ)$-decomposable measures, we need some auxiliary inequalities which we state in the following lemma.

Lemma 3.4.7. *Let $Q \in \mathrm{End}(\mathbb{R}^d)$ and assume $\exp(-tQ) \to 0$ as $t \to \infty$, or equivalently that all eigenvalues of Q have positive real parts. Then there are constants $c_i > 0$, $i = 1, \ldots, 6$, such that*

$$c_1 e^{-c_2 t}\|x\| \leq \|e^{-tQ}x\| \leq c_3 e^{-c_4 t}\|x\| \quad \text{for all } t > 0 \text{ and } x \in \mathbb{R}^d \quad (3.4.6)$$

and

$$c_5 \log(1 + \|x\|^2) \leq \int_0^\infty \frac{\|e^{-tQ}x\|^2}{1 + \|e^{-tQ}x\|^2} \, dt$$
$$\leq c_6 \log(1 + \|x\|^2) \quad \text{for all } x \in \mathbb{R}^d. \quad (3.4.7)$$

Proof. Let $a_1, \ldots, a_p$ denote the eigenvalues of Q and let

$$0 < c_4 < \min_{1 \leq j \leq p} \operatorname{Re}(a_j) \leq \max_{1 \leq j \leq p} \operatorname{Re}(a_j) < c_2 < \infty.$$

Then

$$\lim_{t \to \infty} e^{c_4 t} \exp(-tQ) = 0,$$
$$\lim_{t \to -\infty} e^{c_2 t} \exp(-tQ) = 0.$$

Hence,

$$c_3 := \sup_{t \geq 0} \|e^{c_4 t} \exp(-tQ)\|,$$
$$c_1^{-1} := \sup_{t \leq 0} \|e^{c_2 t} \exp(-tQ)\|$$

are positive and finite. Consequently, for all $t > 0$ and all $x \in \mathbb{R}^d$,

$$\|\exp(-tQ)x\| \leq c_3 e^{-c_4 t} \|x\|.$$

Also, for $t > 0$ and $x \in \mathbb{R}^d$,

$$e^{-c_2 t} \|x\| \leq \|e^{-c_2 t} \exp(tQ)\| \, \|\exp(-tQ)x\| \leq c_1^{-1} \|\exp(-tQ)x\|,$$

which establishes inequality (3.4.6).

Since the function $s \to s^2/(1 + s^2)$ is increasing on $\mathbb{R}^+$, (3.4.6) implies

$$\int_0^\infty \frac{(c_1 e^{-c_2 t} \|x\|)^2}{1 + (c_1 e^{-c_2 t} \|x\|)^2} \, dt \leq \int_0^\infty \frac{\|e^{-tQ}x\|^2}{1 + \|e^{-tQ}x\|^2} \, dt$$
$$\leq \int_0^\infty \frac{(c_3 e^{-c_4 t} \|x\|)^2}{1 + (c_3 e^{-c_4 t} \|x\|)^2} \, dt.$$

But, for a and b in $\mathbb{R}^+$,

$$\int_0^\infty \frac{(ae^{-bt})^2}{1+(ae^{-bt})^2} dt = \frac{1}{2b} \log(1+a^2).$$

Consequently,

$$\frac{1}{2c_2} \log(1 + c_1^2 \|x\|^2) \leq \int_0^\infty \frac{\|e^{-tQ}x\|^2}{1 + \|e^{-tQ}x\|^2} dt$$

$$\leq \frac{1}{2c_4} \log(1 + c_3^2 \|x\|^2).$$

Since the functions $0 < x \to \log(1 + a^2x^2)/\log(1+x^2)$ are bounded above and below by positive constants, depending on a, we obtain (3.4.7) for some positive constants c_5 and c_6. Q.E.D.

From Theorem 3.4.6, we obtain another integral representation of the characteristic functions of $\exp(-tQ)$-decomposable measures in the following corollary.

Corollary 3.4.8. *A measure μ is $\exp(-tQ)$-decomposable for $t > 0$ if and only if there exists a unique infinitely divisible ν having finite logarithmic moment, that is, $\int \log(1 + \|x\|)\nu(dx) < \infty$, such that, for all $y \in \mathbb{R}^d$,*

$$\hat{\mu}(y) = \exp \int_0^\infty \log \hat{\nu}(e^{-tQ^*}y) \, dt. \tag{3.4.8}$$

Proof. The existence of such a ν satisfying (3.4.8) will follow from (3.4.5) if we show that there is $x_0 \in \mathbb{R}^d$ such that

$$\langle y, x_0 \rangle = \int_{\mathbb{R}^d} \langle y, h(x) \rangle G(dx), \tag{3.4.9}$$

where G is the Lévy measure from Theorem 3.4.5 and

$$h(x) := \int_0^\infty e^{-tQ} x \frac{\|x\|^2 - \|e^{-tQ}x\|^2}{(1 + \|e^{-tQ}x\|^2)(1 + \|x\|^2)} dt. \tag{3.4.10}$$

Then taking $a \in \mathbb{R}^d$ such that $Q^{-1}a = b + x_0$, that is, $\int_0^\infty e^{-tQ}a\,dt = b + x_0$, for b in Theorem 3.4.6 and x_0 in (3.4.9), we obtain the desired formula for $\nu = [a, R, G]$. The uniqueness follows from Theorem 3.4.6. From Theorem 1.8.13, we see that ν has a finite logarithmic moment.

To prove (3.4.9), it suffices to show that

$$\|h(x)\| \le K \log(1 + \|x\|^2)$$

from some constant K, because G integrates $\log(1 + \|x\|^2)$ by Theorem 3.4.5. By (3.4.6) and (3.4.7),

$$\int_0^\infty \frac{\|e^{-tQ}x\|^3}{(1 + \|e^{-tQ}x\|^2)(1 + \|x\|^2)}\,dt \le c_3 c_6 \log(1 + \|x\|^2).$$

The function $s \to s/(1 + s^2)$ is increasing for $s \in (0, 1)$. Let $t_0 := \max\{c_4^{-1} \log c_3 \|x\|, 0\}$. Then $t > t_0$ implies that $c_3 \|x\| e^{-c_4 t} < 1$. Since $\|e^{-tQ}x\| \le c_3 \|x\| e^{-c_4 t}$ for all $t > 0$ and all $x \in \mathbb{R}^d$, we have that

$$\int_0^\infty \frac{\|e^{-tQ}x\|}{1 + \|e^{-tQ}x\|^2}\,dt \le \int_0^{t_0} dt + \int_{t_0}^\infty \frac{c_3 e^{-c_4 t}\|x\|}{1 + (c_3 e^{-c_4 t}\|x\|)^2}\,dt$$

$$= t_0 + \frac{1}{c_4} \arctan(\min\{c_3\|x\|, 1\}),$$

since

$$\int_{t_0}^\infty \frac{ae^{-bt}}{1 + (ae^{-bt})^2}\,dt = \frac{1}{b} \arctan(ae^{-bt_0}) \quad \text{for } a, b \in \mathbb{R}^+.$$

Thus,

$$\int_0^\infty \frac{\|e^{-tQ}x\|\,\|x\|^2}{(1 + \|e^{-tQ}x\|^2)(1 + \|x\|^2)}\,dt$$

$$\le \frac{\|x\|^2}{1 + \|x\|^2} \max\left\{\frac{1}{c_4} \log c_3\|x\|, 0\right\} + \frac{\pi}{4c_4}. \quad (3.4.11)$$

Since $y^2/(1 + y^2)\log(1 + y^2)$ is bounded for $y > 0$, there is a constant $K_1 > 0$ such that (3.4.11) is bounded by $K_1 \log(1 + \|x\|^2)$ for $x \in \mathbb{R}^d$ with $\|x\| < c_3^{-1}$. Also, since $y^2 \log c_3 y/(1 + y^2)\log(1 + y^2)$ is

FUNCTIONALS OF exp($-tQ$)-DECOMPOSABLE MEASURES

bounded above for $y > 0$, there is a constant K_2 such that (3.4.11) is bounded by $K_2 \log(1 + \|x\|^2)$ for $x \in \mathbb{R}^d$ with $\|x\| \geq c_3^{-1}$. Letting $K := \max\{K_1, K_2\}$, we obtain the desired bound on $\|h(x)\|$ for all x in $\mathbb{R}^d$, which completes the proof. Q.E.D.

From the integral representation of the Lévy measures corresponding to the exp($-tQ$)-decomposable measures (cf. Theorem 3.4.5), we obtain the following corollaries.

Corollary 3.4.9. *Let M and G be positive Borel measures on $\mathbb{R}^d$, vanishing at $\{0\}$ and such that*

$$M(F) = \int_0^\infty G(e^{tQ}F)\,dt \quad \text{for } F \in \mathscr{B}(\mathbb{R}^d \setminus \{0\}),$$

where Q is a fixed invertible linear operator. Then

(a) *For any $A \in \mathscr{B}(\mathbb{R}^d)$ such that $\exp(tQ)(A) = A$ for all $t \geq 0$, we have that A and $\mathbb{R}^d \setminus A$ have either zero or infinite M-measure; so, in particular, for any Q-invariant subspace V of $\mathbb{R}^d$, either $M(V) = 0$ or $M(V) = \infty$, and either $M(\mathbb{R}^d \setminus V) = 0$ or $M(\mathbb{R}^d \setminus V) = \infty$;*

(b) $M([S_Q, C]) = 0$ *for any countable subset C in $\mathbb{R}^+$; in particular, M has no atoms.*

Proof. (a) is obvious, and (b) follows from the following equality:

$$M([S_Q, C]) = \int_{\mathbb{R}^d \setminus \{0\}} \int_0^\infty 1_{[S_Q, C]}(e^{-sQ}x)\,ds\,G(dx)$$

$$= \int_{\mathbb{R}^d \setminus \{0\}} l_1(\{s: e^{-sQ}x \in [S_Q, C]\})G(dx) = 0,$$

where l_1 denotes the Lebesgue measure on $\mathbb{R}^1$. Q.E.D.

For an arbitrary subset A of $\mathbb{R}^d$, let $\text{lin}(A)$ denote *the smallest linear subspace of $\mathbb{R}^d$* and $\text{lin}_Q(A)$ *the smallest Q-invariant subspace of $\mathbb{R}^d$ containing the set A*, respectively.

Proposition 3.4.10. *For positive measures M and G related as in Corollary 3.4.9, we have*

(a) $\text{supp } M = (\bigcup_{t \geq 0} e^{-tQ}(\text{supp } G))^-$, *where $\bar{A}$ denotes the closure of a set A;*

(b) $\text{lin}(\text{supp } M) = \text{lin}_Q(\text{supp } G).$

Proof

(a) Let $x \in \operatorname{supp} M$ and $x \notin (\bigcup_{t \geq 0} e^{-tQ}(\operatorname{supp} G))^-$. Then there is an open set U containing x and $U \cap (e^{-tQ}(\operatorname{supp} G)) = \emptyset$ for each $t \geq 0$. Hence $e^{tQ}U$ is open and disjoint with $\operatorname{supp} G$, for each $t \geq 0$. Therefore,

$$M(U) = \int_0^\infty G(e^{tQ}U)\, dt = 0,$$

which contradicts $x \in \operatorname{supp} M$. Conversely, let $x \in \bigcup_{t \geq 0}(e^{-tQ}(\operatorname{supp} G))$. Then $x = e^{-t_0 Q} g_0$ for some $t_0 \geq 0$ and $g_0 \in \operatorname{supp} G$. For each open set U containing x, there are an open interval $(t_0 - \delta, t_0 + \delta)$ and an open set V containing g_0 such that $e^{-tQ}g \in U$ for all $t \in (t_0 - \delta, t_0 + \delta)$ and all $g \in V$. Since $e^{tQ}U$ is also open and $g_0 \in e^{tQ}U$, for $t \in (t_0, t_0 + \delta)$ we get

$$M(U) \geq \int_{t_0}^{t_0+\delta} G(e^{tQ}U)\, dt > 0,$$

that is, $x \in \operatorname{supp} M$. Consequently, $(\bigcup_{t \geq 0}(e^{-tQ}(\operatorname{supp} G)))^- \subseteq \operatorname{supp} M$ which completes the proof of (a).

(b) Let $K := \bigcup_{t \geq 0} e^{-tQ}(\operatorname{supp} G)$. Clearly, $e^{-sQ}x \in K$, whenever $x \in K$ and $s \geq 0$. Since $Qx = $ limit of $(x - e^{-tQ}x)/t$ as $t \downarrow 0$, we have $Qx \in \operatorname{lin}(K)$ for any $x \in K$, that is, $\operatorname{lin}(K)$ is a Q-invariant subspace containing $\operatorname{supp} G$. Hence, $\operatorname{lin}_Q(\operatorname{supp} G) \subseteq \operatorname{lin}(\operatorname{supp} M)$.

Conversely, since $Q^k(\operatorname{supp} G) \subseteq \operatorname{lin}_Q(\operatorname{supp} G)$ for $k = 0, 1, 2, \ldots$, we have $e^{-sQ}(\operatorname{supp} G) \subseteq \operatorname{lin}_Q(\operatorname{supp} G)$ for all $s \geq 0$. Finally,

$$\operatorname{lin}\left[\bigcup_{s \geq 0}(e^{-sQ}(\operatorname{supp} G))^-\right] \subseteq \operatorname{lin}(\operatorname{lin}_Q(\operatorname{supp} G))$$
$$= \operatorname{lin}_Q(\operatorname{supp} G),$$

which completes the proof. Q.E.D.

Remark 3.4.11. If $e^{-tQ} \to 0$ as $t \to \infty$ and $\operatorname{supp} G$ is also bounded in $\mathbb{R}^d$, and hence compact, then

$$\left(\bigcup_{t \geq 0} e^{-tQ}(\operatorname{supp} G)\right)^- = \bigcup_{t \geq 0} e^{-tQ}(\operatorname{supp} G) \cup \{0\}.$$

Remark 3.4.12. From Theorem 3.3.5, we see that the formulas (3.4.5) and (3.4.8) as well as Theorem 3.4.5 and all the above corollaries characterize full operator-selfdecomposable measures.

Proposition 3.4.13. *Let R and T be nonnegative self-adjoint operators related by*

$$R = \int_0^\infty e^{-tQ} T e^{-tQ^*} \, dt$$

(*cf. Theorem* 3.4.2). *Then we have*

(a) $\ker R = \bigcap_{t \geq 0} e^{tQ^*}(\ker T)$, *and* $\ker R$ *is a* Q^*-*invariant subspace*;
(b) $R(\mathbb{R}^d) = \lin_Q(T(\mathbb{R}^d))$.

Proof

(a) For a nonnegative selfadjoint operator D, we have $|\langle Dx, y \rangle|^2 \leq \langle Dx, x \rangle \langle Dy, y \rangle$ by the Schwarz inequality. Hence, $\ker D = \{x: \langle Dx, x \rangle = 0\}$. Therefore, $x \in \ker R$ if and only if $\int_0^\infty \langle e^{-tQ} T e^{-tQ^*} x, x \rangle \, dt = 0$ if and only if $\langle e^{-tQ} T e^{-tQ^*} x, x \rangle = 0$ for all $t \geq 0$ if only if $x \in \ker(e^{-tQ} T e^{-tQ^*}) = \ker(T e^{-tQ^*}) = e^{tQ^*}(\ker T)$ for all $t \geq 0$. Thus, $\ker R = \bigcap_{t \geq 0} e^{tQ^*}(\ker T)$. To see that $\ker R$ is Q^*-invariant, note that $e^{-sQ^*}(\ker R) \subseteq \ker R$ for all $s \geq 0$. Hence, for any $x \in \ker R$, $(x - e^{-sQ^*}x)/s \in \ker R$. Consequently, $Q^* x = \lim_{s \downarrow 0} (x - e^{-sQ^*} x)/s \in \ker R$ whenever $x \in \ker R$.

(b) In the following, we use the facts that $(V_1 \cap V_2)^\perp \supseteq V_1^\perp \cup V_2^\perp$ and $(A^*(V_1))^\perp = A^{-1}(V_1^\perp)$, for any subspaces V_1 and V_2, and for any invertible linear operator A. Note that $R(\mathbb{R}^d) = (\ker R)^\perp$ and $T(\mathbb{R}^d) = (\ker T)^\perp$ since R and T are self-adjoint. Using (a), we have

$$R(\mathbb{R}^d) = \left(\bigcap_{t \geq 0} e^{tQ^*}(\ker T) \right)^\perp \supseteq \bigcup_{t \geq 0} (e^{tQ^*}(\ker T))^\perp$$
$$= \bigcup_{t \geq 0} e^{-tQ} T(\mathbb{R}^d).$$

Hence, $R(\mathbb{R}^d) \supseteq T(\mathbb{R}^d)$. Since $\ker R$ is Q^*-invariant, we have $R(\mathbb{R}^d)$ is Q-invariant. Therefore, $R(\mathbb{R}^d) \supseteq \lin_Q(T(\mathbb{R}^d))$.

Now, for any $x \in \mathbb{R}^d$, $e^{-tQ} T e^{-tQ^*} x \in \lin_Q(T(\mathbb{R}^d))$ for all $t \geq 0$. Hence, $Rx = \int_0^\infty e^{-tQ} T e^{-tQ^*} x \, dt \in \lin_Q(T(\mathbb{R}^d))$, that is, $R(\mathbb{R}^d) \subseteq \lin_Q(T(\mathbb{R}^d))$.   Q.E.D.

3.5 GENERATORS OF THE CLASS $L_0(Q)$

We assume the linear operator Q on $\mathbb{R}^d$ is such that $\exp(-tQ) \to 0$ as $t \to \infty$ in the norm topology. By $L_0(Q)$ we denote the class of all $\exp(-tQ)$-decomposable measures, that is,

$$L_0(Q) := \{\mu \in \mathscr{P}: \text{ for each } t > 0, \text{ there is } \nu_t \in \mathscr{P}$$
$$\text{such that } \mu = e^{-tQ}\mu * \nu_t\}, \quad (3.5.1)$$

that is, $\mu \in L_0(Q)$ if and only if its Urbanik decomposability semigroup $\mathbf{D}(\mu)$ contains the one-parameter semigroup $\{e^{-tQ}: t \in \mathbb{R}^+\}$. Note that for $c > 0$, $L_0(cQ) = L_0(Q)$. Thus, without loss of generality, we may assume that

$$\|t^Q\| \leq \beta t^\gamma \quad \text{for all } t \in (0, 1], \text{ where } \gamma > 1/2 \text{ and } \beta > 0. \quad (3.5.2)$$

Hence, for any vector x and any $t \in [a, b]$, $0 < a < b$, we have

$$\|t^Q x\| \leq \beta \left(\frac{t}{b}\right)^\gamma \|b^Q x\|, \quad \beta^{-1}\left(\frac{t}{a}\right)^\gamma \|a^Q x\| \leq \|t^Q x\|,$$

$$\|b^Q x\| \leq \left(\frac{b}{a}\right)^{\|Q\|} \|a^Q x\|$$

and, consequently, for $x \neq 0$ we have

$$g(a, b; x) := \frac{\sup_{t \in [a,b]} \|t^Q x\|}{\inf_{t \in [a,b]} \|t^Q x\|}$$

$$\leq \frac{\beta^2 \|b^Q x\|}{\|a^Q x\|} \leq \beta^3 \left(\frac{b}{a}\right)^{\|Q\|} \quad (3.5.3)$$

From Proposition 1.8.10, we know that the class ID of all infinitely divisible measures can be described as the smallest closed subsemigroup of $\mathscr{P}$ containing the Gaussian and the compound Poisson measures, that is, measures of the form $[x, D, 0]$ with D a symmetric, positive-semidefinite operator and measures of the form $e(\lambda(\delta(y))) * \delta(x)$, where $\lambda > 0$, $x, y \in \mathbb{R}^d$, and $e(m)$, for a finite Borel measure m on $\mathbb{R}^d$, is given by

$$e(m) := e^{-m(\mathbb{R}^d)} \sum_{k=0}^{\infty} \frac{m^{*k}}{k!},$$

$$m^{*0} := \delta(0).$$

In other words, the class ID is generated, in the sense of taking finite convolutions and weak limits, by Gaussian measures and the measures $[x, 0, \lambda\delta(y)]$, $\lambda > 0$, $x, y \in \mathbb{R}^d$.

Proposition 3.5.1. *The class $L_0(Q)$ is a closed subsemigroup of* ID. *Furthermore, if the operator A commutes with Q, then $AL_0(Q) \subseteq L_0(Q)$; in addition, if A is invertible, then $AL_0(Q) = L_0(Q)$. In particular, for $a > 0$, $T_a L_0(Q) = L_0(Q)$.*

Proof. Clearly, if μ_1 and μ_2 satisfy (3.5.1), then $\mu_1 * \mu_2$ also does, that is, $L_0(Q)$ is a semigroup. Assume for each $n \geq 1$ we have $\mu_n \in \mathscr{P}$ such that $\mu_n = e^{-tQ}\mu_n * \nu_{t,n}$ for all $t > 0$ with $\nu_{t,n} \in \mathscr{P}$ and $\mu_n \Rightarrow \mu$. From Theorem 1.7.1, we have that $\{\nu_{t,n}\}$ is conditionally compact, for each $t > 0$. Since $\hat{\mu}_n(y) \neq 0$ for all y (cf. Proposition 1.8.1), we also have that

$$\lim_{n \to \infty} \hat{\nu}_{t,n}(y) = \frac{\hat{\mu}(y)}{\hat{\mu}(e^{-tQ^*}y)}$$

and is a continuous function. Consequently, $\nu_{t,n} \Rightarrow \nu_t$ as $n \to \infty$ for some $\nu_t \in \mathscr{P}$ (cf. Proposition 1.7.6). Thus, $\mu = e^{-tQ}\mu * \nu_t$, that is, $\mu \in L_0(Q)$, which proves the closedness of $L_0(Q)$. The remainder of the lemma is a simple consequence of (3.5.1). Q.E.D.

The fact that $L_0(Q)$ is a closed subsemigroup in ID suggests the problem of finding its generators, analogous to the compound Poisson measures for ID. By Theorem 3.4.4, we know that the Q-Lévy spectral functions L_M of measures belonging to $L_0(Q)$ are characterized by

$$r \to r\frac{\partial}{\partial r}L_M(A, r) \quad \text{is nonincreasing on } \mathbb{R}^+, \text{ for } A \in \mathscr{B}(S_Q).$$

The simplest nonincreasing functions are given by

$$0 < r \to m(A)I_{(0,\alpha]}(r), \quad \text{where } m(A) > 0, \alpha > 0. \quad (3.5.4)$$

For such functions,

$$M([A, [r, \infty)]) = m(A) \int_r^\infty I_{(a,\alpha]}(t) t^{-1} \, dt = m(A) \log \max\left(1, \frac{\alpha}{r}\right),$$

and therefore $m(\cdot)$ must be a finite measure on $\mathscr{B}(S_Q)$. Finally, let

$$M_{\alpha,m}^Q(F) := \int_{S_Q} \int_0^\alpha I_F(s^Q x) s^{-1} \, ds \, m(dx), \quad F \in \mathscr{B}(S_Q). \quad (3.5.5)$$

For ease of notation, we omit the index Q and simply write $M_{\alpha,m}$. Note that $M_{\alpha,m}(F) < \infty$ for every F separated from zero, that is, $0 \notin \overline{F}$, and

$$\int_{[S_Q, (0,1]]} \|x\| M_{\alpha,m}(dx) = \int_{S_Q} \int_0^\alpha \|s^Q x\| s^{-1} \, ds \, m(dx)$$

$$\leq \int_{S_Q} \|x\|_Q m(dx) = m(S_Q) < \infty.$$

Consequently, each $M_{\alpha,m}$ is a Lévy measure (cf. Theorem 1.8.3). Finally, note that

$$M_{\alpha,m} = \Phi(m \times \lambda_\alpha),$$

where Φ is the polar coordinate function (cf. Proposition 3.4.3), m is a finite measure on $\mathscr{B}(S_Q)$, and λ_α is a Borel measure on $\mathbb{R}^+$ with density $t^{-1} I_{(0,\alpha]}(t)$.

Let $\mathscr{K}_Q$ denote the set of all measures of the form $[x, 0, M_{\alpha,m}]$ or of the form $[x, R, 0]$ such that $x \in \mathbb{R}^d$, $\alpha > 0$, m is a finite measure on S_Q, $M_{\alpha,m}$ is given by (3.5.5), and $QR + RQ^*$ is positive-semidefinite.

Lemma 3.5.2. *The set $\mathscr{K}_Q$ is contained in $L_0(Q)$.*

Proof. For $0 < s < 1$ and sets $[F, B] := \Phi(F \times B)$, we have

$$s^Q M_{\alpha,m}([F, B]) = m(F) \int_B I_{(0, s\alpha]}(t) t^{-1} \, dt \leq M_{\alpha,m}([F, B]).$$

Consequently, the above inequality holds for all $A \in \mathscr{B}(\mathbb{R}^d \setminus \{0\})$ (cf. Proposition 3.4.3). Proposition 3.4.1 implies that $[x, 0, M_{\alpha, m}] \in L_0(Q)$. From Proposition 3.4.2, we have $[x, R, 0] \in L_0(Q)$ whenever $QR + RQ^*$ is positive-semidefinite. Q.E.D.

Before we establish the main result of this section, we prove an auxiliary lemma for any function f on a closed interval $[a, b]$ having both right f'_+ and left f'_- derivatives in $[a, b)$ and $(a, b]$, respectively. Let us note that the analog of Rolle's theorem for such a function has the following form: there exists $c \in (a, b)$ such that $f'_+(c) f'_-(c) \leq 0$. Later, f' denotes either the right or the left derivative, possibly different ones at different points.

Lemma 3.5.3. *Let f and g be real-valued functions on $[a, b]$ which have both right and left derivatives on $[a, b)$ and $(a, b]$, respectively. Let h denote either f'_+/g'_+ or f'_-/g'_-. If $g' > 0$ and the function h is nonincreasing on $[a, b]$, then for each $c \in (a, b)$ such that $g(c) \neq g(a)$ and $g(c) \neq g(b)$, we have*

$$\frac{f(c) - f(a)}{g(c) - g(a)} \geq \frac{f(b) - f(c)}{g(b) - g(c)}.$$

Proof. By applying Rolle's theorem to

$$F(x) := f(x) - f(c) - \frac{(f(c) - f(a))(g(x) - g(c))}{g(c) - g(a)}, \quad a \leq x \leq c,$$

we obtain $d \in (a, c)$ such that

$$\left(f'_+(d) - \frac{(f(c) - f(a)) g'_+(d)}{g(c) - g(a)} \right) \left(f'_-(d) - \frac{(f(c) - f(a)) g'_-(d)}{g(c) - g(a)} \right) \leq 0.$$

The analogous inequality is obtained on the interval $[c, b]$ and for some point $e \in (c, b)$. Since $g' > 0$ and h is nonincreasing, we obtain

$$\frac{f(c) - f(a)}{g(c) - g(a)} \geq h(d) \geq h(e) \geq \frac{f(b) - f(c)}{g(b) - g(c)}. \quad \text{Q.E.D.}$$

Lemma 3.5.4. *For each $\mu = [0, 0, M]$ belonging to $L_0(Q)$, there exist a subsequence $\{k_n\}$ of positive integers, positive real numbers α_{nj}, and finite Borel measures m_{nj} on S_Q, $1 \le j \le k_n$, $1 \le n$, such that*

$$\prod_{j=1}^{k_n} [0, 0, M_{\alpha_{nj}, m_{nj}}] \Rightarrow [0, 0, M]. \qquad (3.5.6)$$

Proof. Let L_M be the Q-Lévy spectral function of μ [cf. (3.4.3)]. Then $-L_M(\ , t)$ is a finite Borel measure on S_Q for each $t > 0$. For $n \ge 1$, $1 \le k \le 2^n$ and $F \in \mathcal{B}(S_Q)$, set

$$b_{nk}(F) := L_M\!\left(F, \frac{k}{2^n}\right), \qquad 1 \le k \le n2^n,$$

$$a_{nk}(F) := \frac{b_{nk}(F) - b_{n,k-1}(F)}{\log\!\left(\dfrac{k}{k-1}\right)}, \qquad 2 \le k \le n2^n, \qquad (3.5.7)$$

$$a_{n1}(F) := a_{n2}(F),$$

$$a_{n,n2^n+1}(F) := 0, \qquad b_{n,n2^n+1}(F) := 0.$$

In Lemma 3.5.3, set $f(t) := L_M(F, t)$, $g(t) := \log t$, $a := (k-1)/2^n$, $b := (k+1)/2^n$, $c = k/2^n$. Lemma 3.5.3 applies since the corresponding function h is nonincreasing by Theorem 3.4.4. Thus,

$$a_{n,k+1}(F) \le a_{nk}(F), \qquad 1 \le k \le n2^n.$$

Hence,

$$m_{nk} := a_{nk} - a_{n,k+1}, \qquad 1 \le k \le n2^n, \qquad (3.5.8)$$

are finite Borel measures on S_Q. For the conclusion of the lemma, let

$$\alpha_{nk} := \frac{k}{2^n} \quad \text{and} \quad k_n := n2^n. \qquad (3.5.9)$$

To show that (3.5.6) holds for α_{nk} and m_{nk} constructed above, we will prove that

$$N_n := \sum_{j=1}^{k_n} M_{\alpha_{nj}, m_{nj}} \Rightarrow M \qquad (3.5.10)$$

GENERATORS OF THE CLASS $L_0(Q)$

outside every neighborhood of zero, and that

$$\limsup_{\varepsilon \to 0} \int_{[S_Q,(0,\varepsilon])} \|x\|^2 N_n(dx) = 0 \qquad (3.5.11)$$

(cf. Proposition 1.8.17). We begin the proof of (3.5.10) by noting that it is sufficient to verify it for sets of the form $[F,[t,\infty)]$.

Let H_n be the Q-Lévy spectral function of $[0,0,N_n]$, that is, $H_n(F,r) := -N_n([F;[r,\infty)])$. Taking into account (3.5.5), (3.5.7), and (3.5.8), we obtain for $r > 0$ with $i \in \{2,3,\ldots,k_n\}$ such that $(i-1)/2^n < r \leq i/2^n$,

$$H_n(F,r) = \sum_{k=2}^{k_n} (a_{nk}(F) - a_{n,k+1}(F)) \log\left(\frac{2^n r}{k}\right)$$

$$= \sum_{k=i}^{k_n} (a_{nk}(F) - a_{n,k+1}(F)) \log\left(\frac{2^n r}{k}\right)$$

$$= a_{ni}(F)\log\left(\frac{2^n r}{i}\right) - \sum_{k=i+1}^{k_n} a_{nk}(F)\log\left(\frac{k}{k-1}\right)$$

$$= a_{ni}(F)\log\left(\frac{2^n r}{i}\right) + b_{ni}(F) - b_{n,k_n}(F),$$

where the third equality uses the partial summation formula. For fixed $i \geq 2$, we define the functions $L_{ni}(F,r)$ as follows:

$$L_{ni}(F,r) := a_{ni}(F)\log\left(\frac{2^n r}{i}\right) + b_{ni}(F), \qquad \frac{i-1}{2^n} \leq r < \frac{i}{2^n}.$$

Hence, we have

$$L_{ni}\left(F,\frac{i}{2^n}\right) = L_M\left(F,\frac{i}{2^n}\right),$$

$$L_{ni}\left(F,\frac{i-1}{2^n}\right) = L_M\left(F,\frac{i-1}{2^n}\right)$$

and, consequently,

$$|L_{ni}(F,r) - L_M(F,r)| \leq L_M\left(F,\frac{i}{2^n}\right) - L_M\left(F,\frac{i-1}{2^n}\right).$$

For $\varepsilon > 0$, there exists a positive integer n_0 such that, for all $n \geq n_0$ and all $k = [2^n r] + 1, \ldots, k_n$, we have

$$r_M\left(F, \frac{k}{2^n}\right) - L_M\left(F, \frac{k-1}{2^n}\right) \leq \varepsilon, \qquad -L_M(F, n) \leq \varepsilon.$$

Therefore, $H_n(F, r) \to L_M(F, r)$ as $n \to \infty$, which establishes (3.5.10).

Now, we prove (3.5.11). In view of (3.5.7), (3.5.8), and (3.5.9) and using the partial summation formula,

$$\int_{[S_Q, (0, \varepsilon]]} \|x\|^2 N_n(dx) = \sum_{k=1}^{k_n} \int_{S_Q} \int_0^\varepsilon \|t^Q x\|^2 t^{-1}\, dt\, m_{nk}(dx)$$

$$= \int_{S_Q} \int_0^{2^{-n}} \|t^Q x\|^2 t^{-1}\, dt\, a_{n2}(dx)$$

$$+ \sum_{k=2}^{k_n} \int_{S_Q} \int_{(k-1)/2^n}^{k/2^n} \|t^Q x\|^2 t^{-1}\, dt\, a_{nk}(dx)$$

$$+ \int_{S_Q} \int_{[2^n \varepsilon]/2^n}^{\varepsilon} \|t^Q x\|^2 t^{-1}\, dt\, a_{n,[2^n\varepsilon]+1}(dx),$$

(3.5.12)

where $[b]$ denotes the greatest integer less than or equal to b. We consider the second term in (3.5.12). By the first mean value theorem and by (3.5.3), for $x \in S_Q$ and for the interval $I_{nk} := ((k-1)/2^n, k/2^n]$, with $k = 2, \ldots, [2^n \varepsilon]$, there exists t_{nk} in $\bar{I}_{nk}$ such that

$$\int_{I_{nk}} \|t^Q x\|^2\, d(\log t) = \|t_{nk}^Q x\|^2 \log\left(\frac{k}{k-1}\right)$$

$$\leq g^2\left(\frac{k-1}{2^n}, \frac{k}{2^n}; x\right) \log\left(\frac{k}{k-1}\right) \|s^Q x\|^2$$

$$\leq C \log\left(\frac{k}{k-1}\right) \|s^Q x\|^2$$

for arbitrary s in I_{nk}, where $C = \beta^6 2^{2\|Q\|}$. This together with (3.5.7) yields

$$\int_{S_Q}\int_{I_{nk}} \|t^Q x\|^2 t^{-1}\, dt\, a_{nk}(dx) = \int_{S_Q}\int_{I_{nk}} \|t_{nk}^Q x\|^2 M(\Phi(dx, ds))$$

$$\leq C\int_{S_Q}\int_{I_{nk}} \|s^Q x\|^2 M(\Phi(dx, ds))$$

$$= C\int_{[S_Q, I_{nk}]} \|x\|^2 M(dx).$$

Hence, the second term in (3.5.12) is not greater than

$$C\int_{[S_Q,(0,\varepsilon]]} \|x\|^2 M(dx),$$

which converges to zero as $\varepsilon \to 0$.

By a similar argument, the third term also converges to zero. For the first term, we have

$$\int_{S_Q}\int_0^{2^{-n}} \|t^Q x\|^2 t^{-1}\, dt\, a_{n2}(dx)$$

$$= \int_{S_Q}\int_{2^{-n}}^{2^{-n+1}} \|(t-2^{-n})^Q x\|^2 (t-2^{-n})^{-1}\, dt\, a_{n2}(dx)$$

$$\leq \beta^2 \int_{S_Q}\int_{2^{-n}}^{2^{-n+1}} \left(\frac{t-2^{-n}}{t}\right)^{2\gamma-1} \|t^Q x\|^2 t^{-1}\, dt\, a_{n2}(dx)$$

$$\leq \beta^2 2^{1-2\gamma} \int_{S_Q}\int_{2^{-n}}^{2^{-n+1}} \|t^Q x\|^2 t^{-1}\, dt\, a_{n2}(dx)$$

$$= \beta^2 2^{1-2\gamma} \int_{[S_Q, I_{n2}]} \|x\|^2 M(dx),$$

where the last inequality uses $\gamma > 1/2$ [cf. (3.5.2)]. This bound goes to zero as $n \to \infty$ for each $\varepsilon > 0$. Thus, (3.4.11) is established.   Q.E.D.

From Proposition 3.5.1 and Lemmas 3.5.2 and 3.5.4, we obtain the main result of this section.

Theorem 3.5.5. *The class $L_0(Q)$ is the smallest closed subsemigroup of* ID *containing the generators $\mathcal{K}_Q$.*

Remark 3.5.6. The proof of Lemma 3.5.4 used the assumption (3.5.2) with $\gamma > 1/2$. However, Theorem 3.5.5 holds for any Q with $t^Q \to 0$ as $t \to 0$.

To establish this remark, we need only consider the relationship between $\mathcal{K}_Q$ and $\mathcal{K}_{aQ}$ for $a > 0$. Since

$$\|x\|_{aQ} = \int_0^\infty \|e^{-taQ}x\| \, dt = a^{-1}\|x\|_Q,$$

we have

$$S_{aQ} = aS_Q.$$

Hence, if m is a finite Borel measure on S_{aQ}, then

$$m_a(A) := a^{-1}m(aA), \quad A \in \mathcal{B}(S_Q),$$

is one also on S_Q. Note that

$$c\Phi_Q(t, u) = \Phi_{cQ}(t^{1/c}, cu).$$

Thus, we have

$$M_{\alpha, m}^{aQ}(F) = \int_{S_{aQ}} \int_0^\alpha I_F(s^{aQ}x)s^{-1} \, ds \, m(dx)$$

$$= a^{-1} \int_{S_Q} \int_0^{\alpha^a} I_F(t^Q ay)t^{-1} \, dt \, m(a \, dy)$$

$$= M_{\alpha^a, m_a}^Q(a^{-1}F)$$

for $F \in \mathcal{B}(\mathbb{R}^d \setminus \{0\})$, and this establishes the remark.

3.6 RANDOM INTEGRAL REPRESENTATION

Equation (3.4.1) can be written in terms of random elements as follows: "An $\mathbb{R}^d$-valued r.v. X is $\exp(-tQ)$-decomposable if for each

$t \geq 0$ there exists an X_t independent of X such that

$$X \stackrel{d}{=} e^{-tQ}X + X_t, \tag{3.6.1}$$

where $\stackrel{d}{=}$ means equality in distribution and r.v. means random variable, random vector, or synonymously random element." Since we assume $e^{-tQ} \to 0$ as $t \to \infty$ and since for each $a > 0$ we also have

$$X \stackrel{d}{=} e^{-t(aQ)}X + X_{at},$$

we may assume without loss of generality that

$$\|e^{-Q}\| < 1, \tag{3.6.2}$$

by replacing Q with aQ for sufficiently large $a > 0$. Moreover, from (3.4.1) and its following statement, we see that X is infinitely divisible. Also, (3.6.1) shows that X is the limit in distribution of the stochastic process $\{X_t: t \geq 0\}$ as $t \to \infty$. The main aim of this section is to prove that $X_t \stackrel{d}{=} Z_t$, where Z_t has a precise random integral form. For this purpose, we define the necessary random integrals using the device of formal integration by parts even though our integrand is a deterministic one-parameter operator semigroup. Namely, for the one-parameter semigroup $\{e^{-tQ}: t > 0\}$ and for the $D(\mathbb{R}^d, [0, \infty))$-valued r.v. Y, we set

$$\int_{(a,b]} e^{-sQ} \, dY(s) := e^{-bQ}Y(b) - e^{-aQ}Y(a) - \int_{(a,b]} d(e^{-sQ})Y(s)$$

$$= e^{-bQ}Y(b) - e^{-aQ}Y(a) + \int_{(a,b]} Qe^{-sQ}Y(s) \, ds \tag{3.6.3}$$

for $0 \leq a < b < \infty$, where $D(\mathbb{R}^d, [0, \infty))$-valued means the Skorohod space of all functions from $[0, \infty)$ to $\mathbb{R}^d$ which are right-continuous and have finite left limits (cf. Section 1.9). The last integral in (3.6.3) exists because of the following property of $D(\mathbb{R}^d, [a, b])$.

Lemma 3.6.1. *For each y in $D(\mathbb{R}^d, [a, b])$ and each $\varepsilon > 0$, there exist points $t_0, \ldots, t_r$ such that*

$$a =: t_0 < \cdots < t_r := b,$$
$$w_y[t_{i-1}, t_i) < \varepsilon \quad \text{for } 1 \leq i \leq r, \tag{3.6.4}$$

where

$$w_y[t_{i-1}, t_i) := \sup\{\|y(s) - y(t)\|: t_{i-1} \leq s < t \leq t_i\}.$$

Proof. Let τ be the supremum of those t in $[a, b]$ for which $[a, t)$ can be decomposed into finitely many subintervals $[t_{i-1}, t_i)$ satisfying (3.6.4). Since $y(a) = y(a+)$, we have $\tau > 0$. Since $y(\tau-)$ exists, $(0, \tau]$ can itself be so decomposed. Finally, $\tau < b$ is impossible because $y(\tau) = y(\tau+)$, for $\tau < b$. Q.E.D.

Corollary 3.6.2. *Each y in $D(\mathbb{R}^d, [a, b])$ has at most countably many discontinuities and*

$$\sup\{\|y(t)\|: a \leq t \leq b\} < \infty.$$

Proof. From Lemma 3.6.1, we see that there can be at most finitely many points t at which the jump $\|y(t) - y(t-)\|$ exceeds $1/n$, so y has at most countably many discontinuities. Also, (3.6.4) shows that y is bounded. Q.E.D.

Lemma 3.6.3. *If Y is a $D(\mathbb{R}^d, [0, \infty))$-valued r.v, then*

$$Z(s) := \int_{(0, s]} e^{-tQ}\, dY(t)$$

is also a $D(\mathbb{R}^d, [0, \infty))$-valued r.v. If Y has independent increments, then so has Z.

Proof. Clearly, the sample paths of Z are right-continuous and have left limits. The measurability of $\omega \to Z(\cdot, \omega)$ follows from the following approximation by Riemann–Stieltjes sums. The integral

$$\int_{(a, b]} d(e^{-tQ}) Y(t)$$

can be approximated in norm arbitrarily close by sums of the form

$$\sum_{j=1}^{n} (e^{-t_j Q} - e^{-t_{j-1} Q}) Y(t_{j-1}),$$

with $a = t_0 < \cdots < t_n = b$. Thus,

$$\int_{(a,b]} e^{-tQ} \, dY(t)$$

is also approximated in norm arbitrarily close by sums

$$e^{-bQ}Y(b) - e^{-aQ}Y(a) - \sum_{j=1}^{n} (e^{-t_jQ} - e^{-t_{j-1}Q})Y(t_{j-1})$$

$$= \sum_{j=1}^{n} e^{-t_jQ}\big(Y(t_j) - Y(t_{j-1})\big). \tag{3.6.5}$$

The approximation (3.6.5) also shows that Z has independent increments whenever Y does. Q.E.D.

Lemma 3.6.4. *If Y is a $D(\mathbb{R}^d, [0, \infty))$-valued r.v. with stationary independent increments and $Y(0) = 0$ a.s., then*

$$\hat{\mathscr{L}}\left(\int_{(a,b]} e^{-tQ} \, dY(t)\right)(y) = \exp \int_{(a,b]} \log \hat{\mathscr{L}}(Y(1))(e^{-tQ^*}y) \, dt,$$

where $y \in \mathbb{R}^d$ and $\hat{\mathscr{L}}(\xi)$ denotes the characteristic function of the r.v. ξ.

Proof. Since Y has stationary independent increments with $Y(0) = 0$ a.s., $\hat{\mathscr{L}}(Y(t_j) - Y(t_{j-1})) = \hat{\mathscr{L}}(Y(1))^{t_j - t_{j-1}}$. Approximating $\int_{(a,b]} e^{-tQ} \, dY(t)$ by the last term in (3.6.5), we see that $\log \hat{\mathscr{L}}(\int_{(a,b]} e^{-tQ} \, dY(t))$ is approximated by

$$\sum_{j=1}^{n} (t_j - t_{j-1}) \log \hat{\mathscr{L}}(Y(1))(e^{-t_j Q^*}y).$$

This sum also approximates the integral on the right in the lemma so the lemma is established. Q.E.D.

Let Y be a $D(\mathbb{R}^d, [0, \infty))$-valued r.v. with stationary independent increments and $Y(0) = 0$ a.s. Let $[a, R, M]$ be the infinitely divisible representation of $\mathscr{L}(Y(1))$. Then since the class ID is closed under convolutions, weak limits, and the action of linear operators, that is, if $\mu \in \text{ID}$ and $A \in \text{End}$, then $A\mu \in \text{ID}$ (cf. Section 3.4), we see from

(3.6.5) that $\mathscr{L}(\int_{(0,s]} e^{-tQ} dY(t))$ is also in ID. For each $s > 0$, let

$$[a^s, R^s, M^s] := \mathscr{L}\left(\int_{(0,s]} e^{-tQ} dY(t)\right).$$

From Lemma 3.6.4, we have

$$R^s = \int_0^s e^{-tQ} R e^{-tQ^*} dt, \qquad (3.6.6)$$

$$M^s(F) = \int_0^s M(e^{tQ}F) dt \quad \text{for } F \in \mathscr{B}(\mathbb{R}^d \setminus \{0\}), \qquad (3.6.7)$$

$$a^s = \int_0^s e^{-tQ}(a + b_M(t)) dt, \qquad (3.6.8)$$

where $b_M(t)$ is the vector in $\mathbb{R}^d$ given by

$$b_M(t) := \int_{\mathbb{R}^d \setminus \{0\}} x \frac{\|x\|^2 - \|e^{-tQ}\|^2}{(1 + \|x\|^2)(1 + \|e^{-tQ}x\|^2)} M(dx).$$

This integral exists since Lévy measures integrate $\|x\|^2/(1 + \|x\|^2)$ over $\mathbb{R}^d \setminus \{0\}$. Note that $e^{-tQ}M$ is also a Lévy measure.

When $\int_{(0,s]} e^{-tQ} dY(t)$ converges in distribution as $s \to \infty$, the limit is denoted by $\int_0^\infty e^{-tQ} dY(t)$ and $\int_0^\infty e^{-tQ} dY(t) := [a^\infty, R^\infty, M^\infty]$, where

$$R^\infty = \int_0^\infty e^{-tQ} R e^{-tQ^*} dt, \qquad (3.6.9)$$

$$M^\infty(F) = \int_0^\infty M(e^{tQ}F) dt \quad \text{for } F \in \mathscr{B}(\mathbb{R}^d \setminus \{0\}), \qquad (3.6.10)$$

$$a^\infty = \int_0^\infty e^{-tQ}(a + b_M(t)) dt$$

$$= Q^{-1}a + \int_0^\infty e^{-tQ} b_M(t) dt. \qquad (3.6.11)$$

Our immediate goal is to describe when the convergence in question holds. We start with the following auxiliary lemma on random operator-power series.

Lemma 3.6.5. *Let A be an invertible linear operator with $\|A\| < 1$ and let $\{\xi_n\}$ be a sequence of independently distributed r.v. Then*

$$\sum_{n=1}^{\infty} A^n \xi_n \text{ converges a.s.}$$

if and only if

$$\mathbf{E} \log(1 + \|\xi_1\|) < \infty.$$

Proof. Let $c := \|A\|$. Since A is invertible, we have

$$\|A^{-1}\|^{-n}\|\xi_n\| \leq \|A^n \xi_n\| \leq c^n \|\xi_n\|$$

for all $n \geq 1$. Assume $\sum_n A^n \xi_n$ converges a.s. From Kolomogorov's three-series theorem and the above inequality, we have

$$\sum_n P\big[\|\xi_n\| \geq d\|A^{-1}\|^n\big] \leq \sum_n P\big[\|A^n \xi_n\| \geq d\big] < \infty$$

for each $d > 0$. Consequently,

$$\sum_n P\big[\log^+ \|\xi_1\| > n \log\|A^{-1}\|\big] < \infty,$$

which is equivalent to $E \log^+ \|\xi_1\| < \infty$, so the necessity is established. Conversely, if $\mathbf{E} \log^+ \|\xi_1\| < \infty$, then, for each $d \in (1, 1/c)$,

$$\sum_n P\big[\log^+ \|\xi_1\| \geq n \log d\big] = \sum_n P\big[\|\xi_n\|^{1/n} \geq d\big]$$
$$= \sum_n P\big[\|c^n \xi_n\|^{1/n} \geq cd\big] < \infty.$$

The Borel–Cantelli lemma implies that

$$P\Big[\limsup_{n \to \infty} \|c^n \xi_n\|^{1/n} \geq cd\Big] = 0.$$

Since $cd < 1$, we see that $\sum_n A^n \xi_n$ converges a.s. Q.E.D.

Now, we can give the complete characterization of any $D(\mathbb{R}^d, [0, \infty))$-valued r.v. Y with stationary independent increments such that $\int_0^\infty e^{-tQ} dY(t)$ exists.

Theorem 3.6.6. *Let* $\{\exp(-tQ): t > 0\}$ *be a one-parameter semigroup with* $\lim_{t \to \infty} e^{-tQ} = 0$. *Let* Y *be a* $D(\mathbb{R}^d, [0, \infty))$-*valued r.v. with stationary independent increments,* $Y(0) = 0$ *a.s., and* $\hat{\mathscr{L}}(Y(1)) = [a, R, M]$. *Then the following statements are equivalent.*

(i) $\int_0^s e^{-tQ} \, dY(t)$ *converges in distribution as* $s \to \infty$.
(ii) $\int_0^s e^{-tQ} \, dY(t)$ *converges in norm a.s. as* $s \to \infty$.
(iii) $\mathbf{E} \log(1 + \|\int_0^s e^{-tQ} \, dY(t)\|) < \infty$ *for given* $s > 0$.
(iv) $\mathbf{E} \log(1 + \|Y(1)\|) < \infty$.
(v) $\int_{\|x\| > 1} \log(1 + \|x\|) M(dx) < \infty$.

Proof. In view of (3.6.2) we may assume $\|e^{-Q}\| < 1$ (cf. also Remark 3.6.7).

(iv) $\Leftrightarrow$ (v) This is shown in Proposition 1.8.13.
(ii) $\Rightarrow$ (i) This implication is obvious.
(i) $\Rightarrow$ (ii) Note that if $0 = s_0 < s_n < s_{n+1} \to \infty$, then

$$Z(s_n) = \int_{(0, s_n]} e^{-tQ} \, dY(t) = \sum_{k=1}^n \int_{(s_{k-1}, s_k]} e^{-tQ} \, dY(t)$$

and the summands are independent. Therefore, the convergence in distribution of $\{Z(s_n)\}$ is equivalent to a.s. convergence. Since the sample paths of Z are right-continuous with left limits, the exceptional null set does not depend on the particular sequence $\{s_n\}$ tending to infinity (cf. Remark 1.9.11).

(i) $\Rightarrow$ (v) Let μ be the limit measure in (i). From (3.6.10), we see that the Lévy measure of μ has the form

$$\int_0^\infty M(e^{tQ} \cdot) \, dt.$$

Theorem 3.4.5 gives that M has a logarithmic moment on every open complement of zero.

(v) $\Rightarrow$ (iii) By Proposition 1.8.13 and (3.6.7), it suffices to show that

$$\int_{\|x\| > 1} \log\|x\| M^s(dx) < \infty$$

for all $s > 0$. There is $a > 0$ such that for all $t > 0$, $\|e^{-tQ}\| \leq a$.

Hence,

$$\int_{\|x\|>1} \log\|x\| M^s(dx)$$

$$= \int_0^s \int_{\{x:\,\|e^{-tQ}x\|>1\}} \log\|e^{-tQ}x\| M(dx)\,dt$$

$$\leq s \int_{\{x:\,a\|x\|>1\}} \log a\|x\| M(dx)$$

$$= s \int_{\{x:\,\|x\|>a^{-1}\}} \log\|x\| M(dx) + sM(\|x\|>a^{-1})\log a.$$

The last integral above is finite by (v) and the remaining term is finite since a Lévy measure puts finite mass on any set bounded away from zero.

(iii) $\Rightarrow$ (i) Note that

$$\int_{(0,n]} e^{-tQ}\,dY(t) = \sum_{k=0}^{n-1} e^{-kQ} \int_{(0,1]} e^{-tQ}\,dY(t+k).$$

Let

$$\xi_k := \int_{(0,1]} e^{-tQ}\,dY(t+k).$$

Then

$$\xi_k \stackrel{d}{=} \int_{(0,1]} e^{-tQ}\,dY(t), \qquad \xi_k = e^{kQ} \int_{(k,k+1]} e^{-tQ}\,dY(t).$$

The equality in distribution is due to the stationarity and independence of the increments of Y (cf. Lemma 3.6.4). Hence, $\{\xi_k\}$ are also independent identically distributed random variables (cf. Lemma 3.6.3). By our assumption (iii), each ξ_k has a finite logarithmic moment. Hence, by Lemma 3.6.5,

$$\int_{(0,n]} e^{-tQ}\,dY(t) \to W \quad \text{a.s. as } n \to \infty \qquad (3.6.12)$$

for some random variable W with $\hat{\mathscr{L}}(W) = [a^\infty, R^\infty, M^\infty]$. Furthermore, since the functions $t \to \mathscr{L}(\int_{(0,t]} e^{-sQ}\,dY(s))$ are right-continuous with left limits in $\mathscr{P}$, we see that the set of measures $\{[a^t, R^t, M^t]:$

$0 < t \leq 1\}$ is conditionally compact in $\mathscr{P}$. Hence, we have

$$\int_{(n, n+t_n]} e^{-sQ} \, dY(s) \stackrel{d}{=} e^{-nQ} \int_{(0, t_n]} e^{-sQ} \, dY(s) \stackrel{d}{\to} 0 \quad (3.6.13)$$

for any arbitrary sequence $\{t_n\}$ in $(0, 1]$, as $n \to \infty$. Consequently, (3.6.12) and (3.6.13) imply that

$$\int_{(0, t]} e^{-sQ} \, dY(s) = \int_{(0, [t]]} e^{-sQ} \, dY(s) + \int_{([t], t]} e^{-sQ} \, dY(s) \stackrel{d}{\to} W$$

as $t \to \infty$. Thus, (iii) $\Rightarrow$ (i). Q.E.D.

Remark 3.6.7. In the proof of Theorem 3.6.6, the assumption that $\|e^{-Q}\| < 1$ may be discarded because $\int_{(0, t]} e^{-sQ} \, dY(s)$ converges as $t \to \infty$ if and only if, for arbitrary $a > 0$, $\int_{(0, at]} e^{-sQ} \, dY(s) = \int_{(0, t]} e^{-s(aQ)} \, dY(as)$ converges as $t \to \infty$; in the last integral one can choose $a > 0$ such that $\|e^{-aQ}\| < 1$. Note that $Y(a \cdot)$ is also a $D(\mathbb{R}^d, [0, \infty))$-valued random variable with stationary independent increments and $\mathscr{L}(Y(a)) \in \text{ID}_{\log}$ for all $a > 0$ if and only if $\mathscr{L}(Y(1)) \in \text{ID}_{\log}$, where $\text{ID}_{\log}$ is the class of all ID measures with finite logarithmic moment.

From these preliminary results we are ready to obtain the random integral representation of an $\exp(-tQ)$-decomposable random variable mentioned at the beginning of this section.

Theorem 3.6.8. *Let $\{\exp(-tQ): t > 0\}$ be a one-parameter semigroup with the property that $e^{-tQ} \to 0$ as $t \to \infty$. Then X is an $\exp(-tQ)$-decomposable random variable if and only if there exists a $D(\mathbb{R}^d, [0, \infty))$-valued random variable Y such that Y has stationary independent increments, $Y(0) = 0$ a.s., and*

$$\int_{(0, t]} e^{-sQ} \, dY(s) \stackrel{d}{\to} X \quad \text{as } t \to \infty;$$

in which case, the random variable X_t in (3.6.1) satisfies

$$X_t \stackrel{d}{=} \int_{(0, t]} e^{-sQ} \, dY(s) \quad \text{for each } t > 0. \quad (3.6.14)$$

Proof. For the sufficiency, let $X \stackrel{d}{=} \int_0^\infty e^{-sQ}\, dY(s)$. Then, for each $t > 0$,

$$X \stackrel{d}{=} \int_{(t,\infty]} e^{-sQ}\, dY(s) + \int_{(0,t]} e^{-sQ}\, dY(s)$$

$$\stackrel{d}{=} e^{-tQ} \int_0^\infty e^{-sQ}\, dY(s+t) + \int_{(0,t]} e^{-sQ}\, dY(s).$$

Note that the last two terms are independent by the independent increments of Y and also

$$\int_0^\infty e^{-sQ}\, dY(s+t) \stackrel{d}{=} \int_0^\infty e^{-sQ}\, dY(s) \stackrel{d}{=} X$$

by the stationarity and independence (cf. Lemma 3.6.3). Thus, $X \stackrel{d}{=} e^{-tQ}X + X_t$ with X and X_t independent and with X_t as in (3.6.14). Therefore, X is $\exp(-tQ)$-decomposable. Furthermore, the distribution of X_t is uniquely determined since the characteristic function of X never vanishes; X is infinitely divisible (cf. also Proposition 2.3.7). Therefore, (3.6.14) must hold.

Now, for the necessity, assume X is $\exp(-tQ)$-decomposable. We claim that this implies that there exists an $\mathbb{R}^d$-valued process $\{Z_t : t \geq 0\}$ with independent increments such that $Z_0 = 0$ a.s. and, for $t, u \geq 0$,

$$Z_{t+u} - Z_t \stackrel{d}{=} e^{-tQ}X \quad \text{in } \mathbb{R}^d, \tag{3.6.15}$$

so that, in particular,

$$Z_t \stackrel{d}{=} X_t \quad \text{in } \mathbb{R}^d. \tag{3.6.16}$$

To see this, it suffices to show that if (3.6.15) and (3.6.16) hold for two particular values, say t and u in $[0, \infty)$, and Z_t is independent of $Z_{t+u} - Z_t$, then (3.6.16) holds for $t + u$. From (3.6.1), it follows that

$$e^{-(t+u)Q}X + X_{t+u} \stackrel{d}{=} X \stackrel{d}{=} e^{-tQ}\left(e^{-uQ}X + X_u\right) + X_t$$

$$= e^{-(t+u)Q}X + e^{-tQ}X_u + X_t,$$

with each of X_{t+u}, X_u, and X_t independent of X. Since the characteristic function of X never vanishes, we have

$$X_{t+u} \stackrel{d}{=} e^{-tQ}X_u + X_t \quad \text{in } \mathbb{R}^d.$$

Consequently,

$$Z_{t+u} = (Z_{t+u} - Z_t) + Z_t \stackrel{d}{=} e^{-tQ}X_u + X_t \stackrel{d}{=} X_{t+u} \quad \text{in } \mathbb{R}^d$$

as was to be proved. Let Z be a $D(\mathbb{R}^d, [0, \infty))$-valued version of the process $(Z_t)_{t \in \mathbb{R}^+}$ with independent increments. Now, set

$$Y(t) := \int_{(0,t]} e^{sQ} \, dZ(s) \quad \text{for } t > 0.$$

From Lemma 3.6.3, we see that Y is a $D(\mathbb{R}^d, [0, \infty))$-valued random variable with independent increments. Even more, Y has stationary increments. To see this, note that

$$Z(t + \cdot) - Z(t) \stackrel{d}{=} e^{-tQ}Z(\cdot) \quad \text{in } D(\mathbb{R}^d, [0, \infty)),$$

since both sides have independent increments and the same marginal distributions by (3.6.15) and (3.6.16) (cf. Theorem 1.9.9). Consequently, for fixed $t, u \in \mathbb{R}^+$, we get

$$Y(t + u) - Y(t) = \int_{(t, t+u]} e^{sQ} \, dZ(s) = \int_{(0, u]} e^{sQ} e^{tQ} \, dZ(s + t)$$

$$\stackrel{d}{=} \int_{(0, u]} e^{sQ} \, dZ(s) = Y(u) \quad \text{in } \mathbb{R}^d.$$

Finally, from (3.6.3) and the definition of Y,

$$\int_{(0,t]} e^{-sQ} \, dY(s) = \int_{(0,t]} e^{-sQ} e^{sQ} \, dZ(s) = Z_t.$$

Hence, by (3.6.16) and (3.6.1),

$$\int_{(0,t]} e^{-sQ} \, dY(s) \stackrel{d}{=} X_t \stackrel{d}{\to} X \quad \text{as } t \to \infty,$$

which completes the proof. Q.E.D.

Let $L_0(Q)$ denote the set of all $\exp(-tQ)$-decomposable measures [cf. (3.4.1) and (3.6.1)] and let $\mathrm{ID}_{\log}$ denote the set of all infinitely divisible measures μ with $\int \log(1 + \|x\|)\mu(dx) < \infty$. Theorems 3.6.6 and 3.6.8 suggest the following mapping:

$$\mathcal{T}_Q : \mathrm{ID}_{\log} \to L_0(Q)$$

by means of the formula

$$\mathcal{T}_Q(\mu) := \mathcal{L}\left(\int_0^\infty e^{-tQ} \, dY(t)\right), \qquad (3.6.17)$$

where Y is a $D(\mathbb{R}^d, [0, \infty))$-valued random variable with stationary independent increments, $Y(0) = 0$ a.s., and $\mathcal{L}(Y(1)) = \mu$. In other words, Theorem 3.6.8 gives

$$L_0(Q) = \mathcal{T}_Q(\mathrm{ID}_{\log})$$
$$= \left\{\mathcal{L}\left(\int_0^\infty e^{-tQ} \, dY(t)\right) : Y \text{ is a } D(\mathbb{R}^d, [0, \infty))\text{-valued r.v. with}\right.$$

stationary independent increments,

$$\left. Y(0) = 0 \text{ a.s., and } Y(1) \in \mathrm{ID}_{\log} \right\}. \qquad (3.6.18)$$

If we set $[a^\infty, R^\infty, M^\infty] := \mathcal{T}_Q([a, R, M])$, then a^∞, R^∞, and M^∞ are given by (3.6.9) to (3.6.11), respectively. Consequently, the equality $\mathcal{T}_Q(\mathrm{ID}_{\log}) = L_0(Q)$ together with Corollary 3.4.1 gives the following statements.

Remark 3.6.9

1. A Lévy measure M satisfies the inequality $M \le e^{-tQ}M$ for all $t > 0$ if and only if there exists a Lévy measure G with the property $\int_{\|x\|>1} \log(1 + \|x\|)G(dx) < \infty$ such that

$$M(F) = \int_0^\infty G(e^{tQ}F) \, dt \quad \text{for every } F \in \mathcal{B}(\mathbb{R}^d \setminus \{0\}).$$

2. A Gaussian covariance operator R satisfies the inequality $R \ge e^{-tQ}Re^{-tQ^*}$ for all $t > 0$ if and only if there exists a Gaussian

covariance operator T such that

$$R = \int_0^\infty e^{-tQ} T e^{-tQ^*} \, dt.$$

3. The operator T and the Lévy measure G in (1) and (2) are uniquely determined (cf. Proposition 3.6.10).
4. A measure μ is in $L_0(Q)$ if and only if there exists a unique measure $\nu \in \mathrm{ID}_{\log}$ such that

$$\hat{\mu}(y) = \exp \int_0^\infty \log \hat{\nu}(e^{-tQ^*}y) \, dt.$$

This remark shows that the random integral representation provides an alternative method of solving the operator and the Lévy measure inequalities that describe the class $L_0(Q)$ (cf. Proposition 3.4.2, Theorem 3.4.5, and Corollary 3.4.8).

Our next goal is to describe some algebraic and topological properties of the mapping $\mathcal{T}_Q$ acting between the topological semigroups $L_0(Q)$ and $\mathrm{ID}_{\log}$ (cf. Proposition 3.5.1 and Remark 1.8.14 on $\mathrm{ID}_{\log}$).

Proposition 3.6.10

(a) *The mapping $\mathcal{T}_Q$ is an algebraic isomorphism between the semigroups $L_0(Q)$ and $\mathrm{ID}_{\log}$.*
(b) *$\mathcal{T}_Q(\mu^{*s}) = (\mathcal{T}_Q\mu)^{*s}$ for $s > 0$ and $\mu \in \mathrm{ID}_{\log}$.*
(c) *If the linear operator T commutes with Q, then*

$$\mathcal{T}_Q(T\mu) = T\mathcal{T}_Q(\mu) \quad \text{for } \mu \in \mathrm{ID}_{\log}.$$

(d) *For all $s > 0$, $x \in \mathbb{R}^d$, and $\mu \in \mathrm{ID}_{\log}$,*

$$\mathcal{T}_Q(s^Q \mu * \delta(x)) = s^Q \mathcal{T}_Q(\mu) * \delta(Q^{-1}x).$$

Proof

(a) Theorems 3.6.6 and 3.6.8 show that $\mathcal{T}_Q$ maps $\mathrm{ID}_{\log}$ onto $L_0(Q)$. Next, we show that $\mathcal{T}_Q$ is one-to-one. By the definition of $\mathscr{T}_Q$ and Lemma 3.6.4, we have, for all $s > 0$,

$$\log(\mathscr{T}_Q\mu)\hat{\ }(s^{Q^*}y) = \int_0^\infty \log \hat{\mathscr{L}}(Y(1))(e^{-tQ^*}s^{Q^*}y)\,dt$$
$$= \int_0^s \log \hat{\mathscr{L}}(Y(1))(t^{Q^*}y)t^{-1}\,dt.$$

Hence, for each $y \in \mathbb{R}^d$ and $s > 0$,

$$s\left(\frac{d}{ds}\right)\log(\mathscr{T}_Q\mu)\hat{\ }(s^{Q^*}y) = \log \hat{\mathscr{L}}(Y(1))(s^Q y).$$

Setting $s = 1$, we obtain

$$\hat{\mathscr{L}}(Y(1))(y) = \exp\left\{\left(\frac{d}{ds}\right)\log(\mathscr{T}_Q\mu)\hat{\ }(s^{Q^*}y)\bigg|_{s=1}\right\}.$$

This implies that $\mathscr{L}(Y(1))$ is uniquely determined by $\mathscr{T}_Q\mu$, that is, $\mathscr{T}_Q$ is one-to-one.

To see that $\mathscr{T}_Q$ is a homeomorphism, let $\mu, \nu \in \mathrm{ID}_{\log}$ with $\mu = [a, R, M]$ and $\nu = [b, S, N]$. From (3.6.9) to (3.6.11), we see that

$$(M+N)^\infty = M^\infty + N^\infty, \quad (R+S)^\infty = R^\infty + S^\infty,$$
$$(a+b)^\infty = a^\infty + b^\infty.$$

Therefore, $\mathscr{T}_Q$ is a homeomorphism.

(b) Note that $\mu \in \mathrm{ID}_{\log}$ if and only if $\mu^{*s} \in \mathrm{ID}_{\log}$ for every $s > 0$. From (3.6.9) to (3.6.11), we obtain the desired equality in (b).

(c) Since T commutes with Q,

$$T\mathscr{L}\left(\int_0^\infty e^{-tQ}\,dY(t)\right) = \mathscr{L}\left(\int_0^\infty Te^{-tQ}\,dY(t)\right)$$
$$= \mathscr{L}\left(\int_0^\infty e^{-tQ}\,d(TY(t))\right).$$

Note that TY is a $D(\mathbb{R}^d, [0, \infty))$-valued r.v. with stationary independent increments and $TY(0) = 0$ a.s. Also, $\mathscr{L}(TY(1)) \in \text{ID}_{\log}$ because

$$\int \log(1 + \|x\|)(T\mu)(dx) \le \log(1 + \|T\|)$$
$$+ \int \log(1 + \|x\|)\mu(dx) < \infty.$$

Hence, (c) is established.

(d) Since $\int_0^\infty e^{-tQ} a\, dt = Q^{-1}a$ for all $a \in \mathbb{R}^d$, (d) follows from (a) and (c). Q.E.D.

Corollary 3.6.11. *Let A be invertible and $\mu \in \text{ID}_{\log}$. Then*

$$\mathscr{T}_{AQA^{-1}}(\mu) = A\mathscr{T}_Q(A^{-1}\mu);$$

consequently,

$$L_0(AQA^{-1}) = AL_0(Q).$$

Proof. We have

$$\mathscr{T}_{AQA^{-1}}(\mu) = \mathscr{L}\left(A\int_0^\infty e^{-tQ}\, d(A^{-1}Y(t))\right) = A\mathscr{T}_Q(A^{-1}\mu).$$

Since, for invertible A, $A\mu \in \text{ID}_{\log}$ if and only if $\mu \in \text{ID}_{\log}$, this first equality and (3.6.18) implies the second equality, that is, $L_0(AQA^{-1}) = AL_0(Q)$. Q.E.D.

Now, after establishing the algebraic properties of the mapping $\mathscr{T}_Q$, we wish to describe its topological properties, with the topology of weak convergence on $\mathscr{P}$. First, note that $\int \log(1 + \|x\|)\gamma(dx) < \infty$ if and only if $\int \log(1 + \|x\|^2)\gamma(dx) < \infty$ for all $\gamma \in \mathscr{P}$. In particular, for the Lévy measure M of a $\mu \in \text{ID}_{\log}$, we have that, for all $\varepsilon > 0$,

$$\int_{\|x\|>\varepsilon} \log(1 + \|x\|^2)M(dx) < \infty$$

by Theorem 3.6.6. Thus, we may define, for $F \in \mathscr{B}(\mathbb{R}^d \setminus \{0\})$,

$$\tilde{M}(F) := \int_F \log(1 + \|x\|^2)M(dx). \tag{3.6.19}$$

Then $\tilde{M}$ is a finite measure. The following lemma concerns the weak convergence of a sequence $\{M_n\}$ of Lévy measures with this integral condition and of the corresponding sequence $\{\tilde{M}_n\}$.

Lemma 3.6.12. *Let M and M_n, $n \geq 1$, be Lévy measures with finite logarithmic moments outside every neighborhood of zero and let $\tilde{M}$ and $\tilde{M}_n$ be given by (3.6.19). Then*

$$M_n \Rightarrow M \text{ outside very neighborhood of zero}$$

and

$$\lim_{s \to \infty} \sup_{n \geq 1} \int_{\|x\| > s} \log(1 + \|x\|^2) M_n(dx) = 0$$

if and only if

$$\tilde{M}_n \Rightarrow \tilde{M} \text{ outside every neighborhood of zero}.$$

Proof. First, we consider the "only if" part. Let $f: \mathbb{R}^d \to \mathbb{R}^1$ be continuous, bounded by $K > 0$, and vanishes in some neighborhood of zero. Let $\varepsilon > 0$ be given and select $s > 0$ so that $M(\{x: \|x\| = s\}) = 0$ and

$$\sup_{n \geq 1} \int_{\|x\| > s} \log(1 + \|x\|^2) M_n(dx) < \frac{\varepsilon}{K}.$$

Since $f(x)\log(1 + \|x\|^2) I_{\{\|x\| \leq s\}}(x)$ is bounded and vanishes in a neighborhood of zero and its set of discontinuity points has M-measure zero, we have

$$\lim_{n \to \infty} \int_{\|x\| \leq s} f(x) \tilde{M}_n(dx) = \lim_{n \to \infty} \int_{\|x\| \leq s} f(x) \log(1 + \|x\|^2) M_n(dx)$$

$$= \int_{\|x\| \leq s} f(x) \log(1 + \|x\|^2) M(dx)$$

$$= \int_{\|x\| \leq s} f(x) \tilde{M}(dx).$$

Also,

$$\sup_{n\geq 1}\left|\int_{\|x\|\geq s} f(x)\tilde{M}_n(dx)\right| \leq \sup_{n\geq 1}\int_{\|x\|>s} K\log(1+\|x\|^2)M_n(dx) < \varepsilon.$$

Hence,

$$\lim_{n\to\infty}\int f(x)\tilde{M}_n(dx) = \int f(x)\tilde{M}(dx),$$

which shows that $\tilde{M}_n \Rightarrow \tilde{M}$ outside every neighborhood of zero.

Now, for the converse, note that $\tilde{M}_n \Rightarrow \tilde{M}$ outside every neighborhood of zero implies that the sequence $\{\tilde{M}_n\}$ is tight. Thus,

$$0 = \lim_{s\to\infty}\sup_{n\geq 1}\tilde{M}_n(\|x\|>s) = \lim_{s\to\infty}\sup_{n\geq 1}\int_{\|x\|>s}\log(1+\|x\|^2)M_n(dx).$$

Furthermore, if f is continuous, bounded, and vanishes in some neighborhood of zero, so does $g(x) := f(x)/\log(1+\|x\|^2)$. Hence,

$$\lim_{n\to\infty}\int f(x)M_n(dx) = \lim_{n\to\infty}\int g(x)\tilde{M}_n(dx)$$

$$= \int g(x)\tilde{M}(dx) = \int f(x)M(dx).$$

Therefore, $M_n \Rightarrow M$ outside every neighborhood of zero. Q.E.D.

Theorem 3.6.13. *Let $\mu_n = [a_n, R_n, M_n]$ and $\mu = [a, R, M]$ belong to $\mathrm{ID}_{\log}$ for $n \geq 1$. Then*

$$\mu_n \Rightarrow \mu \quad \text{and} \quad \lim_{s\to\infty}\sup_{n\geq 1}\int_{\|x\|>s}\log(1+\|x\|^2)M_n(dx) = 0$$

if and only if

$$\mathcal{T}_Q(\mu_n) \Rightarrow \mathcal{T}_Q(\mu).$$

Proof. Let

$$[a_n^\infty, R_n^\infty, M_n^\infty] := \mathcal{T}_Q(\mu_n), \quad [a^\infty, R^\infty, M^\infty] := \mathcal{T}_Q(\mu)$$

[cf. (3.6.9) to (3.6.11) and (3.6.17)]. Also, for a Lévy measure M and

$\varepsilon > 0$, let $T_{M,\varepsilon}$ be the operator defined by

$$\langle T_{M,\varepsilon} y, y \rangle := \int_{B_\varepsilon} \langle y, x \rangle^2 M(dx) \quad \text{for } y \in \mathbb{R}^d, \qquad (3.6.20)$$

where $B_\varepsilon := \{x \in \mathbb{R}^d : \|x\| < \varepsilon\}$.

Sufficiency

From Corollary 1.8.16, we need to show that the following conditions:

(a) $\lim_{n \to \infty} \int f(x) M_n(dx) = \int f(x) M(dx)$ for every f which is continuous, bounded, and vanishes in some neighborhood of zero, and

$$\lim_{s \to \infty} \sup_{n \geq 1} \int_{B_s^c} \log(1 + \|x\|^2) M_n(dx) = 0;$$

(b) $\lim_{\varepsilon \downarrow 0} \limsup_{n \to \infty} \left(\langle R_n y, y \rangle + \langle T_{M_n, \varepsilon} y, y \rangle \right)$

$$= \lim_{\varepsilon \downarrow 0} \liminf_{n \to \infty} \left(\langle R_n y, y \rangle + \langle T_{M_n, \varepsilon} y, y \rangle \right)$$

$$= \langle R y, y \rangle \quad \text{for all } y \in \mathbb{R}^d;$$

(c) $\lim_{n \to \infty} a_n = a$

imply that

(a′) $\lim_{n \to \infty} \int f(x) M_n^\infty(dx) = \int f(x) M^\infty(dx)$ for every f which is continuous, bounded, and vanishes in some neighborhood of zero;

(b′) $\lim_{\varepsilon \downarrow 0} \limsup_{n \to \infty} \left(\langle R_n^\infty y, y \rangle + \langle T_{M_n^\infty, \varepsilon} y, y \rangle \right)$

$$= \lim_{\varepsilon \downarrow 0} \liminf_{n \to \infty} \left(\langle R_n^\infty y, y \rangle + \langle T_{M_n^\infty, \varepsilon} y, y \rangle \right)$$

$$\langle R^\infty y, y \rangle \quad \text{for all } y \in \mathbb{R}^d;$$

(c′) $\lim_{n \to \infty} a_n^\infty = a^\infty$,

where M_n^∞, M^∞, R_n^∞, a_n^∞, and a^∞ are given by (3.6.9) to (3.6.11).

Let f be as given in (a'). Then

$$g(x) := \int_0^\infty f(e^{-tQ}x)\, dt, \qquad x \in \mathbb{R}^d,$$

is continuous because $f(e^{-tQ}x)$ is zero for all sufficiently large t so, for each x, $g(x)$ is obtained by integrating a bounded continuous function of (t, x) over a bounded t-interval. Since f vanishes on some ball B_ε about zero and, for some $a > 0$, $\|e^{-tQ}\| \leq a$ for all $t \geq 0$, we see that g must also vanish on the ball $B_{\varepsilon/a}$ about zero. Also, there is $K > 0$ such that, for all x,

$$|f(x)| \leq \frac{K\|x\|^2}{1 + \|x\|^2},$$

because f is bounded and vanishes around zero. Thus, by Lemma 3.4.7 for all x,

$$|g(x)| \leq Kc_6 \log(1 + \|x\|^2).$$

Therefore, using Lemma 3.6.12,

$$\lim_{n\to\infty} \int f(x) M_n^\infty(dx) = \lim_{n\to\infty} \int \int_0^\infty f(e^{-tQ}x)\, dt\, M_n(dx)$$

$$= \lim_{n\to\infty} \int \frac{g(x)}{\log(1 + \|x\|^2)} \tilde{M}_n(dx)$$

$$= \int \frac{g(x)}{\log(1 + \|x\|^2)} \tilde{M}(dx) = \int f(x) M^\infty(dx).$$

Thus, (a') holds.

Let

$$S_{n,\varepsilon} := R_n + T_{M_n,\varepsilon},$$

$$S_{n,\varepsilon}^\infty := \int_0^\infty e^{-tQ} S_{n,\varepsilon} e^{-tQ^*}\, dt,$$

$$I_{n,\varepsilon}(y) := \int_0^\infty \int \bigl(I_{B_\varepsilon}(e^{-tQ}x) - I_{B_\varepsilon}(x)\bigr)\langle e^{-tQ^*}y, x\rangle^2 M_n(dx)\, dt$$

for $y \in \mathbb{R}^d$. Since

$$\int g(x) M^\infty(dx) = \int_0^\infty \int g(e^{-tQ}x) M(dx)\, dt,$$

from (3.6.20) we see that

$$\langle T_{M^\infty, \varepsilon} y, y \rangle = \int_0^\infty \int I_{B_\varepsilon}(e^{-tQ}x) \langle e^{-tQ^*}y, x \rangle^2 M(dx)\, dt.$$

Hence,

$$\begin{aligned}\langle R_n^\infty y, y \rangle + \langle T_{M_n^\infty, \varepsilon} y, y \rangle &= \int_0^\infty \langle e^{-tQ} S_{n,\varepsilon} e^{-tQ^*} y, y \rangle\, dt + I_{n,\varepsilon}(y) \\ &= \langle S_{n,\varepsilon}^\infty y, y \rangle + I_{n,\varepsilon}(y). \quad (3.6.21)\end{aligned}$$

We will show $I_{n,\varepsilon}(y) \to 0$ as $n \to \infty$ and then $\varepsilon \downarrow 0$. We have $|I_{n,\varepsilon}(y)| \leq I'_{n,\varepsilon}(y) + I''_{n,\varepsilon}(y)$, where

$$I'_{n,\varepsilon}(y) := \int_0^\infty \int I_{B_\varepsilon}(e^{-tQ}x) I_{B_\varepsilon^c}(x) \langle e^{-tQ^*}y, x \rangle^2 M_n(dx)\, dt,$$

$$I''_{n,\varepsilon}(y) := \int_0^\infty \int I_{B_\varepsilon^c}(e^{-tQ}x) I_{B_\varepsilon}(x) \langle e^{-tQ^*}y, x \rangle^2 M_n(dx)\, dt.$$

Let

$$k(x, t) := \frac{\|e^{-tQ}x\|^2}{1 + \|e^{-tQ}x\|^2} \quad \text{for } x \in \mathbb{R}^d \text{ and } t \in \mathbb{R}^+.$$

Then

$$I'_{n,\varepsilon}(y) \leq (1 + \varepsilon^2) \|y\|^2 \int_0^\infty \int I_{B_\varepsilon^c \cap (e^{tQ} B_\varepsilon)}(x) k(x, t) M_n(dx)\, dt.$$

Let

$$B_I := \left\{ x : \|x\| < \frac{\varepsilon e^{c_2 t}}{c_1} \right\} = B_{\varepsilon c_1^{-1} e^{c_2 t}}.$$

By (3.4.6),
$$B_\varepsilon^c \cap B_I \supseteq B_\varepsilon^c \cap (e^{tQ} B_\varepsilon),$$
where c_1 and c_2 are positive constants. Hence,

$$I'_{n,\varepsilon}(y) \le (1+\varepsilon^2)\|y\|^2 \int_0^\infty \int I_{B_\varepsilon^c \cap B_I}(x) k(x,t) M_n(dx)\, dt$$

$$\le (1+\varepsilon^2)\|y\|^2 \Bigg\{ \int_0^\infty \int_{\varepsilon c_1^{-1} \le \|x\| < \varepsilon c_1^{-1} e^{c_2 t}} k(x,t) M_n(dx)\, dt$$

$$+ \int_0^\infty \int_{\varepsilon \le \|x\| < \varepsilon c_1^{-1}} k(x,t) M_n(dx)\, dt \Bigg\}$$

$$\le (1+\varepsilon^2)\|y\|^2 \Bigg\{ c_6 \int_{\varepsilon \le \|x\| < \varepsilon c_1^{-1}} \log(1+\|x\|^2) M_n(dx)$$

$$+ \int g_\varepsilon(x) M_n(dx) \Bigg\},$$

where
$$g_\varepsilon(x) := I_{B_{\varepsilon c_1^{-1}}^c}(x) \int_{\log(\varepsilon^{-1} c_1 \|x\|)}^\infty k(x,t)\, dt,$$

and the inequalities above use (3.4.7). From Lemma 3.4.7, we have $0 \le g_\varepsilon(x) \le c_6 I_{B_{\varepsilon c_1^{-1}}^c}(x) \log(1+\|x\|^2)$. Also, for all x, $g_\varepsilon(x) \to 0$ as $\varepsilon \downarrow 0$, and for all $\varepsilon > 0$, g_ε vanishes around zero. Note that the set of discontinuity points of g_ε is contained in $\{x: \|x\| = c_1^{-1}\varepsilon\}$ which we may assume has M-measure zero. Thus, by (a), Lemma 3.6.12, and Corollary 1.6.5, we have $\lim_{\varepsilon \downarrow 0} \limsup_{n \to \infty} I'_{n,\varepsilon}(y) = 0$.

For $I''_{n,\varepsilon}(y)$, let $B_{II} := \{x: \|x\| < \varepsilon e^{c_4 t}/c_3\} = B_{c_3^{-1}\varepsilon e^{c_4 t}}$, where c_3 and c_4 are from (3.4.6). Then

$$I''_{n,\varepsilon}(y) \le \int_0^\infty \int I_{B_\varepsilon \cap B_{II}^c}(x) \|y\|^2 c_3^2 e^{-2c_4 t} \|x\|^2 M_n(dx)\, dt$$

$$\le c_3^2 \|y\|^2 \int_{c_3^{-1}\varepsilon \le \|x\| < \varepsilon} \|x\|^2 \int_0^{c_4^{-1} \log(c_3 \|x\|/\varepsilon)} e^{-2c_4 t}\, dt\, M_n(dx)$$

$$= \frac{c_3^2 \|y\|^2}{2c_4} \int_{c_3^{-1}\varepsilon \le \|x\| < \varepsilon} \left(\|x\|^2 - \frac{\varepsilon^2}{c_3^2} \right) M_n(dx).$$

Therefore, from (a), $\lim_{\varepsilon \downarrow 0} \limsup_{n \to \infty} I''_{n,\varepsilon}(y) = 0$. Consequently, $\lim_{\varepsilon \downarrow 0} \limsup_{n \to \infty} I_{n,\varepsilon}(y) = 0$. Together with (b) and (3.6.21), we have

$$\lim_{\varepsilon \downarrow 0} \limsup_{n \to \infty} (R_n^\infty + T_{M_n^\infty,\varepsilon}) \leq R^\infty.$$

Using (b), the definition of $S_{n,\varepsilon}$, and Fatou's lemma, we also have

$$\langle R^\infty y, y \rangle = \int_0^\infty \lim_{\varepsilon \downarrow 0} \sup \inf_{n \to \infty} \langle e^{-tQ^*} S_{n,\varepsilon} e^{-tQ^*} y, y \rangle \, dt$$

$$\leq \lim_{\varepsilon \downarrow 0} \liminf_{n \to \infty} \left(\langle R_n^\infty y, y \rangle + \langle T_{M_{n,\varepsilon}^\infty} y, y \rangle \right) \leq \langle R^\infty y, y \rangle.$$

Therefore, (b') is established.

It remains to prove (c'). First, note that

$$a_n^\infty = Q^{-1} a_n + \int h(x) M_n(dx),$$

where, for $x \in \mathbb{R}^d$,

$$h(x) := \int_0^\infty e^{-tQ} x \left(\frac{\|x\|^2 - \|e^{-tQ}x\|^2}{(1 + \|e^{-tQ}x\|^2)(1 + \|x\|^2)} \right) dt.$$

Since $a_n \to a$ by (c), it is sufficient to show that

$$\lim_{n \to \infty} \int \langle y, h(x) \rangle M_n(dx) = \int \langle y, h(x) \rangle M(dx) \quad (3.6.22)$$

for all $y \in \mathbb{R}^d$. The function h is the same as in (3.4.10); therefore, $\|h(x)\| \leq K \log(1 + \|x\|^2)$ for some constant K. Thus, by Lemma 3.6.12, for $y \in \mathbb{R}^d$ and B_ε a continuity set of M,

$$\lim_{n \to \infty} \int_{B_\varepsilon^c} \langle y, h(x) \rangle M_n(dx) = \int_{B_\varepsilon^c} \langle y, h(x) \rangle M(dx).$$

On the other hand, using the inequalities from Lemma 3.4.7, we obtain

$$\left| \int_{B_\varepsilon} \langle y, h(x) \rangle M_n(dx) \right|$$

$$\leq \|y\| \left\{ \int_{B_\varepsilon} \int_0^\infty \|e^{-tQ}x\| \frac{\|x\|^2}{(1 + \|e^{-tQ}x\|^2)(1 + \|x\|^2)} \, dt \, M_n(dx) \right.$$

$$\left. + \int_{B_\varepsilon} \int_0^\infty \|e^{-tQ}x\| \frac{\|e^{-tQ}x\|^2}{(1 + \|e^{-tQ}x\|^2)(1 + \|x\|^2)} \, dt \, M_n(dx) \right\}$$

$$\leq \|y\| \left\{ c_3 \varepsilon \int_{B_\varepsilon} \int_0^\infty e^{-c_4 t} \frac{\|x\|^2}{1 + \|x\|^2} \, dt \, M_n(dx) \right.$$

$$\left. + \varepsilon c_3 \int_{B_\varepsilon} \int_0^\infty e^{-c_4 t} \frac{c_3^2 \|x\|^2}{1 + c_3^2 \|x\|^2} \, dt \, M_n(dx) \right\}$$

$$\leq \|y\| \frac{c_3 \varepsilon}{c_4} \left\{ \int_{B_\varepsilon} \frac{\|x\|^2}{1 + \|x\|^2} M_n(dx) + c_3^2 \int_{B_\varepsilon} \frac{\|x\|^2}{1 + c_3^2 \|x\|^2} M_n(dx) \right\}$$

$$\leq C \|y\| \varepsilon \int \frac{\|x\|^2}{1 + \|x\|^2} M_n(dx)$$

$$\leq C \|y\| \varepsilon \sup_{n \geq 1} \int (1 + \|x\|^2) M_n(dx) < \infty,$$

where $C := c_4^{-1} c_3 (1 + c_3^2)$ and the last supremum is finite because by (a) $\{M_n\}$ is weakly convergent on B_1^c and by (b) $\sup_{n \in \mathbb{N}} \|T_{M_{n,1}}\| < \infty$. Consequently,

$$\lim_{\varepsilon \downarrow 0} \sup_{n \geq 1} \int_{B_\varepsilon} \langle y, h(x) \rangle M_n(dx) = 0,$$

which proves (3.6.22) and completes the proof of the sufficiency.

Necessity

Assume $[a_n^\infty, R_n^\infty, M_n^\infty] \Rightarrow [a^\infty, R^\infty, M^\infty]$.

From Corollary 1.8.16, we have

$$\sup_{n \geq 1} \left\{ \|a_n^\infty\|, \|R_n^\infty\|, \int \frac{\|x\|^2}{1 + \|x\|^2} M_n^\infty(dx) \right\} < \infty \qquad (3.6.23)$$

and $\{M_n^\infty\}$ is tight on every set bounded away from zero in $\mathbb{R}^d$.

$$(3.6.24)$$

By Proposition 3.4.2 and (3.6.9), we have $QR_n^\infty + R_n^\infty Q = R_n$. Consequently, from (3.6.23), we obtain

$$\sup_{n \geq 1} \|R_n\| \leq 2\|Q\| \sup_{n \geq 1} \|R_n^\infty\| < \infty.$$

For $\eta > 0$, select $r > 0$ and $t_0 > 0$ so that

$$B_\eta^c \subseteq \left(e^{tQ} B_r\right)^c \text{ for every } t \in [0, t_0].$$

From (3.6.24), we see that, for $\varepsilon > 0$, there exists a compact set $K_\varepsilon \subseteq B_r^c$ such that $M_n^\infty(B_r^c \setminus K_\varepsilon) \leq \varepsilon$. Note that

$$\tilde{K}_\varepsilon := \left(\bigcup_{0 \leq t \leq r_0} e^{tQ} K_\varepsilon \right) \cap B_\eta^c$$

is contained in B_η^c and is compact. Also,

$$\varepsilon \geq M_n^\infty(B_r^c \setminus K_\varepsilon) \geq \int_0^{t_0} M_n\left(\left(e^{tQ} B_r\right)^c \setminus e^{tQ} K_\varepsilon\right) dt$$

$$\geq \int_0^{t_0} M_n\left(\left(e^{tQ} B_r\right)^c \cap B_\eta^c \setminus \left(e^{tQ} K_\varepsilon\right) \cap B_\eta^c\right) dt$$

$$\geq t_0 M_n\left(B_\eta^c \setminus \tilde{K}_\varepsilon\right)$$

for all $n \geq 1$. Thus, we see that $\{M_n\}$ is tight on any set bounded away from zero in $\mathbb{R}^d$.

From Lemma 3.4.7 and formulas (3.6.10) and (3.6.23), we obtain

$$\sup_{n\geq 1} \int \log(1 + \|x\|^2) M_n(dx) \leq c_5^{-1} \sup_{n\geq 1} \int_0^\infty \int k(x,t) M_n(dx)\, dt$$

$$= c_5^{-1} \sup_{n\geq 1} \int \frac{\|z\|^2}{1+\|z\|^2} M_n^\infty(dz) < \infty. \quad (3.6.25)$$

Note that

$$a_n^\infty = Q^{-1} a_n + \int h(x) M_n(dx),$$

where h is given by (3.4.10). Hence, for some $K > 0$,

$$\|h(x)\| \leq K \log(1 + \|x\|^2)$$

(cf. the proof of Corollary 3.4.8). Hence, (3.6.25) implies that

$$\sup_{n\geq 1} \int \|h(x)\| M_n(dx) < \infty$$

and consequently we obtain $\sup_{n\geq 1} \|a_n\| < \infty$. Summarizing these facts, we have

$$\{[a_n, R_n, M_n]\}_{n\in\mathbb{N}} \text{ is conditional compact in ID}(\mathbb{R}^d). \quad (3.6.26)$$

The next step is to prove that

$$\lim_{s\to\infty} \sup_{n\geq 1} \int_{\|x\|>s} \log(1+\|x\|^2) M_n(dx) = 0. \quad (3.6.27)$$

Clearly, one may replace the norm $\|\ \|$ by the norm $\|\ \|_Q$ given by (3.4.3) (cf. Proposition 3.4.3). Let

$$\tilde{f}_n(t) := M_n(\{x: \|x\|_Q > t\})$$

for $t > 0$. Hence,

$$\int_{\|x\|_Q \geq s} \log\|x\|_Q M_n(dx) = \tilde{f}_n(s) \log s + \int_s^\infty \tilde{f}_n(t) t^{-1}\, dt.$$

Since, for $s > 1$,

$$2\int_{\sqrt{s}}^{\infty} \tilde{f}_n(t)t^{-1}\,dt \geq 2\int_{\sqrt{s}}^{s} \tilde{f}_n(t)t^{-1}\,dt \geq \tilde{f}_n(s)\log s,$$

we see that

$$\limsup_{n\to\infty} \int_{\|x\|_Q \geq s} \log\|x\|_Q M_n(dx) = 0 \text{ iff } \limsup_{s\to\infty} \int_s^{\infty} \tilde{f}_n(t)t^{-1}\,dt = 0. \tag{3.6.28}$$

Note that from (3.6.10) and the properties of polar coordinates

$$M_n^{\infty}\!\left([S_Q, [t, \infty)]\right) = \alpha \int_{t^{1/\alpha}}^{\infty} M_n\!\left([S_Q, [r^{\alpha}, \infty)]\right) r^{-1}\,dr$$

and

$$\tilde{f}_n(r) \leq M_n\!\left([S_Q, [r^{\alpha}, \infty)]\right)$$

for $r > 1$ and $\alpha := \|Q\|_Q^{-1}$ (cf. the last part of the proof of Theorem 3.4.5). Hence,

$$\limsup_{s\to\infty} \int_s^{\infty} \tilde{f}_n(t)t^{-1}\,dt \leq \limsup_{s\to\infty} \int_s^{\infty} M_n\!\left([S_Q, [t^{\alpha}, \infty)]\right) t^{-1}\,dt$$

$$= \alpha^{-1} \limsup_{s\to\infty} M_n^{\infty}\!\left([S_Q, [s^{\alpha}, \infty)]\right) = 0,$$

because of (3.6.24). This together with (3.6.28) gives (3.6.27). Finally, for any strictly increasing sequences $\{n'\}$ of positive integers, there is a subsequence $\{n''\}$ and $[a', R', M']$ in ID($\mathbb{R}^d$) such that

$$[a_{n''}, R_{n''}, M_{n''}] \Rightarrow [a', R', M']$$

[cf. (3.6.26)], and also by (3.6.27)

$$\limsup_{s\to\infty} \int_{\|x\|>s} \log(1 + \|x\|^2) M_{n''}(dx) = 0.$$

Thus, the first part of this theorem gives

$$\mathscr{T}_Q([a_{n''}, R_{n''}, M_{n''}]) \Rightarrow \mathscr{T}_Q([a', R', M']) = \mathscr{T}_Q([a, R, M])$$

and Proposition 3.6.10 implies that $[a', R', M'] = [a, R, M]$. Consequently, $[a_n, R_n, M_n] \Rightarrow [a, R, M]$ (cf. Corollary 1.6.3). Q.E.D.

Remark 3.6.14. In the preceding section, we found the generators for the class $L_0(Q)$. From Proposition 1.8.10, we have that the class ID($\mathbb{R}^d$) is generated by the Gaussian and Poisson measures. Since $L_0(Q) \subseteq$ ID, it is natural to consider how these generators are transformed by the mapping $\mathscr{T}_Q$. First, consider $a \in \mathbb{R}^d \setminus \{0\}$ and look at δ_a^∞. There are unique $z \in S_Q$ and $s > 0$ such that $\Phi(z, s) = a$. From (3.6.10), for $r > 0$ and $A \in \mathscr{B}(S_Q)$,

$$\delta_a^\infty([A, [r, \infty)]) = \int_0^\infty \delta_{\Phi(z,s)}([A, [e^t r, \infty)]) \, dt$$

$$= \delta_z(A) \int_{\{t > 0: \, s > e^t r\}} dt = \delta_z(A) \log\left(\max\left(1, \frac{s}{r}\right)\right)$$

$$= \delta_z(A) \int_r^\infty I_{(0, s]}(t) t^{-1} \, dt.$$

Hence, for $\lambda > 0$ and $[r, u) \subseteq (0, \infty)$,

$$(\lambda \delta_{\Phi(z,s)})^\infty([A, [r, u)]) = \lambda \delta_z(A) \int_r^u I_{(0, s]}(t) t^{-1} \, dt,$$

so, for arbitrary $F \in \mathscr{B}(\mathbb{R}^d \setminus \{0\})$,

$$(\lambda \delta_{\Phi(z,s)})^\infty(F) = \int_{S_Q} \int_0^s I_F(t^Q x) t^{-1} \, dt \, \lambda \delta_z(dx)$$

[cf. (3.5.5) and Theorem 3.5.5].

Second, for a covariance operator R, we have $R^\infty = \int_0^\infty e^{-tQ} R e^{-tQ^*} \, dt$ by (3.6.9), so $R = QR^\infty + R^\infty Q^*$ by Theorem 3.4.2. Therefore, these facts raise the possibility that all of $L_0(Q)$ is generated by these simple Poisson measures $[x, 0, (\lambda \delta_a)^\infty]$, $\lambda > 0$, $x \in \mathbb{R}^d$, and by Gaussian measures $[x, R, 0]$ such that $0 \leq QR + RQ^*$ (cf. Theorem 3.5.5).

Proposition 3.6.10 and Corollary 3.6.11 give algebraic properties of $L_0(Q)$ with respect to some linear operators, either commuting with Q

or invertible. From the random integral representation of measures in $L_0(Q)$, although it is possible to use the definition of $L_0(Q)$ directly, we will see how this class is unchanged under some idempotents. Note that in Sections 3.1 to 3.4, we worked under the very essential assumption that μ in $L_0(Q)$ is full on $\mathbb{R}^d$. We now introduce the following notation. For a subspace W of $\mathbb{R}^d$, let

$$L_Q(Q;W) := \{\mu \in \mathscr{P}(\mathbb{R}^d): W = \text{lin}(\text{supp } \mu), \text{ and, for all } t \geq 0,$$
$$\mu = e^{-tQ}\mu * \nu_t \text{ for some } \nu_t \in \mathscr{P}(\mathbb{R}^d) \text{ with supp } \nu_t \subseteq W\}.$$

Theorem 3.6.15. *If $P \in \text{End}(\mathbb{R}^d)$ is an idempotent, that is, $P^2 = P$, and the space $\ker P$ is Q-invariant, then for any $\mu \in L_0(Q;\mathbb{R}^d)$ we have $P\mu \in L_0(PQ|W;W)$, where $W := P(\mathbb{R}^d)$ and $PQ|W$ is the restriction of PQ to W.*

Proof. Note that $P|\ker P = 0$ and $\ker P = (I - P)(\mathbb{R}^d)$. Since $\ker P$ is Q-invariant, we obtain $Pe^{-tQ}(I - P) = 0$ for every $t \in \mathbb{R}^1$, and $PQ(I - P) = 0$. Hence, $PQ = PQP$ and by induction $PQ^kP = (PQ)^kP$ since $PQ^{k+1}P = (PQ)(Q^kP) = (PQP)(Q^kP) = (PQ)(PQ^kP) = (PQ)(PQ)^kP = (PQ)^{k+1}P$. Consequently, $Pe^{-tQ}P = e^{-tPQ}P$, and e^{-tPQ} and Pe^{-tQ} coincide on $W := P(\mathbb{R}^d)$. Since $\mu \in L_0(Q;\mathbb{R}^d)$, we have $\mu = \mathscr{L}(\int_0^\infty e^{-tQ}\,dY(t))$ for some $D(\mathbb{R}^\infty,[0,\infty))$-valued r.v. Y with stationary independent increments, $\mathscr{L}(Y(1)) \in \text{ID}_{\log}(\mathbb{R}^d)$ and $Y(0) = 0$ a.s. [cf. (3.6.18)]. Hence,

$$P\mu = \mathscr{L}\left(\int_0^\infty Pe^{-tQ}\,d(PY(t)) + \int_0^\infty Pe^{-tQ}\,d((I-P)Y(t))\right)$$
$$= \mathscr{L}\left(\int_0^\infty e^{-tPQ}\,d(PY(t))\right) \in L_0(PQ|W;W),$$

because $Pe^{-tQ} = e^{-tPQ}$ on W, $Pe^{-tQ}(I - P) = 0$, and $\mathscr{L}(PY(1)) \in \text{ID}_{\log}(W)$. Since

$$\|e^{-tPQ}\|_W \leq \sup\{\|e^{-tPQ}Px\|: x \in \mathbb{R}^d \text{ with } \|x\| \leq 1\}$$
$$= \sup\{\|Pe^{-tQ}Px\|: x \in \mathbb{R}^d \text{ with } \|x\| \leq 1\} \leq \|e^{-tQ}\|\|P^2\|,$$

we also have that $e^{-tPQ} \to 0$ as $t \to \infty$ in the norm $\|\cdot\|_W$ on the subspace W. Q.E.D.

Corollary 3.6.16. *If P is an idempotent on $\mathbb{R}^d$ whose kernel is Q-invariant, then $P(L_0(Q; \mathbb{R}^d)) \subseteq L_0(PQ|W; W)$, where W is the range of P.*

3.7 INFINITESIMAL GENERATORS

From the previous section, we know that $\exp(-tQ)$-decomposable measures are limits of some stochastic processes as time tends to infinity. These processes are given by random integrals. Here, we show that they are stationary Markov processes, so a natural problem is to characterize the infinitesimal generators associated with a Markov semigroup. This is the aim of the present section which is divided into two subsections. The first collects some basic facts and tools for one-parameter semigroups on $C_0(\mathbb{R}^d)$, and the second gives the complete description of the infinitesimal generators associated with $\exp(-tQ)$-decomposable measures.

3.7.1 One-Parameter Semigroups on $C_0(\mathbb{R}^d)$

Let $C_0(\mathbb{R}^d)$ denote the real Banach space of all real-valued continuous functions on $\mathbb{R}^d$ vanishing at infinity with norm $\|f\| := \max\{|f(x)|: x \in \mathbb{R}^d\}$; ∞ is treated as the one-point compactification of $\mathbb{R}^d$. By a *one-parameter contraction semigroup of operators on $C_0(\mathbb{R}^d)$*, we mean a family of operators U_t, $t \geq 0$, on $C_0(\mathbb{R}^d)$ such that

(i) $U_t U_s = U_{t+s}$ for all $t, s \geq 0$, and $U_0 = I$, the identity operator;
(ii) for $f \in C_0(\mathbb{R}^d)$ and $x \in \mathbb{R}^d$, $\lim_{t \downarrow 0}(U_t f)(x) = f(x)$;
(iii) $\|U_t\| \leq 1$ for all $t \geq 0$.

From (i) and (ii), we see that the functions $t \to U_t f(x)$ are right-continuous on $[0, \infty)$. Consequently, one can define, for $\lambda > 0$, $f \in C_0(\mathbb{R}^d)$, and $x \in \mathbb{R}^d$,

$$(R_\lambda f)(x) := \int_0^\infty e^{-\lambda t} U_t f(x)\, dt.$$

The Lebesgue dominated convergence theorem implies that $R_\lambda f \in C_0(\mathbb{R}^d)$ for each $\lambda > 0$. Clearly, R_λ is a bounded linear operator and $\|R_\lambda f\| \leq \lambda^{-1}\|f\|$. Other important properties of R_λ are stated in the following proposition.

INFINITESIMAL GENERATORS

Proposition 3.7.1. *For λ, λ_1, and λ_2 positive constants, we have*

(a) $R_{\lambda_1} - R_{\lambda_2} + (\lambda_1 - \lambda_2) R_{\lambda_1} R_{\lambda_2} = 0$, *the resolvent equation;*
(b) $R_{\lambda_1} R_{\lambda_2} = R_{\lambda_2} R_{\lambda_1}$;
(c) $R_{\lambda_1} f = 0$ *if and only if* $R_{\lambda_2} f = 0$;
(d) $R_\lambda f = 0$ *if and only if* $f = 0$;
(e) $\mathscr{R} := R_\lambda(C_0(\mathbb{R}^d))$ *does not depend on* λ;
(f) *On* $\mathscr{R}$, $\Gamma := I - R_1^{-1}$ *is well defined and* $R_\lambda^{-1} = \lambda I - \Gamma$.

Proof

(a) Assume $\lambda_1 \neq \lambda_2$ and $f \in C_0(\mathbb{R}^d)$. Then

$$R_{\lambda_1} R_{\lambda_2} f = \int_0^\infty e^{-\lambda_1 t} U_t \left(\int_0^\infty e^{-\lambda_2 s} U_s f \, ds \right) dt$$

$$= \int_0^\infty \int_0^\infty e^{-\lambda_1 t - \lambda_2 s} U_{t+s} f \, ds \, dt = \int_0^\infty \int_t^\infty e^{-\lambda_1 t - \lambda_2 (r-t)} U_r f \, dr \, dt$$

$$= \int_0^\infty \int_0^r e^{-t(\lambda_1 - \lambda_2) - \lambda_2 r} U_r f \, dt \, dr = \int_0^\infty \frac{e^{-\lambda_2 r} - e^{-\lambda_1 r}}{\lambda_1 - \lambda_2} U_r f \, dr$$

$$= (\lambda_1 - \lambda_2)^{-1} (R_{\lambda_2} f - R_{\lambda_1} f).$$

Now, rearrange this to obtain (a).

(b) Note $R_{\lambda_1} R_{\lambda_2} = (\lambda_1 - \lambda_2)^{-1}(R_{\lambda_2} - R_{\lambda_1}) = (\lambda_2 - \lambda_1)^{-1}(R_{\lambda_1} - R_{\lambda_2}) = R_{\lambda_2} R_{\lambda_1}$.
(c) $R_{\lambda_2} f = 0$ implies $R_{\lambda_1} f = (R_{\lambda_2} - (\lambda_1 - \lambda_2) R_{\lambda_1} R_{\lambda_2}) f = 0$ by (a).
(d) We need only prove that $R_\lambda f = 0$ implies $f = 0$. From the definition of R_λ, we have

$$|\lambda R_\lambda f(x) - f(x)| \leq \int_0^\infty \lambda e^{-\lambda t} |U_t f(x) - f(x)| \, dt$$

$$= \int_0^\infty e^{-s} |U_{s/\lambda} f(x) - f(x)| \, ds. \quad (3.7.1)$$

By (ii) and the Lebesgue dominated convergence theorem, $\lambda R_\lambda f(x) \to f(x)$ as $\lambda \to \infty$ for each x. Since $R_\lambda f = 0$ for some λ, by (c), it is zero for all $\lambda > 0$. Hence, $f = 0$.

(e) Let $f \in R_{\lambda_1}(C_0(\mathbb{R}^d))$, that is, $f = R_{\lambda_1} g$ for some $g \in C_0(\mathbb{R}^d)$. Using (a) and (b), we have

$$f = R_{\lambda_1} g = R_{\lambda_2}\bigl(g + (\lambda_2 - \lambda_1) R_{\lambda_1} g\bigr) = R_{\lambda_2}\bigl(g + (\lambda_2 - \lambda_1) f\bigr),$$

so $f \in R_{\lambda_2}(C_0(\mathbb{R}^d))$ for all $\lambda_2 > 0$.

(f) On $\mathscr{R}$, the operator R_λ^{-1} exists by linearity and (d). Applying R_1^{-1} to the resolvent equation with $\lambda_1 = 1$ and $\lambda_2 = \lambda$, we obtain $(R_1^{-1} + (\lambda - 1)I)R_\lambda = I$, that is, $R_\lambda^{-1} = \lambda I - \Gamma$.

Q.E.D.

Let

$$\mathscr{X}^\infty := \{ f \in C_0(\mathbb{R}^d) : \lambda R_\lambda f \to f \text{ in } C_0(\mathbb{R}^d) \text{ as } \lambda \to \infty \}.$$

Clearly, $\mathscr{X}^\infty$ is a linear subspace of $C_0(\mathbb{R}^d)$. From the inequality

$$\|f - \lambda R_\lambda f\| \leq \|f - f_n\| + \|f_n - \lambda R_\lambda f_n\| + \|\lambda R_\lambda(f_n - f)\|$$
$$\leq 2\|f - f_n\| + \|f_n - \lambda R_\lambda f_n\|,$$

we see that $\mathscr{X}^\infty$ is a closed subspace of $C_0(\mathbb{R}^d)$. Furthermore, if $f = R_s g$ for some $s > 0$ and $g \in C_0(\mathbb{R}^d)$, then from the resolvent equation we have

$$\|f - \lambda R_\lambda f\| = |\lambda - s|^{-1} \|\lambda R_\lambda g - s R_s g\| \leq \frac{2\|g\|}{|\lambda - s|},$$

so $f \in \mathscr{X}^\infty$, that is, $R_s(C_0(\mathbb{R}^d)) \subseteq \mathscr{X}^\infty$ or $\mathscr{R} \subseteq \mathscr{X}^\infty$. Since $\mathscr{X}^\infty \subset \overline{\mathscr{R}}$, the closure of $\mathscr{R}$ in the topology of $C_0(\mathbb{R}^d)$, we have $\mathscr{X}^\infty = \overline{\mathscr{R}}$. Actually, more than this is true.

Proposition 3.7.2. $C_0(\mathbb{R}^d) = \mathscr{X}^\infty = \overline{\mathscr{R}}$.

Proof. Suppose that $\mathscr{X}^\infty$ is a proper subset of $C_0(\mathbb{R}^d)$. Since $\mathscr{X}^\infty$ is a closed subset of $C_0(\mathbb{R}^d)$, by the Riesz theorem there is a bounded signed measure μ such that $\int f d\mu = 0$ for all $f \in \mathscr{X}^\infty$ and μ is not identically zero. Since $R_\lambda(C_0(\mathbb{R}^d)) = \mathscr{X}^\infty$, we have

$$\int \lambda R_\lambda f \, d\mu = 0 \quad \text{for all } \lambda > 0 \text{ and all } f \in C_0(\mathbb{R}^d). \quad (3.7.2)$$

On the other hand, $\lambda R_\lambda f(x) \to f(x)$ as $\lambda \to \infty$ [cf. (3.7.1)], so by the bounded convergence theorem and (3.7.2) we obtain $\int f \, d\mu = 0$ for each $f \in C_0(\mathbb{R}^d)$. This implies that $\mu = 0$ which contradicts the supposition. Q.E.D.

Now, let
$$\mathscr{X}^0 := \{f \in C_0(\mathbb{R}^d) : U_t f \to f \text{ in } C_0(\mathbb{R}^d) \text{ as } t \to 0\}.$$

Arguing as in the case of $\mathscr{X}^\infty$ we see that $\mathscr{X}^0$ is a closed subspace of $C_0(\mathbb{R}^d)$. Even more can be said.

Proposition 3.7.3. *We have $C_0(\mathbb{R}^d) = \mathscr{X}^0$, that is, $\{U_t : t \geq 0\}$ forms a one-parameter strongly continuous contraction semigroup (cf. Section 1.4).*

Proof. Note that, for $f = R_\lambda g$ with $g \in C_0(\mathbb{R}^d)$, we have

$$|U_t f(x) - f(x)| = \left| (e^{\lambda t} - 1) \int_t^\infty e^{-\lambda r} U_r g(x) \, dr - \int_0^t e^{-\lambda r} U_r g(x) \, dr \right|$$

$$\leq \frac{(e^{\lambda t} - 1)\|g\|}{\lambda} + \frac{\|g\|(1 - e^{-\lambda t})}{\lambda}.$$

Hence, $R_\lambda(C_0(\mathbb{R}^d)) \subseteq \mathscr{X}^0$ so $\mathscr{R} \subseteq \mathscr{X}^0$. By Propositions 3.7.1 and 3.7.2, $C_0(\mathbb{R}^d) = \overline{\mathscr{R}} \subseteq \mathscr{X}^0$. Q.E.D.

With the same proofs all of this can be generalized as follows.

Remark 3.7.4. For an arbitrary locally compact space K with a countable base, each one-parameter contraction semigroup $\{U_t : t \geq 0\}$ on $C_0(K)$, the space of all continuous functions on K which vanish at infinity ∞, the one-point compactification, satisfying

$$\lim_{t \to 0} U_t f(x) = f(x) \quad \text{for each } f \in C_0(K) \text{ and each } x \in K,$$

also satisfies

$$\lim_{t \to 0} U_t f = f \quad \text{in } C_0(K) \text{ for each } f \in C_0(K),$$

that is, it is a strongly continuous one-parameter semigroup.

With the semigroup $\{U_t, t \geq 0\}$ on $C_0(\mathbb{R}^d)$ [cf. (i), (ii), and (iii)], we associate the operator Γ defined on the dense subset $\mathscr{R}$ of $C_0(\mathbb{R}^d)$

[cf. Propositions 3.7.1(f) and 3.7.2]. Denoting by $\Gamma_t := t^{-1}(U_t - I)$ for $t > 0$, we define two other operators Γ_0 and Γ_{00} on the sets $\mathscr{R}_0$ and $\mathscr{R}_{00}$, respectively. Namely,

$$\mathscr{R}_0 := \Big\{ f \in C_0(\mathbb{R}^d) : \text{there is a } g_f \in C_0(\mathbb{R}^d) \text{ such that}$$

$$\lim_{t \to 0} |\Gamma_t f(x) - g_f(x)| = 0 \text{ for each } x \in \mathbb{R}^d \text{ and } \sup_{t > 0} \|\Gamma_t f\| < \infty \Big\},$$

$$\mathscr{R}_{00} := \Big\{ f \in C_0(\mathbb{R}^d) : \text{there is an } h_f \in C_0(\mathbb{R}^d) \text{ such that}$$

$$\lim_{t \to 0} \|\Gamma_t f - h_f\| = 0 \Big\},$$

$\Gamma_0 f := g_f$ and $\Gamma_{00} f := h_f$.

The relations among these operators are given in the following proposition.

Proposition 3.7.5. *We have $\mathscr{R} = \mathscr{R}_0 = \mathscr{R}_{00}$ and $\Gamma = \Gamma_0 = \Gamma_{00}$, so $\Gamma f = \lim_{t \to 0} t^{-1}(U_t f - f)$ in $C_0(\mathbb{R}^d)$.*

Proof. Obviously, $\mathscr{R}_{00} \subseteq \mathscr{R}_0$ and $\Gamma_{00} = \Gamma_0$ on $\mathscr{R}_{00}$. Now, let $f \in \mathscr{R}$, so $f = R_\lambda g$ for some $g \in C_0(\mathbb{R}^d)$ and $\lambda > 0$. Since, by Proposition 3.7.1(f),

$$\Gamma f = \lambda f - R_\lambda^{-1} f = \lambda R_\lambda g - g,$$

we obtain

$$\Gamma_t f(x) - \Gamma f(x)$$

$$= t^{-1}\bigg[(e^{\lambda t} - 1 - \lambda t)\int_t^\infty e^{-\lambda r} U_r g(x)\, dr$$

$$+ \int_0^t (g(x) - e^{-\lambda r} U_r g(x))\, dr - \lambda t \int_0^t e^{-\lambda r} U_r g(x)\, dr\bigg]$$

$$= t^{-1}\bigg[(e^{\lambda t} - 1 - \lambda t)\int_t^\infty e^{-\lambda r} U_r g(x)\, dr + \int_0^t e^{-\lambda r}(g(x) - U_r g(x))\, dr$$

$$+ g(x)\int_0^t (1 - e^{-\lambda r})\, dr - \lambda t \int_0^t e^{-\lambda r} U_r g(x)\, dr\bigg].$$

Hence,

$$\|\Gamma_t f - \Gamma f\| \leq \frac{(e^{\lambda t} - 1 - \lambda t)\|g\|}{\lambda t} + t^{-1}\int_0^t e^{-\lambda r}\|g - U_r g\|\, dr$$

$$+ \|g\| t^{-1}\int_0^t (1 - e^{-\lambda r})\, dr + \lambda \|g\|\int_0^t e^{-\lambda r}\, dr.$$

INFINITESIMAL GENERATORS

By Proposition 3.7.3, we see that the second term goes to zero as $t \to 0$. Therefore, $\Gamma_t f \to \Gamma f$ in $C_0(\mathbb{R}^d)$ as $t \to 0$, that is, $\mathscr{R} \subseteq \mathscr{R}_{00}$ and $\Gamma = \Gamma_{00}$ on $\mathscr{R}$.

Assume now that $f \in \mathscr{R}_0$. Then $\Gamma_0 f \in C_0(\mathbb{R}^d)$ and by the bounded convergence theorem we have

$$R_1(I - \Gamma_0)f(x) = R_1 f(x) - \lim_{t \to 0} R_1 \Gamma_t f(x)$$

$$= R_1 f(x) - \lim_{t \to 0} t^{-1}\left[(e^t - 1)\int_0^\infty e^{-r} U_r f(x)\, dr\right.$$

$$\left. - e^t \int_0^t e^{-r} U_r f(x)\, dr\right] = f(x).$$

Thus, $f = R_1(I - \Gamma_0)f \in R_1(C_0(\mathbb{R}^d)) = \mathscr{R}$, that is, $\mathscr{R}_0 \subseteq \mathscr{R}$ and $\Gamma_0 = I - R_1^{-1} = \Gamma$ on $\mathscr{R}_0$.   Q.E.D.

Remark 3.7.6. Proposition 3.7.5 holds with the same proof on $C_0(K)$, where K is locally compact with a countable base (cf. Remark 3.7.4).

The operator Γ, or Γ_0 or Γ_{00} in the case $C_0(K)$, is referred to as the *infinitesimal generator* of the one-parameter semigroup $\{U_t: t \geq 0\}$. It must be emphasized that Γ with its domain $\mathscr{R}$ completely characterizes the semigroup $\{U_t: t \geq 0\}$ (cf. Corollary 1.4.3).

3.7.2 Infinitesimal Generators of Ornstein–Uhlenbeck Type Processes

Let Y be a $D(\mathbb{R}^d, [0, \infty))$-valued random variable with stationary independent increments and $Y(0) = 0$ a.s. For $t \geq 0$ and $x \in \mathbb{R}^d$, let

$$Z_x(t) := e^{-tQ} x + \int_{(0, t]} e^{-sQ}\, dY(s) \qquad (3.7.3)$$

We refer to Z_x as an *Ornstein–Uhlenbeck type process* since, when Y is Brownian motion, Z_x is called the Ornstein–Uhlenbeck process. Our assumption on Q requires $e^{-tQ} \to 0$ as $t \to \infty$, so we see that a measure μ is in $L_0(Q)$ if and only if it is the limit in distribution as $t \to \infty$ of some Ornstein–Uhlenbeck type process. Moreover, this limit exists if and only if $\mathscr{L}(Y(1)) \in \mathrm{ID}_{\log}$ (cf. Theorem 3.6.6). The stochastic properties of the Z_x process are given in the following proposition.

Proposition 3.7.7. *The Ornstein–Uhlenbeck type processes are stationary Markov processes with independent increments and with paths in $D(\mathbb{R}^d, [0, \infty))$.*

Proof. From Lemma 3.6.3, we see that Z_x has independent increments and paths in $D(\mathbb{R}^d, [0, \infty))$. For $t \geq 0$, $x \in \mathbb{R}^d$, and $A \in \mathcal{B}(\mathbb{R}^d)$, let

$$P(t, x, A) := \mathcal{L}(Z_x(t))(A).$$

From (3.7.3), we have $P(0, x, A) = \delta_x(A)$. Clearly, $P(t, x, \cdot)$ is a probability on $\mathcal{B}(\mathbb{R}^d)$ and $P(t, \cdot, A)$ is a Borel-measurable function. To prove the proposition, we show that, for $t \geq 0$, $s \geq 0$, $x \in \mathbb{R}^d$, and $A \in \mathcal{B}(\mathbb{R}^d)$,

$$P(t + s, x, A) = \int P(t, y, A) P(s, x, dy). \tag{3.7.4}$$

For the moment let ν denote the measure given by the right side of (3.7.4). From (3.7.3) and Lemma 3.6.4, we have

$$\hat{P}(t, x, \cdot)(z) = \exp\left\{i\langle e^{-tQ^*}z, x\rangle + \int_0^t \log \hat{\mathcal{L}}(Y(1))(e^{-rQ^*}z)\, dr\right\}.$$

Consequently, for $z \in \mathbb{R}^d$,

$$\hat{\nu}(z) = \int\int e^{i\langle z, v\rangle} P(t, y, dv) P(s, x, dy)$$

$$= \int \hat{P}(t, y, \cdot)(z) P(s, x, dy)$$

$$= \int e^{i\langle e^{tQ^*}z, y\rangle} \exp\left\{\int_0^t \log \hat{\mathcal{L}}(Y(1))(e^{rQ^*}z)\, dr\right\} P(s, x, dy)$$

$$= \hat{P}(s, x, \cdot)(e^{-tQ^*}z)\exp\left\{\int_0^t \log \hat{\mathcal{L}}(Y(1))(e^{-rQ^*}z)\, dr\right\}$$

$$= \exp\left\{i\langle e^{-(s+t)Q^*}z, x\rangle + \int_0^{s+t} \log \hat{\mathcal{L}}(Y(1))(e^{-rQ^*}z)\, dr\right\}$$

$$= \hat{P}(t + s, x, \cdot)(z),$$

which establishes (3.7.4). Q.E.D.

Before we begin the search for the infinitesimal generator associated with Z_x, note that as $x_n \to x$ and $t_n \to t > 0$ with $t_n \geq t$, we have $Z_{x_n}(t_n) \to Z_x(t)$. Therefore, under these conditions

$$P(t_n, x_n, \cdot) \Rightarrow P(t, x, \cdot). \tag{3.7.5}$$

For any real-valued bounded Borel-measurable function f on $\mathbb{R}^d$, let

$$(U_t f)(x) := \mathbf{E}f(Z_x(t)) = \int f(y) P(t, x, dy)$$
$$= \int f(e^{-tQ}x + y) P(t, 0, dy) \tag{3.7.6}$$

for $t \geq 0$ and $x \in \mathbb{R}^d$. From (3.7.4), we have that $\mathbb{U} := \{U_t : t \geq 0\}$ is a one-parameter semigroup of operators, that is, $U_t(U_s f) = U_{t+s} f$, and $P(0, x, A) = \delta_x(A)$ implies that $U_0 f = f$, and $U_0 = I$, the identity operator. Furthermore, if $f \in C_0(\mathbb{R}^d)$, that is, f is continuous and vanishes at infinity, from the last inequality in (3.7.6) and the Lebesgue bounded convergence theorem, we see that $U_t f \in C_0(\mathbb{R}^d)$; thus,

$$U_t : C_0(\mathbb{R}^d) \to C_0(\mathbb{R}^d), \quad t \geq 0,$$

are bounded linear operators on the Banach space $C_0(\mathbb{R}^d)$ with the supremum norm. Clearly, if $f \geq 0$, then $U_t f \geq 0$. Since $P(t, x, \cdot)$ are tight measures, their supports are σ-compact, so one can find an increasing sequence $\{K_n\}$ of compact sets such that $P(t, x, K_n) \to 1$. Selecting an $f_n \in C_0(\mathbb{R}^d)$ such that $0 \leq f_n \leq 1$ and $f_n(y) = 1$ for all $y \in K_n$, we see that

$$\|U_t\| \geq \int f_n(y) P(t, x, dy) \geq P(t, x, K_n).$$

Clearly, $U_t \leq 1$, so this implies that $U_t = 1$ for all $t \geq 0$. Since Z_x is $D(\mathbb{R}^d, [0, \infty))$-valued, the functions $t \to \mathscr{L}(Z_x(t))$ are $D(\mathscr{P}(\mathbb{R}^d), [0, \infty))$-valued, and, in particular, for any real-valued bounded continuous f on $\mathbb{R}^d$,

$$\lim_{t \to 0} (U_t f)(x) = f(x) \quad \text{for each } x \in \mathbb{R}^d. \tag{3.7.7}$$

In fact, the functions $t \to (U_t f)(x)$ are $D(\mathbb{R}, [0, \infty))$-valued. For $f \in$

$C_0(\mathbb{R}^d)$, Proposition 3.7.3 gives that (3.7.7) is equivalent to

$$\lim_{t \to 0} U_t f = f \quad \text{in } C_0(\mathbb{R}^d). \tag{3.7.8}$$

Summarizing the above, we have $\mathbb{U} = \{U_t: t \geq 0\}$, with U_t given by (3.7.6), is a strongly continuous one-parameter semigroup of operators on $C_0(\mathbb{R}^d)$ with $\|U_t\| = 1$ and $0 \leq f \leq 1$ implies $0 \leq U_t f \leq 1$. Such semigroups $\mathbb{U}$ are called Feller–Dynkin semigroups. Our aim now is to find the infinitesimal generator of $\mathbb{U}$, that is, an operator Γ and its dense domain $\mathscr{D}(\Gamma)$ in $C_0(\mathbb{R}^d)$ such that, for $f \in \mathscr{D}(\Gamma)$,

$$\Gamma f := \lim_{t \to 0} t^{-1}(U_t f - f) \quad \text{in } C_0(\mathbb{R}^d). \tag{3.7.9}$$

In view of Proposition 3.7.5, we see that this is equivalent to the following conditions:

$$\sup_{t > 0} t^{-1}\|U_t f - f\| < \infty \quad \text{and} \quad \lim_{t \to 0} \left| t^{-1}[U_t f(x) - f(x)] - \Gamma f(x) \right| = 0$$

$$\text{for } x \in \mathbb{R}^d. \tag{3.7.10}$$

Now, we formulate our main result of this section. Let $C_K^2(\mathbb{R}^d)$ denote the space of all twice continuously differentiable real-valued functions on $\mathbb{R}^d$ with compact support. Let $Df(x)$ and $D^2 f(x)$ denote the first and second derivatives, respectively, of f at x, and tr A denote the trace of the linear operator A on $\mathbb{R}^d$.

Theorem 3.7.8. *For $x \in \mathbb{R}^d$, let $\{Z_x(t): t \geq 0\}$ be the process given by the random integral (3.7.3), where Y is a $D(\mathbb{R}^d, [0, \infty))$-valued random variable with stationary independent increments, $Y(0) = 0$ a.s. and $\mathscr{L}(Y(1)) = [a, R, M]$. Let $\mathbb{U} = \{U_t: t \geq 0\}$ be the one-parameter semigroup on $C_0(\mathbb{R}^d)$ defined by $(U_t f)(x) := \mathbf{E} f(Z_x(t))$. Then the infinitesimal generator Γ of $\mathbb{U}$ is the smallest closed extension of the operator G on $C_K^2(\mathbb{R}^d)$ given by*

$$(Gf)(x) := \langle a - Qx, Df(x) \rangle + \tfrac{1}{2} \operatorname{tr} R D^2 f(x)$$

$$+ \int_{\mathbb{R}^d \setminus \{0\}} \left[f(x + y) - f(x) - \langle y, Df(x) \rangle (1 + \|y\|^2)^{-1} \right] M(dy).$$

INFINITESIMAL GENERATORS 153

Proof. First, note that applying the Taylor formula to $f \in C_K^2(\mathbb{R}^d)$, we have

$$\left| f(x+y) - f(x) - \langle y, Df(x) \rangle \left(1 + \|y\|^2\right)^{-1} \right| \leq c_f \|y\|^2 \left(1 + \|y\|^2\right)^{-1} \tag{3.7.11}$$

for some positive constant c_f. Therefore, Gf is well defined since M integrates $\|y\|^2(1 + \|y^2\|)^{-1}$ over $\mathbb{R}^d$, and $Gf \in C_0(\mathbb{R}^d)$.

Step 1. Γ is an extension of G, that is, $C_K^2(\mathbb{R}^d) \subseteq \mathscr{D}(\Gamma)$ and $\Gamma = G$ on $C_K^2(\mathbb{R}^d)$.

We show that for $f \in C_K^2(\mathbb{R}^d)$ condition (3.7.10) holds for Γ replaced by G, which is equivalent to (3.7.9) (cf. Proposition 3.7.3). From (3.7.3) and Lemma 3.6.4, we see that

$$t^{-1}\left[e^{-i\langle z, x\rangle} \hat{\mathscr{L}}(Z_x(t))(z) - 1\right] = t^{-1}[\exp h_t(x, z) - 1],$$

where

$$h_t(x, z) := i\langle z, e^{-tQ}x - x\rangle + \int_0^t \log \hat{\mathscr{L}}(Y(1))(e^{-sQ^*}z)\, ds.$$

Since $\lim_{t \to 0} t^{-1} h_t(x, z) = -i\langle z, Qx\rangle + \log \hat{\mathscr{L}}(Y(1))(z)$, we see that

$$\lim_{t \to 0} \exp t^{-1}\left[e^{i\langle z, -x\rangle}\hat{\mathscr{L}}(Z_x(t))(z) - 1\right] = \exp\{i\langle z, -Qx\rangle\}\hat{\mathscr{L}}(Y(1))(z)$$

$$= \hat{\mathscr{L}}(Y(1) - Qx)(z).$$

The last characteristic function above corresponds to the measure $[a - Qx, R, M]$. But, rewriting the above expression on the left without the limit, we have

$$\exp t^{-1}\left[e^{i\langle z, -x\rangle}\hat{\mathscr{L}}(Z_x(t))(z) - 1\right]$$

$$= \exp\left[\int (e^{i\langle z, u-x\rangle} - 1) t^{-1} P(t, x, du)\right]$$

$$= \exp\left\{i\langle z, a_{x,t}\rangle + \int_{\mathbb{R}^d\setminus\{0\}} \left[e^{i\langle z, u\rangle} - 1 - \frac{i\langle z, u\rangle}{1 + \|u\|^2}\right] N(t, x, du)\right\},$$

where

$$N(t, x, A) := t^{-1} P(t, x, A + x) \quad \text{for } A \in \mathscr{B}(\mathbb{R}^d \setminus \{0\}),$$

$$\langle z, a_{x,t} \rangle := \int \frac{\langle z, u \rangle}{1 + \|u\|^2} N(t, x, du) \quad \text{for } z \in \mathbb{R}^d.$$

Consequently, the above convergence of characteristic functions is equivalent to the statement that

$$[a_{x,t}, 0, N(t, x, \cdot)] \Rightarrow [a - Qx, R, M] \quad \text{as } t \to 0. \quad (3.7.12)$$

Therefore, for any sequence $0 < t_n \to 0$, we have

(α) $$\lim_{n \to \infty} \int h(y) N(t_n, x, dy) = \int h(y) M(dy)$$

for every h bounded, vanishing around zero, and with discontinuity set of M-measure zero;

(β) $$\lim_{\varepsilon \to 0} \liminf_{n \to \infty} \int_{B_\varepsilon} \langle u, y \rangle^2 N(t_n, x, dy)$$

$$= \lim_{\varepsilon \to 0} \limsup_{n \to \infty} \int_{B_\varepsilon} \langle u, y \rangle^2 N(t_n, x, dy) = \langle Ru, u \rangle$$

for all $u \in \mathbb{R}^d$;

(γ) $$\lim_{n \to \infty} a_{x, t_n} = a - Qx$$

(cf. Corollary 1.8.16). For $f \in C_K^2(\mathbb{R}^d)$ and $x \in \mathbb{R}^d$, set

$$g_{f,x}(y) := f(y + x) - f(x) - \langle y, Df(x) \rangle \left(1 + \|y\|^2\right)^{-1}$$

for $y \in \mathbb{R}^d$. We see that $g_{f,x}$ is continuous and bounded, and from the Taylor formula we obtain

$$g_{f,x}(y) = \langle y, Df(x) \rangle \|y\|^2 \left(1 + \|y\|^2\right)^{-1}$$
$$+ \tfrac{1}{2} \langle y, D^2 f(x) y \rangle + O(\|y\|^3).$$

INFINITESIMAL GENERATORS

Furthermore, for $\varepsilon > 0$ we have

$$t_n^{-1}[U_{t_n}f(x) - f(x)]$$
$$= \langle a_{x,t_n}, Df(x) \rangle + \int_{B_\varepsilon^c} g_{f,x}(y) N(t_n, x, dy)$$
$$+ \frac{1}{2}\int_{B_\varepsilon} \langle y, D^2f(x)y \rangle N(t_n, x, dy)$$
$$+ \int_{B_\varepsilon} \langle y, Df(x) \rangle \|y\|^2 (1 + \|y\|^2)^{-1} N(t_n, x, dy)$$
$$+ \int_{B_\varepsilon} O(\|y\|^3) N(t_n, x, dy). \tag{3.7.13}$$

The absolute value of the last summand is bounded by

$$C\varepsilon \int_{\|y\| \leq \varepsilon} \|y\|^2 N(t_n, x, dy)$$

for some positive constant C, whereas

$$\|Df(x)\|^2 \varepsilon^2 \sup_{n \geq 1} \int (1 + \|y\|^2)^{-1} N(t_n, x, dy)$$

bounds the absolute value of the next to the last summand. These bounds together with (α) and (β) imply that these summands tend to zero if we take lim sup over n and then limit as $\varepsilon \to 0$. Next, note that in terms of coordinates

$$\int_{B_\varepsilon} \langle y, D^2f(x)y \rangle N(t, x, dy) = \sum_{k,\ell=1}^{d} \int_{B_\varepsilon} y_k y_\ell \, \partial_{k,\ell} f(x) \, N(t, x, dy),$$

where $y_k := \langle y, e_k \rangle$, $\{e_1, \ldots, e_d\}$ is the standard base for $\mathbb{R}^d$, and $\partial_{k,\ell} f(x)$ denotes the second-order partial derivative of f with respect to the kth and ℓth coordinates evaluated at x. Hence, from (β) we obtain

$$\lim_{\varepsilon \to 0} \limsup_{n \to \infty} \int_{B_\varepsilon} \langle y, D^2f(x)y\rangle N(t_n, x, dy)$$

$$= \sum_{k,\ell=1}^{d} R_{k\ell} \partial_{k,\ell} f(x) = \operatorname{tr} RD^2f(x),$$

where $(R_{k\ell})$ is the matrix representation of the operator R. Consequently, (3.7.13) with (α), (β), and (γ) gives

$$\lim_{t \to 0} t^{-1}[U_t f(x) - f(x)]$$

$$= \lim_{\varepsilon \to 0} \lim_{t \to 0} \left\{ \langle a_{x,t}, Df(x)\rangle + \int_{B_\varepsilon^c} g_{f,x}(y) N(t, x, dy) \right.$$

$$\left. + \frac{1}{2}\int_{B_\varepsilon} \langle y, D^2f(x)y\rangle N(t, x, dy) \right\}$$

$$= \langle a - Qx, Df(x)\rangle + \int_{\mathbb{R}^d \setminus \{0\}} g_{f,x}(y) M(dy)$$

$$+ \frac{1}{2} \operatorname{tr} RD^2f(x) = (Gf)(x).$$

This establishes the second part of (3.7.10).

To prove the first part of (3.7.10), it suffices to show that $\sup_{0 < t \le \varepsilon} t^{-1}\|U_t f - f\| < \infty$ for some $\varepsilon > 0$. Choose and fix $\varepsilon > 0$ so small that $\|e^{-tQ} - I\| < 1/2$ for $0 < t \le \varepsilon$. Since f has compact support, we can choose a positive integer k such that $f(x) = 0$ for $\|x\| > k$. Then for $0 < t \le \varepsilon$ we have

$$\sup_{\|x\|>4k} |t^{-1}(U_t f(x) - f(x))| = \sup_{\|x\|>4k} \left| \int f(e^{-tQ}x + y) t^{-1} P(t, 0, dy) \right|$$

$$\le \sup_{\|x\|>4k} \left| \int_{\|y\|>k} f(e^{-tQ}x + y) N(t, 0, dy) \right|$$

$$\le \|f\| \sup_{0 < t \le \varepsilon} N(t, 0, (\|y\| > k)) < \infty,$$

INFINITESIMAL GENERATORS

since $N(t, 0, \cdot)$ is weakly convergent as $t \to 0$ outside very neighborhood of zero [cf. (α)]. For $\|x\| \leq 4k$, using (3.7.11), we have

$$t^{-1}[U_t f(x) - f(x)]$$

$$\leq \|a_{x,t}\| \|Df(x)\| + c_f \int \frac{\|y\|}{1 + \|y\|^2} N(t, x, dy)$$

$$\leq \text{constant} \left\{ \left\| \int \frac{\|y\|}{1 + \|y\|^2} N(t, x, dy) \right\| \right.$$

$$\left. + \int_{B_1} \|y\|^2 N(t, x, dy) + \int_{B_1^c} N(t, x, dy) \right\}$$

$$\leq \text{constant} \left\{ \int_{B_1} \|y\| N(t, x, dy) + \int_{B_1} \|y\|^2 N(t, x, dy) \right.$$

$$\left. + 2 \int_{B_1^c} N(t, x, dy) \right\}.$$

Without loss of generality, we may assume that $\|e^{-tQ}x - x\| \leq 1/2$ for $0 \leq t \leq \varepsilon$ and $\|x\| \leq 4k$. Then for the third integral, using (3.7.6) and (α), we have

$$N(t, x, B_1^c) = t^{-1} P(t, 0, B_1^c + x - e^{-tQ}x) \leq N(t, 0, B_{1/2}^c)$$

and hence $\sup_{0 < t \leq \varepsilon} \sup_{\|x\| \leq 4k} N(t, x, B_1^c) < \infty$. For the second integral, we have

$$\int_{B_1} \|y\|^2 N(t, x, dy)$$

$$= \int I_B(e^{-tQ}x - x + z) \|e^{-tQ}x - x + z\|^2 N(t, 0, dz)$$

$$\leq 2 \int_{B_{3/2}} (\|e^{-tQ}x - x\|^2 + \|z\|^2) t^{-1} P(t, 0, dz)$$

$$\leq 2 \sup_{0 < t \leq \varepsilon} \sup_{\|x\| \leq 4k} \left\| \frac{e^{-tQ}x - x}{t} \right\|^2 + 2 \sup_{0 < t \leq \varepsilon} \int_{B_{3/2}} \|z\|^2 N(t, 0, dz).$$

Therefore, from (β), we obtain

$$\sup_{0 \le t \le \varepsilon} \sup_{\|x\| \le 4k} \int_{B_1} \|y\|^2 N(t, x, dy) < \infty.$$

Finally, for the first integral, we have

$$\int_{B_1} \|y\| N(t, x, dy)$$

$$= \int I_{B_1}(e^{-tQ}x - x + y) \|e^{-tQ}x - x + y\| t^{-1} P(t, 0, dy)$$

$$\le t^{-1} \|e^{-tQ}x - x\| + \int_{B_{3/2}} \|y\| N(t, 0, dy)$$

$$\le t^{-1} \|e^{-tQ}x - x\| + \text{constant} \int \frac{\|y\|}{1 + \|y\|^2} N(t, 0, dy)$$

$$\le \sup_{0 < t \le \varepsilon} t^{-1} \|e^{-tQ} - I\|(4k) + \text{constant} \sup_{0 < t \le \varepsilon} \|a_{0, t}\|.$$

Consequently, from (γ) and the fact that the mapping $(x, t) \to a_{x, t}$ is continuous [cf. (3.7.5)], we obtain

$$\sup_{0 < t \le \varepsilon} \sup_{\|x\| \le 4k} \int_{B_1} \|y\| N(t, x, dy) < \infty,$$

which establishes the first part of (3.7.10) and therefore Step 1 is done.

Since the infinitesimal generator Γ is always a closed operator (cf. Proposition 1.4.1), from Step 1 and Theorem 1.2.1 we see that $\overline{G}$ exists, where $\overline{G}$ is the smallest closed extension of G. Let $C^2(\mathbb{R}^d)$ denote the space of all twice continuously differentiable real-valued functions and let $\partial_k f$ and $\partial_{k, \ell} f$ stand for the partial derivatives of the first and second order, respectively. Then $Df(x)$ has the vector representation with coordinates $\partial_k f(x)$, and $D^2 f(x)$ has the matrix representation with elements $\partial_{k, \ell} f(x)$, $1 \le k, \ell \le d$ (cf. Proposition 1.3.3). Let us define $F_1(\mathbb{R}^d)$ as follows:

$$F_1(\mathbb{R}^d) := \{f \in C_0(\mathbb{R}^d) : f \in C^2(\mathbb{R}^d), \|x\| \partial_k f(x) \to 0,$$

$$\partial_{k, \ell} f(x) \to 0 \text{ as } \|x\| \to \infty \text{ for } 1 \le k, \ell \le d\}.$$

Furthermore, let G_1 be defined on $F_1(\mathbb{R}^d)$ by the same formula as G in the statement of the theorem. Since (3.7.11) also holds for $f \in F_1(\mathbb{R}^d)$, we see that $G_1 f \in C_0(\mathbb{R}^d)$.

Step 2. $\overline{G}$ is an extension of G_1.

It suffices to show that given $f \in F_1(\mathbb{R}^d)$, one can find a sequence $\{f_n\}$ in $C_K^2(\mathbb{R}^d)$ such that $\|f_n - f\| \to 0$ and $\|Gf_n - G_1 f\| \to 0$ in the norm of $C_0(\mathbb{R}^d)$. Select $h \in C^2(\mathbb{R}^d)$ such that $0 \le h \le 1$, $h(x) = 1$ for $\|x\| \le 1$, and $h(x) = 0$ for $\|x\| \ge 2$. Setting $f_n(x) := f(x) h(x/n)$, we have $f_n \in C_K^2(\mathbb{R}^d)$ and $\|f_n - f\| \le \sup_{\|x\| \ge n} \|f(x)\| \to 0$ as $n \to \infty$. Furthermore, from (3.7.11), we have $\lim_{n \to \infty} \|Gf_n - G_1 f\| = 0$, which completes Step 2.

Step 3. If $\int \|x\| \mu(dx) < \infty$, equivalently $\int_{\|x\| > 1} \|x\| M(dx) < \infty$, then $\Gamma = \overline{G}$.

First, we prove that $U_t \colon F_1(\mathbb{R}^d) \to F_1(\mathbb{R}^d)$. Let M^t be the Lévy measure of $\mathscr{L}(Z_x(t))$. From (3.7.3), Lemma 3.6.4, and (3.6.7), we have

$$M^t(A) = \int_0^t M(e^{sQ}A)\, ds \quad \text{for } A \in \mathscr{B}(\mathbb{R}^d \setminus \{0\}).$$

Setting $c_t := \sup_{0 \le s \le t} \|e^{-sQ}\|$, we obtain

$$\int_{B_1^c} \|y\| M^t(dy) = \int_0^t \int I_{B_1^c}(e^{-sQ}z) \|z\| M(dz)\, ds$$

$$\le c_t \int_0^t \int I_{(e^s B_1)^c}(z) \|z\| M(dz)\, ds$$

$$\le c_t t \int_{B_{1/c_t}^c} \|z\| M(dz) < \infty,$$

because of the additional assumption on M. Hence (cf. Proposition 1.8.13), we have

$$E\|Z_x(t)\| = \int \|e^{-tQ}x + y\| P(t, 0, dy) < \infty. \tag{3.7.14}$$

If $f \in F_1(\mathbb{R}^d)$, from (3.7.14) and the equality

$$(U_t f)(x) = Ef(Z_x(t)) = \int f(e^{-tQ}x + y) P(t, 0, dy),$$

we see the processes of differentiation and integration may be interchanged. Consequently, from the Lebesgue bounded convergence theorem, we obtain $\partial_k U_t f$ and $\partial_{k,\ell} U_t f$ belong to $C_0(\mathbb{R}^d)$. Furthermore,

$$\|x\| |\partial_j U_t f(x)|$$

$$\leq \|e^{-tQ}\| \|x\| \sum_{k=1}^{d} \int |\partial_k f(e^{-tQ}x + y)| P(t, 0, dy)$$

$$\leq \|e^{-tQ}\| \|e^{tQ}\| \sum_{k=1}^{d} \left\{ \int \|e^{-tQ}x + y\| |\partial_k f(e^{-tQ}x + y)| P(t, 0, dy) \right.$$

$$\left. + \int \|y\| |\partial_k f(e^{-tQ}x + y)| P(t, 0, dy) \right\}.$$

Since $f \in F_1(\mathbb{R}^d)$, all of these summands go to zero as $\|x\| \to \infty$. Therefore, $F_1(\mathbb{R}^d)$ is U_t-invariant.

This invariance with Step 2 and the Watanabe lemma (cf. Lemma 1.4.4) completes the proof of Step 3.

Step 4. Let us define, for $f \in F_1(\mathbb{R}^d)$,

$$(G_0 f)(x)$$
$$:= \langle a - Qx, Df(x) \rangle + \tfrac{1}{2} \operatorname{tr} R D^2 f(x)$$
$$+ \int_{\|y\| \leq 1} \left[f(y + x) - f(x) - \langle y, Df(x) \rangle (1 + \|y\|^2)^{-1} \right] M(dy)$$
$$- \int_{\|y\| > 1} \langle y, Df(x) \rangle (1 + \|y\|^2)^{-1} M(dy)$$

and, for $f \in C_0(\mathbb{R}^d)$,

$$(Vf)(x) := \int_{\|y\| > 1} [f(y + x) - f(x)] M(dy).$$

Clearly, V is a bounded linear operator on $C_0(\mathbb{R}^d)$ and $G_1 = G_0 + V$ on $F_1(\mathbb{R}^d)$. From Step 3, we conclude that the smallest closed extension $\overline{G}_0$ of G_0 is an infinitesimal generator. Consequently, $\overline{G}_1 = \overline{G}_0 + V = \Gamma$, which yields the theorem. Q.E.D.

INFINITESIMAL GENERATORS

Corollary 3.7.9. *The operator G in Theorem 3.7.8 corresponds to an exp(tQ)-decomposable measure if and only if its Lévy measure M has finite logarithmic moment outside every neighborhood of zero.*

Let V_t, $t \geq 0$, be a one-parameter semigroup of bounded linear operators on the Banach space $C_b(S)$ of bounded measurable functions and let ν be a probability on the measurable space $(S, \mathscr{S})$. We way that V_t, $t \geq 0$, is ν-stationary if

$$\int_S (V_t f)(x) \nu(dx) = \int_S f(x) \nu(dx)$$

for all $t \geq 0$ and for all $f \in C_b(S)$.

Proposition 3.7.10. *Let U_t, $t \geq 0$, be the one-parameter semigroup on $C_b(\mathbb{R}^d)$ associated with the Ornstein–Uhlenbeck type process $Z_x(t)$ given by (3.7.3) with $\mathscr{L}(Y(1)) \in \mathrm{ID}_{\log}$. Then the limit measure μ of $\mathscr{L}(Z_x(t))$ as $t \to \infty$ is the unique measure for which U_t is μ-stationary.*

Proof. The limit measure μ exists since $\mathscr{L}(Y(1)) \in \mathrm{ID}_{\log}$ (cf. Theorem 3.6.6.). Letting $P(t, x, \cdot) := \mathscr{L}(Z_x(t))$, we have, for $f \in C_b(\mathbb{R}^d)$,

$$\int_{\mathbb{R}^d} f(y) P(t+s, x, dy) = \int_{\mathbb{R}^d} \int_{\mathbb{R}^d} f(z) P(t, y, dz) P(s, x, dy)$$

(cf. Proposition 3.7.7). Letting $s \to \infty$, we obtain

$$\int_{\mathbb{R}^d} f(y) \mu(dy) = \int_{\mathbb{R}^d} (U_t f)(y) \mu(dy),$$

which shows that U_t is μ-stationary. Suppose there is another measure ν such that U_t is ν-stationary. Then, for all $t > 0$ and for all $f \in C_0(\mathbb{R}^d)$,

$$\int_{\mathbb{R}^d} f(y) \nu(dy) = \int_{\mathbb{R}^d} \int_{\mathbb{R}^d} f(z) P(t, y, dz) \nu(dy).$$

Letting $t \to \infty$, we obtain

$$\int_{\mathbb{R}^d} f(y) \nu(dy) = \int_{\mathbb{R}^d} \int_{\mathbb{R}^d} f(z) \mu(dz) \nu(dy) = \int_{\mathbb{R}^d} f(z) \mu(dz)$$

for all $f \in C_0(\mathbb{R}^d)$. Therefore, $\nu = \mu$. Q.E.D.

3.8 THE ABSOLUTE CONTINUITY OF $\exp(-tQ)$-DECOMPOSABLE MEASURES

For two measures ρ_1, ρ_2, we say that ρ_1 is *absolutely continuous* with respect to ρ_2, $\rho_1 \ll \rho_2$, if the class of sets with ρ_2-measure zero is contained in the class of sets with ρ_1-measure zero, that is, $\rho_2(F) = 0$ implies $\rho_1(F) = 0$. If, in addition, $\rho_2 \ll \rho_1$, we say that ρ_1 and ρ_2 are *equivalent*.

Our main goal in this section is to show that all full $\exp(-tQ)$-decomposable measures on $\mathbb{R}^d$ are absolutely continuous with respect to the d-dimensional Lebesgue measure. Consequently, by the Radon-Nikodym theorem, for the full $\exp(-tQ)$-decomposable measure μ, there is a Borel-measurable function g on $\mathbb{R}^d$ such that, for all $F \in \mathcal{B}(\mathbb{R}^d)$,

$$\mu(F) = \int_F g(x)\,dx.$$

The function g is called the Radon-Nikodym derivative, or density, of μ with respect to the Lebesgue measure. Notationally, $g = d\mu/dx$ or $\mu(dx) = g(x)\,dx$. We will also relate g to the infinitesimal generator Γ (cf. Theorem 3.7.8).

Let l_d denote the d-dimensional Lebesgue measure. When $\rho \ll l_d$, we simply say that ρ is absolutely continuous. We begin with the following simple observations:

$$\nu \ll l_d \text{ implies } \nu * \rho \ll l_d \text{ for all } \rho \in \mathscr{P}(\mathbb{R}^d), \quad (3.8.1)$$

$$\nu \ll l_d \text{ if and only if } A\nu \ll l_d \text{ for all } A \in \text{Aut}(\mathbb{R}^d), \quad (3.8.2)$$

$$\nu \ll l_d \text{ if and only if } \nu * \delta(a) \ll l_d \text{ for all } a \in \mathbb{R}^d. \quad (3.8.3)$$

These observations are simple consequences of the following properties of l_d:

$$l_d(AF) = |\det A|\, l_d(F) \quad \text{and} \quad l_d(F + x) = l_d(F)$$

for all $A \in \text{Aut}(\mathbb{R}^d)$, $F \in \mathcal{B}(\mathbb{R}^d)$, and $x \in \mathbb{R}^d$. A further observation is that if γ is a full Gaussian measure on $\mathbb{R}^d$ with expected value $m \in \mathbb{R}^d$ and covariance operator D, then $D \in \text{Aut}(\mathbb{R}^d)$ and

$$\frac{d\gamma}{dx}(x) = (2\pi)^{-d/2} \exp\left(-\frac{1}{2}\langle D^{-1}(x-m), x-m \rangle\right). \quad (3.8.4)$$

The proof that a full measure $\mu \in L_0(Q)$ is absolutely continuous is rather technical and long. Part of it depends on the dimension d, requiring an induction-type argument. Hence, we first examine the class $L_0(I)$ on $\mathbb{R}^1$. Another part uses a sufficient condition for the absolute continuity of infinitely divisible measures on $\mathbb{R}^d$ (cf. Theorem 3.8.1). Finally, crucial to our consideration is the form of the Lévy measure of $\mu \in L_0(Q)$ (cf. Proposition 3.4.1 and Theorem 3.4.5). We begin with a sufficient condition for the absolute continuity of a measure $\mu \in \mathrm{ID}(\mathbb{R}^d)$.

Theorem 3.8.1. *Let μ be infinitely divisible on $\mathbb{R}^d$ with an infinite Lévy measure M. If, for some $n \geq 1$, $\tilde{M}^n$ is absolutely continuous, then μ is absolutely continuous. [Here, $\tilde{M}(F) := \int_F \|x\|^2/(1+\|x\|^2) M(dx)$ is a finite measure, the so-called Khintchine spectral measure of μ.]*

Proof. We use the following notation. $M|B$ is the restriction of M to $B \in \mathscr{B}(\mathbb{R}^d \setminus \{0\})$; B_k^c is the complement of the closed sphere centered at the origin with radius $1/k$; for a finite measure m on $\mathscr{B}(\mathbb{R}^d)$,

$$e(m) := \exp(-m(\mathbb{R}^d)) \sum_{j=0}^{\infty} (j!)^{-1} m^j,$$

where $m^0 := \delta(0)$ [cf. (1.8.1)].

For $k \geq 1$, let μ_k be the compound Poisson measure on $\mathbb{R}^d$ given by $\mu_k := e(M|B_k^c)$. Letting $c_k := M(B_k^c)$ and $\nu_k := c_k^{-1} M|B_k^c$, we obtain

$$\mu_k = e^{-c_k} \sum_{j=0}^{\infty} (j!)^{-1} c_k^j \nu_k^j. \tag{3.8.5}$$

Set

$$a_{k1} := e^{-c_k} \sum_{j=0}^{n-1} (j!)^{-1} c_k^j \nu_k^j(B_k^c),$$

$$a_{k2} := e^{-c_k} \sum_{j=n}^{\infty} (j!)^{-1} c_k^j \nu_k^j(B_k^c),$$

where n is such that $\tilde{M}^n \ll l_d$. Then

$$\mu_{k1} := a_{k1}^{-1} e^{-c_k} \sum_{j=0}^{n-1} (j!)^{-1} c_k^j \nu_k^j,$$

$$\mu_{k2} := a_{k2}^{-1} e^{-c_k} \sum_{j=n}^{\infty} (j!)^{-1} c_k^j \nu_k^j$$

are such that $\mu_{k1}, \mu_{k2} \in \mathscr{P}(\mathbb{R}^d)$ and $\mu_k = a_{k1}\mu_{k1} + a_{k2}\mu_{k2}$. Since M has infinite mass, we see that $c_k \to \infty$, so $a_{k1} \to 0$ as $k \to \infty$. Furthermore, since M and $\tilde{M}$ are equivalent measures and $\tilde{M}^n \ll l_d$, we have $\nu_k^n \ll l_d$. This with (3.8.1) implies that $\mu_{k2} \ll l_d$. Let $\rho_k \in \mathrm{ID}(\mathbb{R}^d)$ be such that $\mu = \mu_k * \rho_k$ and let $\mu = b_1 \mu^{(1)} + b_2 \mu^{(2)}$ be the Lebesgue decomposition of μ into a convex combination of a singular measure $\mu^{(1)}$ and an absolutely continuous measure $\mu^{(2)}$. On the other hand, $\mu = a_{k1}(\mu_{k1} * \rho_k) + a_{k2}(\mu_{k2} * \rho_k)$ with the last term being absolutely continuous. Hence, $b_1 \leq a_{k1}$. Since $a_{k1} \to 0$, we have $b_1 = 0$. Therefore, $\mu = \mu^{(2)}$, that is, μ is absolutely continuous. Q.E.D.

We know that $L_0(Q) = L_0(cQ)$ for any $c > 0$. This shows that when $d = 1$, $L_0(Q) = L_0(I)$, that is, there is exactly one class of selfdecomposable measures on $\mathbb{R}^1$. Theorems 3.4.5 and 3.8.1 combine to give a simple proof that each selfdecomposable measure on $\mathbb{R}^1$ is absolutely continuous.

Corollary 3.8.2. *All nondegenerate selfdecomposable measures on $\mathbb{R}^1$ are absolutely continuous.*

Proof. By (3.8.1) and (3.8.4), if μ has a nondegenerate Gaussian component, it is absolutely continuous. So, assume that $\mu = [0, 0, M] \in L_0(I)$ is nondegenerate. By Theorem 3.4.5, for any $F \in \mathscr{B}(\mathbb{R}^1 \setminus \{0\})$, we have

$$M(F) = \int_{\mathbb{R}^1 \setminus \{0\}} \int_0^{\infty} I_F(e^{-s}x) \, ds \, G(dx)$$

$$= \int_{\mathbb{R}^1 \setminus \{0\}} \int_0^1 s^{-1} I_{x^{-1}F}(s) \, ds \, G(dx).$$

Since $l_1(F) = 0$ implies $l_1(x^{-1}F) = 0$ for all $x \neq 0$, we see that $M \ll l_1$. By Corollary 3.4.9 and Theorem 3.8.1, we have $\mu \ll l_1$. Q.E.D.

For results when $d > 1$, we need to examine the absolute continuity of μ on $\mathbb{R}^d$ given information concerning the convolution components of μ. Let V be an r-dimensional subspace of $\mathbb{R}^d$ with $r < d$. We may think of V as a "copy" of $\mathbb{R}^r$ and r-dimensional Lebesgue measure l_r defined on the Borel subsets of V. For $\mu \in \mathscr{P}(\mathbb{R}^d)$ with $\mathrm{lin}(\mathrm{supp}\,\mu) \subseteq V$, we may consider $\mu \in \mathscr{P}(V)$. To say that such a measure μ is absolutely continuous on V means that for all $B \in \mathscr{B}(V)$, $l_r(B) = 0$ implies $\mu(B) = 0$.

Lemma 3.8.3. *Let V_1 be an r-dimensional subspace of $\mathbb{R}^d$ with $r < d$ and let V_2 be its orthogonal complement. Let P_j be the orthogonal projection of $\mathbb{R}^d$ onto V_j for $j = 1, 2$. Assume that $\mu_1 \in \mathscr{P}(\mathbb{R}^d)$ with $\mathrm{lin}(\mathrm{supp}\,\mu_1) \subseteq V_1$ and μ_1 is absolutely continuous on V_1. Also, assume that $\mu_2 \in \mathscr{P}(\mathbb{R}^d)$ and $P_2\mu_2$ is absolutely continuous on V_2. Then $\mu := \mu_1 * \mu_2 \in \mathscr{P}(\mathbb{R}^d)$ is absolutely continuous on $\mathbb{R}^d$.*

Proof. Identify V_1 and V_2 with $\mathbb{R}^r$ and $\mathbb{R}^{d-r}$, respectively, and consider $\mathbb{R}^d = V_1 \oplus V_2$. For $x \in \mathbb{R}^d$, let $x_j := P_j x$ for $j = 1, 2$, and identify x with the pair (x_1, x_2). Let $B \in \mathscr{B}(\mathbb{R}^d)$ have l_d-measure zero. We will show that $\mu(B) = 0$.

Since $l_d(B) = 0$, we have

$$\int_{\mathbb{R}^{d-r}} \int_{\mathbb{R}^r} I_B((x_1, x_2)) l_r(dx_1) l_{d-r}(dx_2) = 0.$$

Consequently, there is a $B_2 \in \mathscr{B}(\mathbb{R}^{d-r})$ of l_{d-r}-measure zero such that, for all $x_2 \notin B_2$,

$$\int_{\mathbb{R}^r} I_B((x_1, x_2)) l_r(dx_1) = 0. \tag{3.8.6}$$

Now, define the function h for $(y_1, y_2) \in V_1 \oplus V_2$ by

$$h(y_1, y_2) := \int_{\mathbb{R}^r} I_B((x_1 + y_1, y_2)) \mu_1(dx_1).$$

Then h is Borel-measurable. Since μ_1 is absolutely continuous on V_1, there is a function $f: V_1 \to \mathbb{R}^1$ which is Borel-measurable such that $\mu_1(dx_1) = f(x_1) l_r(dx_1)$. Hence,

$$h(y_1, y_2) = \int_{\mathbb{R}^r} I_B((x_1, y_2)) f(x_1 - y_1) l_r(dx_1).$$

From (3.8.6), we obtain

$$h(y_1, y_2) = I_{B_2}(y_2)h(y_1, y_2). \tag{3.8.7}$$

Let ξ be an $\mathbb{R}^d$-valued r.v. such that $\mathscr{L}(\xi) = \mu_2$ and let $\nu(dy_1|y_2)$ be the conditional probability of $P_1\xi$ given $P_2\xi = y_2$, that is, for all $F \in \mathscr{B}(\mathbb{R}^r)$,

$$\nu(F|y_2) := \mathrm{Prob}\{P_1\xi \in F | P_2\xi = y_2\}.$$

Note that

$$\mu(B) = \int_{\mathbb{R}^d}\int_{\mathbb{R}^d} I_{B-y}(x)\mu_1(dx)\mu_2(dy)$$

$$= \int_{\mathbb{R}^d}\int_{\mathbb{R}^r} I_B((x_1 + y_1, y_2))\mu_1(dx_1)\mu_2(dy)$$

$$= \int_{\mathbb{R}^d} h(y_1, y_2)\mu_2(dy).$$

This last integral is the expected value of $h(P_1\xi, P_2\xi)$. Therefore,

$$\mu(B) = \int_{\mathbb{R}^{d-r}}\int_{\mathbb{R}^r} h(y_1, y_2)\nu(dy_1|y_2)(P_2\mu_2)(dy_2)$$

$$= \int_{B_2}\int_{\mathbb{R}^r} h(y_1, y_2)\nu(dy_1|y_2)(P_2\mu_2)(dy_2) = 0$$

by (3.8.7), and since $P_2\mu_2 \ll l_{d-r}$, $l_{d-r}(B_2) = 0$. Q.E.D.

Before proceeding to the proof of the main result, we need two more technical lemmas. The first one relies on the fact that two analytic functions on $\mathbb{R}^1$ are equal everywhere whenever they are equal on a set containing an accumulation point; in other words, the set of zeros of an analytic function on $\mathbb{R}^1$ is either discrete or the whole line. The second lemma provides a decomposition of $\mathbb{R}^d$ into a direct sum of subspaces related to the operator Q.

Lemma 3.8.4. *Let f be a real analytic function on $\mathbb{R}^d$ which is not identically equal to zero. Then the set $Z := \{t \in \mathbb{R}^d: f(t) = 0\}$ of zeros of f has d-dimensional Lebesgue measure zero.*

Proof. For $d = 1$, this result is well known. Let $d > 1$. For $x \in \mathbb{R}^{d-1}$, let $Z_x := \{s \in \mathbb{R}^1 : (s, x) \in Z\}$. With fixed $x \in \mathbb{R}^{d-1}$, the function $s \mapsto f(s, x)$ is a real analytic function on $\mathbb{R}^1$. So, if $l_1(Z_x) > 0$, we have $f(\cdot, x) \equiv 0$. Therefore,

$$D := \{x \in \mathbb{R}^{d-1} : l_1(Z_x) > 0\}$$
$$= \{x \in \mathbb{R}^{d-1} : f(s, x) = 0 \text{ for each } s \in \mathbb{R}^1\}.$$

Since f is not identically equal to zero, there is $s \in \mathbb{R}^1$ such that $\tilde{Z}_s := \{x \in \mathbb{R}^{d-1} : f(s, x) = 0\} \neq \mathbb{R}^{d-1}$. Note that $\tilde{Z}_s$ is the set of zeros of a real analytic function on $\mathbb{R}^{d-1}$ which is not identically equal to zero. Furthermore, since we have

$$l_d(Z) = \int_{\mathbb{R}^{d-1}} l_1(Z_x) l_{d-1}(dx)$$
$$= \int_D l_1(Z_x) l_{d-1}(dx) \leq \int_{\tilde{Z}_s} l_1(Z_x) l_{d-1}(dx),$$

it suffices to show that $l_{d-1}(\tilde{Z}_s) = 0$. Since the lemma is true for $d = 1$, by induction, we see that it holds for any $d \geq 1$. Q.E.D.

Lemma 3.8.5. *Let Q be an invertible linear operator on $\mathbb{R}^d$ and let V be a k-dimensional subspace of $\mathbb{R}^d$ with $k < d$ such that $\lin_Q(V)$, the smallest Q-invariant subspace of $\mathbb{R}^d$ containing V, coincides with $\mathbb{R}^d$. Then there are $r \geq 2$ positive integers $k = j_1 \geq j_2 \geq \cdots \geq j_r$ and subspaces $V_1, \ldots, V_r$ with $V_1 = V$ such that*

(i) $\mathbb{R}^d = V_1 \oplus \cdots \oplus V_r$, $\dim V_m = j_m$ for $1 \leq m \leq r$;
(ii) V_m is a subspace of QV_{m-1} for $2 \leq m \leq r$;
(iii) $Q^{m-1}V$ is a subspace of $V_1 \oplus \cdots \oplus V_{m-1}$ for $2 \leq m \leq r$.

Proof. Let $\{e_1, \ldots, e_k\}$ be an orthogonal basis for V and let $V(1) := \lin\{e_1, \ldots, e_k, Qe_1, \ldots, Qe_k\}$. Then V is a proper subspace of $V(1)$, because otherwise V would be Q-invariant so $V = \lin_Q(V) = \mathbb{R}^d$ by assumption. So, renumbering $e_1, \ldots, e_k$ if necessary, we have that there is a positive integer j_2 such that $j_2 \leq j_1 := k$ such that $e_1, \ldots, e_{j_1}, Qe_1, \ldots, Qe_{j_2}$ are linearly independent, span the subspace $V(1)$, and $Qe_n \in V(1)$ for all $j_2 + 1 \leq n \leq k$.

Repeating the above procedure on

$$V(2) := \text{lin}\{V(1), QV(1)\}$$
$$= \text{lin}\{e_1, \ldots, e_{j_1}, Qe_1, \ldots, Qe_{j_2}, Q^2e_1, \ldots, Q^2e_{j_2}\},$$

we obtain $j_3 \leq j_2$ such that, after renumbering if necessary, the vectors $e_1, \ldots, e_{j_1}, Qe_1, \ldots, Qe_{j_2}, Q^2e_1, \ldots, Q^2e_{j_3}$ are linearly independent, span $V(2)$, and $Q^2e_n \in V(2)$ for all $j_3 + 1 \leq n \leq k$ [note that for $j_2 + 1 \leq n \leq k$, we have $Q^2e_n \in QV(1)$]. After finitely many similar steps, we obtain r positive integers $k = j_1 \geq \cdots \geq j_r > 0$ such that $j_1 + \cdots + j_r = d$ and

(a) the vectors $e_1, \ldots, e_{j_1}, Qe_1, \ldots, Qe_{j_2}, \ldots, Q^{r-1}e_1, \ldots, Q^{r-1}e_{j_r}$ are linearly independent;
(b) for every $2 \leq m \leq r$, the vectors $Q^{m-1}e_{j_m+1}, \ldots, Q^{m-1}e_k$ are contained in $\text{lin}\{e_1, \ldots, e_{j_1}, Qe_1, \ldots, Qe_{j_2}, \ldots, Q^{m-1}e_1, \ldots, Q^{m-1}e_{j_m}\}$.

Now, set $V_m := \text{lin}\{Q^{m-1}e_1, \ldots, Q^{m-1}e_{j_m}\}$ for $1 \leq m \leq r$. Then (a) implies (i). The definition of V_m with $j_{m-1} \geq j_m$ gives (ii). Finally, (b) implies (iii). Q.E.D.

The most crucial fact needed to prove our main result is given in the following proposition.

Proposition 3.8.6. *Let μ be an $\exp(-tQ)$-decomposable measure with $\mu = \mathcal{T}_Q[a, T, G]$ for $[a, T, G] \in \text{ID}_{\log}$. If there is a subspace V of $\mathbb{R}^d$ containing the supp G with the property that $\text{lin}_Q(V) = \mathbb{R}^d$ and $G(W) = 0$ for every proper subspace W of V, then μ is absolutely continuous.*

Proof. Let M be the Lévy measure of μ. By (3.6.10), for $B \in \mathcal{B}(\mathbb{R}^d \setminus \{0\})$,

$$M(B) = \int_0^\infty G(e^{tQ}B)\, dt.$$

Hence, by Corollary 3.4.9, either $M(\mathbb{R}^d) = 0$ or $M(\mathbb{R}^d) = \infty$. When $M(\mathbb{R}^d) = 0$, μ is Gaussian, so it is absolutely continuous. Hence, assume $M(\mathbb{R}^d) = \infty$. Therefore, it suffices to show that $\tilde{M}^d \ll l_d$ (cf. Theorem 3.8.1), where

$$\tilde{M}(B) := \int_B \frac{\|x\|^2}{1 + \|x\|^2} M(dx).$$

Since $V \supseteq \operatorname{supp} G$, we have

$$\tilde{M}(B) := \int_0^\infty \int_{V \setminus \{0\}} I_B(e^{-tQ}x) \frac{\|e^{-tQ}x\|^2}{1 + \|e^{-tQ}x\|^2} G(dx) \, dt.$$

For any set A, let $(A)^d$ denote the Cartesian product of A with itself d times. Then

$$\tilde{M}^d(B) = \int_{((0,\infty))^d} \int_{(V \setminus \{0\})^d} I_B(e^{-t_1 Q} x_1 + \cdots + e^{-t_d Q} x_d)$$

$$\times \prod_{j=1}^d \frac{\|e^{-t_j Q} x_j\|^2}{1 + \|e^{-t_j Q} x_j\|^2} G(dx_1) \cdots G(dx_d) \, dt_1 \cdots dt_d.$$

(3.8.8)

To show that $l_d(B) = 0$ implies $\tilde{M}^d(B) = 0$, we will divide $(V \setminus \{0\})^d$ into a countable union of subsets and show that the (inner) integral in (3.8.8) vanishes on each of these subsets.

Let $V_1, \ldots, V_r$ be the subspaces of $\mathbb{R}^d$ given in Lemma 3.8.5, with $r = 1$ when $V = \mathbb{R}^d$, and let P_m be the projection from $\mathbb{R}^d$ onto V_m, $1 \le m \le r$. Since V_m is a subspace of $Q^{m-1}(V)$, we have $P_m Q^{m-1}(V) = V_m$, $1 \le m \le r$. Let $N_m := \{x \in V : P_m Q^{m-1} x = 0\}$, that is, N_m is the kernel of $P_m Q^{m-1}$ restricted to V. Then $\dim N_m = k - j_m$. Furthermore, if $W_m(x_1, \ldots, x_s) := \operatorname{lin}\{N_m, \{x_1, \ldots, x_s\}\}$, then $k - j_m \le \dim W_m(x_1, \ldots, x_s) \le k - j_m + s$ and $W_m(x_1, \ldots, x_s)$ is a subspace of V whenever each $x_i \in V$. Let $n_0 := 0$ and $n_m := \sum_{s=1}^m j_s$ for $1 \le m \le r$. Set

$$L := \Big\{ (x_1, \ldots, x_d) \in (V \setminus \{0\})^d : \dim W_m(x_{n_{m-1}+1}, \ldots, x_{n_m}) = k$$

$$\text{for each } 1 \le m \le r \Big\},$$

and, for $1 \le m \le r$, $0 \le s \le j_{m-1}$, set

$$K_{ms} := \left\{(x_1, \ldots, x_d) \in (V \setminus \{0\})^d \setminus L : \dim W_m(x_{n_{m-1}+1}, \ldots, x_{n_m}) = k - j_m + s\right\}.$$

Note that K_{m0} consists of those nonzero vectors $(x_1, \ldots, x_d)$ such that $x_{n_{m-1}+1}, \ldots, x_{n_m}$ are in N_m. In particular, K_{10} is empty since $N_1 = \{0\}$. For $s \ge 1$ and integers $i_1, \ldots, i_s$ such that $n_{m-1} < i_1 < \cdots < i_s \le n_m$, set

$$K_m(i_1, \ldots, i_s)$$
$$:= \left\{(x_1, \ldots, x_d) \in K_{ms} : \dim W_m(x_{i_1}, \ldots, x_{i_s}) = k - j_m + s\right\}.$$

Hence, we decompose $(V \setminus \{0\})^d$ into

$$(V \setminus \{0\})^d = L \cup \left(\bigcup_{m=1}^{r} K_m\right), \tag{3.8.9}$$

where

$$K_m := K_{m0} \cup \left(\bigcup_{s=1}^{j_m - 1} \bigcup_{n_{m-1}+1 \le i_1 < \cdots < i_s \le n_m} K_m(i_1, \ldots, i_s)\right),$$

since $K_m(i_1, \ldots, i_s)$ consists of all those $(x_1, \ldots, x_d) \in (V \setminus \{0\})^d$ such that (i) $\dim W_m(x_{i_1}, \ldots, x_{i_s}) = k - j_m + s$; (ii) for $n_{m-1} < j \le n_m$ and $j \ne i_1, \ldots, i_s$, $x_j \in W_m(x_{i_1}, \ldots, x_{i_s})$; and (iii) for $1 \le j \le n_{m-1}$ and $n_m < j \le d$, $x_j \in V \setminus \{0\}$.

Let G_i denote the product measure on $(V \setminus \{0\})^i$ obtained by producting G with itself. Thus, we see that $G_d(K_m(i_1, \ldots, i_s))$ is given by

$$\int_{(V \setminus \{0\})^s} G_{d-s}(F(x_{i_1}, \ldots, x_{i_s})) G_s(dx_{i_1}, \ldots, dx_{i_s}),$$

where

$$F(x_{i_1}, \ldots, x_{i_s}) := \left\{(\xi_j)_{j \ne i_1, \ldots, i_s} \in V^{d-s} : (\xi_1, \ldots, \xi_d) \in K_m(i_1, \ldots, i_s)\right.$$
$$\left. \text{and } \xi_{i_j} = x_{i_j} \text{ for all } 1 \le j \le s\right\}.$$

Then $F(x_{i_1}, \ldots, x_{i_s})$ is also equal to

$$\Big\{(\xi_j)_{j \neq i_1, \ldots, i_s} \in V^{d-s}: \dim W_m(x_{i_1}, \ldots, x_{i_s}) = k - j_m + s;$$
$$\xi_j \in W_m(x_{i_1}, \ldots, x_{i_s}) \text{ for } m_{n-1} < j \leq n_m \text{ with } j \neq i_1, \ldots, i_s;$$
$$\xi_j \in V \text{ for } 1 \leq j \leq n_{m-1} \text{ and } n_m < j \leq d\Big\}.$$

Hence, $G_d(K_m(i_1, \ldots, i_s))$ is bounded above by

$$\int_{(V \setminus \{0\})^s} [G(V)]^{d-j_m} \Big[G\big(W_m(x_{i_1}, \ldots, x_{i_s})\big)\Big]^{j_m-s} G_s(dx_{i_1}, \ldots, dx_{i_s}).$$

This bound is equal to zero because $W_m(x_{i_1}, \ldots, x_{i_s})$ is a proper subspace of V, so by assumption it has G-measure zero. Similarly, we obtain $G_d(K_{m0}) = [G(N_m)]^{j_m}[G(V)]^{d-j_m} = 0$, since N_m is also a proper subspace of V. This shows that the inner integral in (3.8.8) is equal to zero when taken over the sets K_m [cf. (3.8.9)].

Now, fix $(x_1, \ldots, x_d) \in L$. Since $\dim N_m = k - j_m$, for each $1 \leq m \leq r$, the vectors $x_{n_{m-1}+1}, \ldots, x_{n_m}$ are linearly independent and so are $Q^{m-1}x_{n_{m-1}+1}, \ldots, Q^{m-1}x_{n_m}$. Also, by Lemma 3.8.5, $\lin\{Q^{m-1}x_{n_{m-1}+1}, \ldots, Q^{m-1}x_{n_m}\} \subseteq Q^{m-1}V \subseteq V_1 \oplus \cdots \oplus V_m$. We claim that

$$x_1, \ldots, x_{n_1+1}, Qx_{n_1+1}, \ldots, Qx_{n_2}, \ldots, Q^{r-1}x_{n_{r-1}+1}, \ldots, Q^{r-1}x_{n_r}$$
(3.8.10)

are linearly independent.
Suppose that

$$a_1 x_1 + \cdots + a_{n_1} x_{n_1} + \cdots + a_{n_{r-1}} Q^{r-1}x_{n_{r-1}+1} + \cdots + a_{n_r} Q^{r-1}x_{n_r} = 0.$$
(3.8.11)

Since $\lin\{x_1, \ldots, Q^{r-2}x_{n_{r-1}}\} \subseteq V_1 \oplus \cdots \oplus V_{r-1}$, applying P_r to (3.8.11) we see that

$$a_{n_{r-1}+1} x_{n_{r-1}+1} + \cdots + a_{n_r} x_{n_r} \in N_r \cap \lin\{x_{n_{r-1}+1}, \ldots, x_{n_r}\} = \{0\}.$$

Therefore, $a_{n_{r-1}+1}, \ldots, a_{n_r}$ are all equal to zero. By consecutively applying $P_{r-1}, \ldots, P_1$ to (3.8.11), we see that all of the a's are equal to zero, so (3.8.10) is established.

For $(t_1, \ldots, t_d) \in \mathbb{R}^d$, set

$$D(t_1, \ldots, t_d) := \det(e^{-t_1 Q} x_1, \ldots, e^{-t_d Q} x_d).$$

Then D is a real analytic function on $\mathbb{R}^d$ and it is not identically equal to zero, because from (3.8.10) its $(j_2 + 2j_3 + \cdots + (r-1)j_r)$-order derivative is not identically equal to zero, that is,

$$\partial t_{n_1+1} \cdots \partial t_{n_2} \partial^2 t_{n_2+1} \cdots \partial^2 t_{n_3} \cdots \partial^{r-1} t_{n_r} D(0, \ldots, 0)$$
$$= \det(x_1, \ldots, x_{n_1}, Qx_{n_1+1}, \ldots, Qx_{n_2}, \ldots, Q^{r-1} x_{n_r}) \neq 0.$$

Note that the Jacobian, $J\psi$, of the mapping

$$\psi(t_1, \ldots, t_d) := e^{-t_1 Q} x_1 + \cdots + e^{-t_d Q} x_d$$

has the form

$$(-1)^d \det(Q) \det(e^{-t_1 Q} x_1, \ldots, e^{-t_d Q} x_d).$$

Hence, the Jacobian is nonzero outside of some closed set N of l_d-measure zero (cf. Lemma 3.8.4). Therefore, for any $u \in \mathbb{R}^d \setminus N$, there is an open set U containing u such that ψ^{-1} exists on $\psi(U)$ and is differentiable. Hence, by change of variables

$$\int_U I_B(\psi(t_1, \ldots, t_d)) l_d(dt_1, \ldots, dt_d)$$
$$= \int_{\psi(U)} I_B(s_1, \ldots, s_d) |J\psi^{-1}(s_1, \ldots, s_d)| l_d(ds_1, \ldots, ds_d),$$

which is equal to zero since $l_d(B) = 0$. Thus,

$$\int_{\mathbb{R}^d} I_B(\psi(t_1, \ldots, t_d)) l_d(dt_1, \ldots, dt_d)$$
$$= \int_N I_B(\psi(t_1, \ldots, t_d)) l_d(dt_1, \ldots, dt_d) = 0,$$

since $l_d(N) = 0$. Therefore, the inner integral in (3.8.8) is also zero when taken over L. Consequently, the integral (3.8.8) is zero, so $\tilde{M}^d(B) = 0$. Thus, we finally have $\tilde{M}^d \ll l_d$. Q.E.D.

Remark 3.8.7. For any nonzero measure m on $\mathbb{R}^d$ with $m(\{0\}) = 0$, there is a subspace W such that $m(W) > 0$ and $m(\tilde{W}) = 0$ for all proper subspaces $\tilde{W}$ of W.

Proof. Suppose not. Then for each subspace W we could find a strictly decreasing sequence of proper subspaces $W_1, \ldots, W_r$ such that $\dim W_r = 1$ and $m(W_r) > 0$. Consequently, $m(\{0\}) > 0$, a contradiction. Q.E.D.

Remark 3.8.8. If $[b, T, G] \in \mathrm{ID}_{\log}$ and $G|B$ is the restriction of G to the set B, then $[b, T, G|B] \in \mathrm{ID}_{\log}$ and $\mathcal{T}_Q[b, T, G|B] \in L_0(Q)$. Furthermore, if V is a Q-invariant subspace of $\mathbb{R}^d$ and $[a, R, M] = \mathcal{T}_Q[b, T, G]$, then $\mathcal{T}_Q[b, T, G|V]$ has Lévy measure $M|V$ [cf. Theorem 3.6.8 and (3.6.17) and (3.6.10)].

We are now prepared to state and prove the main result of this section.

Theorem 3.8.9. *Each full $\exp(-tQ)$-decomposable measure on $\mathbb{R}^d$ is absolutely continuous.*

Proof. Let $\mu = [a, R, M] = \mathcal{T}_Q[b, T, G] \in L_0(Q)$ be full, with $[b, T, G] \in \mathrm{ID}_{\log}$ (cf. Proposition 3.6.10). By (3.6.10), we have, for $B \in \mathcal{B}(\mathbb{R}^d \setminus \{0\})$,

$$M(B) = \int_0^\infty G(e^{tQ}B)\, dt. \quad (3.8.12)$$

Consider the following three cases.

(i) $G = 0$.
(ii) G is nonzero, but $G(V) = 0$ for every property Q-invariant subspace V of $\mathbb{R}^d$.
(iii) $G(V) > 0$ for some proper Q-invariant subspace V of $\mathbb{R}^d$.

In the case of (i), μ is a full Gaussian measure on $\mathbb{R}^d$, so $\mu \ll l_d$ [cf. (3.8.4)].

In the case of (ii), let W be a subspace of $\mathbb{R}^d$ such that $G(W) > 0$ and $G(\tilde{W}) = 0$ for every proper subspace $\tilde{W}$ of W (cf. Remark 3.8.7). Let $\mu_1 := \mathcal{T}_Q[0, 0, G|W]$, where $G|W$ is the restriction of G to W. Then $\mu_1 \in L_0(Q)$. Clearly, $\mathrm{supp}(G|W) \subseteq W$. Suppose $\lin_Q W$ is a proper subspace of $\mathbb{R}^d$. Then $G(\lin_Q W) = 0$ by (ii), so $G(\tilde{W}) = 0$, a contradiction. Hence, $\lin_Q W = \mathbb{R}^d$. By our selection of W, $(G|W)(\tilde{W}) = G(\tilde{W}) = 0$ for every proper subspace $\tilde{W}$ of W. Thus, by Proposition 3.8.6, $\mu_1 \ll l_d$. Since $\mu = \mu_1 * \mathcal{T}_Q[b, T, G|W^c]$, we have $\mu \ll l_d$, by (3.8.1).

In the case of (iii), we will prove that μ is absolutely continuous by induction on the dimension d together with Lemma 3.8.3 and Proposition 3.8.6. For $d = 1$, Corollary 3.8.2 states that μ is absolutely continuous. So, assume that $d \geq 2$ and Theorem 3.8.9 holds for any full $\exp(-tQ)$-decomposable measure on a k-dimensional space for all $k < d$. Let V be as in (iii), that is, V is a proper Q-invariant subspace of $\mathbb{R}^d$ with $G(V) > 0$. Since $QV = V$, we have $e^{tQ}V = V$ for all $t \geq 0$, so by (3.8.12), $M(V) = \infty$. Let $V_1 := \lin_Q(\supp(G|V))$ and $\mu_1 := \mathcal{T}_Q[0, 0, G|V_1]$. Note that $V \supset V_1 \supset \supp(G|V)$, so that $G|V_1 = G|V$. We have that V_1 is a Q-invariant subspace of V, $\dim V_1 < d$, $\mu_1 \in L_0(Q|V_1)$, and the Lévy measure of μ_1 is $M|V_1$ by Remark 3.8.8. Then $\mu_1 = [c, 0, M|V_1]$ for some c in V_1 by (3.6.11). From Proposition 3.4.10,

$$\lin(\supp \mu_1) = \lin(\supp(\mu_1 * \delta(-c))) = \lin(\supp(M|V_1))$$
$$= \lin_Q(\supp(G|V_1)) = V_1.$$

Hence, μ_1 is full on V_1. Considering μ_1 as in $\mathscr{P}(V_1)$, we have that μ_1 is full and $\exp(-t(Q|V_1))$-decomposable on the r-dimensional space V_1 with $r < d$. So, by the induction hypothesis, we see that $\mu_1 \ll l_r$ on the space V_1.

Now, let $\mu_2 := \mathcal{T}_Q[b, T, G|V_1^c]$. Then $\mu_2 \in L_0(Q)$ on $\mathbb{R}^d$ and $\mu = \mu_1 * \mu_2$. Let $V_2 := V_1^\perp$ and let P_j be the orthogonal projection of $\mathbb{R}^d$ onto V_j for $j = 1, 2$. Since $\mu_2 \in L_0(Q; \mathbb{R}^d)$ and $\ker P_2 = V_1$ is Q-invariant, from Corollary 3.6.16 we have $P_2\mu_2 \in L_0(P_2Q|V_2; V_2)$.

Suppose that the $\supp P_2\mu_2$ is contained in a proper hyperplane $v + K$ of V_2, so $\dim K \leq d - r - 1$. Let ξ and ζ be independent $\mathbb{R}^d$-valued r.v.'s such that $\mathscr{L}(\xi) = \mu_1$ and $\mathscr{L}(\zeta) = \mu_2$. Since $\mathscr{L}(P_2\zeta) = P_2\mu_2$ and $\mathscr{L}(P_1\xi) = \mu_1$, we see that $\xi + \zeta = \xi + P_1\zeta + P_2\zeta \in V_1 + a + K$, which is a proper hyperplane of $\mathbb{R}^d$. Since $\mu = \mathscr{L}(\xi + \zeta)$, this implies that μ is not full, a contradiction. Hence, $\lin(\supp P_2\mu_2) = V_2$. Therefore, $P_2\mu_2 \in \mathscr{P}(V_2)$ is full and is in $L_0(P_2Q|V_2; V_2)$. Since $\dim V_2 = d - r$, by the induction hypothesis, $P_2\mu_2 \ll l_{d-r}$ on V_2.

Finally, applying Lemma 3.8.3 to $\mu = \mu_1 * \mu_2$, we have that μ is absolutely continuous on $\mathbb{R}^d$.   Q.E.D.

Except for the Gaussian densities already known, we have not obtained any information about the density for an arbitrary full

$\mu \in L_0(Q)$. Our next aim is to relate the density of a full $\mu = [a, R, M] \in L_0(Q)$ with the operator G (note: G is now an operator, not a Lévy measure), whose smallest closed extension Γ is the infinitesimal generator of the semigroup $(U_t f)(x) := \mathbf{E} f(Z_x(t))$, where $f \in C_0(\mathbb{R}^d)$,

$$Z_x(t) := e^{-tQ}x + \int_0^t e^{-sQ} \, dY(s)$$

and Y is a $D(\mathbb{R}^d, [0, \infty))$-valued r.v. with stationary independent increments, $Y(0) = 0$ a.s., and $\mathscr{L}(Y(1)) = [a, R, M]$ (cf. Theorem 3.7.8). Furthermore, from Theorem 3.6.8, we have

$$L_0(Q) = \{\mathscr{L}(Z_x(\infty)): \mathscr{L}(Y(1)) \in \mathrm{ID}_{\log}\}$$

[cf. (3.6.18)].

Theorem 3.8.10. *Let G be the linear operator on $C_K^2(\mathbb{R}^d)$ given in Theorem 3.7.8. Then g is the density of the full $\exp(-tQ)$-decomposable measure $\mu = [a, R, M]$ on $\mathbb{R}^d$ if and only if g is the unique weak solution of*

$$G^*g = 0 \quad \text{with } g \geq 0 \quad \text{and} \quad \int_{\mathbb{R}^d} g(x) l_d(dx) = 1. \quad (3.8.13)$$

Here, G^* is the formal adjoint of G, that is,

$$(G^*f)(x) := \langle Qx - a, Df(x) \rangle + \mathrm{tr}(Qf(x))$$
$$+ \frac{1}{2} \mathrm{tr}(RD^2 f(x))$$
$$+ \int_{\mathbb{R}^d \setminus \{0\}} \left(f(x - y) - f(x) + \frac{\langle y, Df(x) \rangle}{1 + \|y\|^2} \right) M(dy),$$
$$(3.8.14)$$

with Df and $D^2 f$ denoting derivatives of f and tr denoting the trace.

Proof. Assume that g is the density of $\mu = [a, R, M] \in L_0(Q)$. From Proposition 3.7.10, we have, for all $t \geq 0$,

$$\int_{\mathbb{R}^d} (U_t f)(x) \mu(dx) = \int_{\mathbb{R}^d} f(x) \mu(dx)$$

for all $f \in C_0(\mathbb{R}^d)$, that is, the semigroup U_t, $t \geq 0$, is μ-stationary. Hence, from $\mu(dx) = g(x)l_d(dx)$,

$$\int_{\mathbb{R}^d} (Gf)(x)g(x)l_d(dx) = 0, \quad f \in \mathscr{D}(G), \quad (3.8.15)$$

that is, g is a weak solution of (3.8.13).

Conversely, assume that g is a weak solution of (3.8.13), that is, (3.8.15) holds. Since Γ is a closed extension of G, we have, for all $f \in \mathscr{D}(\Gamma)$,

$$\int_{\mathbb{R}^d} \Gamma f \cdot g l_d(dx) = 0.$$

Furthermore, since $\mathscr{D}(\Gamma) = R_\lambda(C_0(\mathbb{R}^d))$ and $(\lambda I - \Gamma)^{-1} = R_\lambda$ (cf. Propositions 3.7.1 and 3.7.5), we have $\Gamma R_\lambda = R_\lambda \Gamma$ on $\mathscr{D}(\Gamma)$. Hence, for $f \in \mathscr{D}(\Gamma)$,

$$\int_{\mathbb{R}^d} (\Gamma R_\lambda f)(x) \cdot g(x) l_d(dx) = \int_{\mathbb{R}^d} (R_\lambda \Gamma f)(x) \cdot g(x) l_d(dx) = 0.$$

Therefore, from $R_\lambda(\lambda \Gamma - I) = I$ on $\mathscr{D}(\Gamma)$, we obtain

$$\alpha := \int_{\mathbb{R}^d} f(x) g(x) l_d(dx) = \int_{\mathbb{R}^d} R_\lambda(\lambda \Gamma - I) f(x) \cdot g(x) l_d(dx)$$

$$= \lambda \int_{\mathbb{R}^d} R_\lambda f \cdot g \, dl_d = \int_0^\infty \lambda e^{-\lambda t} \left(\int_{\mathbb{R}^d} U_t f \cdot g \, dl_d \right) dt$$

for each $\lambda > 0$. Thus, the Laplace transform of the function $t \to \int_{\mathbb{R}^d} U_t f \cdot g l_d$ is equal to the constant α, so, for all $f \in \mathscr{D}(\Gamma)$,

$$\int_{\mathbb{R}^d} f \cdot g l_d = \int_{\mathbb{R}^d} U_t f \cdot g l_d.$$

Since this equality extends to $C_0(\mathbb{R}^d)$, we see that the semigroup U_t, $t \geq 0$, is ν-stationary, where $\nu(dx) := g(x)l_d(dx)$. By the uniqueness of Proposition 3.7.10, we see that ν is $\exp(-tQ)$-decomposable with g for a density.

Finally, to see that G^* is of the form (3.8.13), it suffices to check that

$$\int_{\mathbb{R}^d} Gh_1 \cdot h_2 \, dx = \int_{\mathbb{R}^d} h_1 \cdot G^* h_2 \, dx$$

for $h_1, h_2 \in C_K^2(\mathbb{R}^d)$. The above is easy to see when both G and G^* are expressed by coordinate partial derivatives. Q.E.D.

3.9 MULTIVARIATE SELFDECOMPOSABLE MEASURES

A probability measure μ (on a Banach space) is said to be *multivariate selfdecomposable*, or simply *selfdecomposable*, if there exist a sequence $\{a_n\}$ of positive numbers, a sequence $\{\nu_n\}$ of probability measures on the Banach sphere in question, and a sequence $\{v_n\}$ of vectors in the same Banach space such that the triangular array of measures

$$\{T_{a_n}\nu_j : 1 \le j \le n\} \text{ is uniformly infinitesimal} \qquad (3.9.1)$$

and

$$T_{a_n}(\nu_1 * \cdots * \nu_n) * \delta(v_n) \Rightarrow \mu. \qquad (3.9.2)$$

Later, L_0 denotes the class of all selfdecomposable measures and it is often called the *Lévy class* of probability measures (not to be confused with Lévy spectral measures of infinitely divisible measures; cf. Section 1.8). A quick look at Section 3.1 leads us to observe that selfdecomposable measures are operator-selfdecomposable ones with normings by $A_n := a_n I$, $a_n > 0$. Thus, all characterizations and properties discussed so far can be specialized to this class, but often only for *full* measures. In fact, fullness can be replaced by nondegeneracy, because of the very particular form of A_n's and the convergence of types theorems discussed in Section 2.7. In particular, in case of nondegenerate limits, Corollary 2.7.3 is used in the same fashion as Theorem 2.2.10 was utilized for operator-selfdecomposability in Section 3.2. It might be worthwhile here to call attention to Remark 2.7.5. It implies that allowing $a_n < 0$ in (3.9.2) would lead to limit measures which are shifts of symmetric measures; thus, restricting the class L_0.

Proposition 3.9.1. *If μ is nondegenerate and selfdecomposable, then in (3.9.2) we have $a_n \to 0$ and $a_{n+1}/a_n \to 1$ as $n \to \infty$.*

Proof. Proceed along the lines of the proofs of Proposition 3.2.2 and (3.2.2) using convergence of types theorems from Section 2.7. Of course, some of the steps in this case are trivial. Q.E.D.

Theorem 3.9.2. *A measure μ on X is selfdecomposable, that is, $\mu \in L_0$ if and only if, for each $0 < c < 1$, there exists $\mu_c \in \mathscr{P}$ such that*

$$\mu = T_c\mu * \mu_c. \tag{3.9.3}$$

Proof. Since degenerate measures are selfdecomposable and satisfy (3.9.3), we assume that μ is nondegenerate. From Proposition 3.9.1, for each $0 < c < 1$, there exists a sequence $k_n \geq n$ such that $a_{k_n}/a_n \to c$ as $n \to \infty$. Setting $\rho_n := T_{a_n}(\nu_1 * \cdots * \nu_n) * \delta(v_n)$, we have that $\rho_n \Rightarrow \mu$ and

$$\rho_{k_n} = T_{a_{k_n}/a_n}\rho_n * \left[T_{a_{k_n}}(\nu_{n+1} * \cdots * \nu_{k_n}) * \delta(w_n)\right] \tag{3.9.4}$$

for some $w_n \in X$. With Theorem 1.7.1(a), this implies that the second factor in (3.9.4) is conditionally compact. Denoting its limit point by μ_c, we get the decomposition $\mu = T_c\mu * \mu_c$ which proves the necessity of (3.9.3). [In fact, the second factor in (3.9.3) converges to μ_c.]

For the sufficiency of (3.9.3), let us set $\nu_1 := \mu$, $\nu_k := \mu_{(k-1)/k}$, $k \geq 2$, using (3.9.3) with $c = (k-1)/k$. Taking $a_n := n^{-1}$, we obtain

$$T_{a_n}(\nu_1 * \nu_2 * \cdots * \nu_n) = \mu.$$

[Note that $\hat{\nu}_k = \hat{\mu}/(T_{(k-1)/k}\mu)\hat{\ }$ for $k \geq 2$ and use Fourier transforms to show the above equation.] Q.E.D.

The factorization, or decomposition, in (3.9.3) is the justification for the term "selfdecomposable" measure. Also, writing it in the form

$$\mu = T_{e^{-t}}\mu * \nu_t, \quad \text{for all } t \geq 0, \tag{3.9.5}$$

we see that Theorem 3.9.2 is analogous to the fundamental characterization of operator-selfdecomposable measures given in Theorem 3.5.5, due to Urbanik. It is very interesting how much deeper analysis was needed to establish Theorem 3.5.5, while Theorem 3.9.2 is a straightforward calculation.

Theorem 3.9.3 (Random Integral Representation). *A measure μ is selfdecomposable if and only if there exists a $D(X,[0,\infty))$-valued random variable Y such that Y has stationary independent increments, $Y(0) = 0$*

a.s., $\mathbf{E}\log(1 + \|Y(1)\|) < \infty$, and

$$\mathscr{L}\left(\int_0^\infty e^{-s}\, dY(s)\right) = \mu.$$

Proof. The proof goes by the same steps as the proof of Theorem 3.6.8. Q.E.D.

Corollary 3.9.4. *A measure μ on a Banach space X is selfdecomposable if and only if, for $x^* \in X^*$,*

$$\hat{\mu}(x^*) = \exp\{i\langle x^*, a\rangle - 4^{-1}\langle x^*, Rx^*\rangle$$
$$+ \int_{X\setminus\{0\}}\left(\int_{(0,1)} \frac{e^{iw\langle x^*, x\rangle} - 1}{w}\, dw - i\langle x^*, x\rangle I_B(x)\right)M(dx),$$
(3.9.6)

where the triple $a \in X$, R is a Gaussian covariance operator, and M is a Lévy spectral measure satisfying the condition $\int_{\|x\|>1} \log(1 + \|x\|)M(dx) < \infty$ and they are uniquely determined.

Proof. Using Lemma 3.6.4, compute $\hat{\mu}$ from the random integral representation in Theorem 3.9.3, with $\mathscr{L}(Y(1)) = [\alpha_0, R, M]$. Because of Proposition 3.6.10, $\mathscr{L}(Y(1))$ is uniquely determined. Q.E.D.

Besides the above, other properties of selfdecomposable measures can be derived from those of operator-selfdecomposable ones. However, let us focus on the problem of distinguishing selfdecomposable measures among those which are operator-selfdecomposable. Note that (3.9.5) characterizes the class L_0 as

$$\mu = e^{-tI}\mu * \nu_t \quad \text{for all } t \geq 0.$$

In other words, I is a U-exponent of the selfdecomposable measure μ, that is, $e^{-tI} \in \mathbf{D}(\mu)$ for $t \geq 0$ (cf. the second part of Section 3.3). The rest of this section deals only with $X = \mathbb{R}^d$, a finite-dimensional linear space.

Proposition 3.9.5. *Let μ be a measure on $\mathbb{R}^d$ such that $\{0\}$ and $\mathbb{R}^d$ are the only invariant subspaces with respect to the symmetry semigroup $\mathbf{A}(\mu)$. If μ is full and $\mathscr{T}(\mathbf{D}(\mu)) \setminus \mathscr{T}(\mathbf{A}(\mu)) \neq \varnothing$, then μ is selfdecom-*

posable. Moreover, for a nondegenerate selfdecomposable μ, we have $-I \in \mathscr{T}(\mathbf{D}(\mu)) \setminus \mathscr{T}(\mathbf{A}(\mu))$.

Proof. Since μ is full, $\mathbf{A}(\mu)$ is a compact subgroup of $\text{Aut}(\mathbb{R}^d)$ (cf. Corollary 2.3.2). Furthermore, by Corollary 2.4.2, there exist a subgroup $\mathscr{O}_0$ of the orthogonal group $\mathscr{O} = \mathscr{O}(\mathbb{R}^d)$ and a positive-definite symmetric W such that $\mathbf{A}(\mu) = W\mathscr{O}_0 W^{-1}$. Since $\mathbf{A}(W^{-1}\mu) = \mathscr{O}_0$, $\mathbf{D}(W^{-1}\mu) = W^{-1}\mathbf{D}(\mu)W$, and $W^{-1}\mu$ is full, we may assume without loss of generality that $\mathbf{A}(\mu) = \mathscr{O}_0$ (cf. Lemma 2.3.3 and Corollary 2.1.4). Let $Q \in \mathscr{T}(\mathbf{D}(\mu)) \setminus \mathscr{T}(\mathbf{A}(\mu))$ and let Q_c be the commuting exponent of μ, from Proposition 3.3.8(b). For $A \in \mathbf{A}(\mu)$, $A^* = A^{-1}$, so Q_c^* also commutes with $\mathbf{A}(\mu)$ as well as $Q_c Q_c^*$. By Schur's lemma [cf. Lange (1975)], we conclude $Q_c Q_c^* = \lambda^2 I$ and $Q_c^* Q_c = \rho^2 I$ for some real λ and ρ. Hence, $\lambda^2 Q_c = (Q_c Q_c^*)Q_c = Q_c(Q_c^* Q_c) = \rho^2 Q_c$, so $(\lambda^2 - \rho^2)Q_c = 0$. Consequently, $Q_c Q_c^* = Q_c^* Q_c$ and both Q_c, Q_c^* commute with $\mathbf{A}(\mu)$. Once again by Schur's lemma, we obtain $Q_c = \lambda I$. Since $\mathbf{D}(\mu)$ is a compact subgroup of $\text{End}(\mathbb{R}^d)$ (cf. Theorem 2.3.1) and $\exp(-\lambda t I) \in \mathbf{D}(\mu)$ for $t \geq 0$, we see that $\lambda \geq 0$. Furthermore, from

$$d\lambda = \text{tr}(Q_c) = \int_{\mathbf{A}(\mu)} \text{tr}(g^{-1}Qg)H(dg) = \text{tr}(Q),$$

we see that $\lambda > 0$, because otherwise $\det(e^{-tQ}) = e^{-t(\text{tr } Q)} = 1$, $e^{-tQ} \in \mathbf{D}(\mu)$, and by Proposition 2.3.5 we conclude that $e^{-tQ} \in \mathbf{A}(\mu)$ and consequently $Q \notin \mathscr{T}(\mathbf{D}(\mu)) \setminus \mathscr{T}(\mathbf{A}(\mu))$ which contradicts our assumption. Thus, $\exp(-\lambda t I) \in \mathbf{D}(\mu)$ for each $t \geq 0$, where $\lambda > 0$, that is, μ is selfdecomposable.

Conversely, if μ is selfdecomposable, then I is its Urbanik exponent, that is, $-I \in \mathscr{T}(\mathbf{D}(\mu))$. If also $-I \in \mathscr{T}(\mathbf{A}(\mu))$, then $e^{-t}I \in \mathbf{A}(\mu)$ and consequently $|\hat{\mu}(x^*)| = |\hat{\mu}(e^{-t}x^*)| \to 1$ as $t \to \infty$, which implies that μ is degenerate. Q.E.D.

Let us conclude this section with a discussion of measures on $\mathbb{R}^d$ that have "large" symmetry semigroups. In fact, we will say that μ is *elliptically symmetric* if its symmetry semigroup $\mathbf{A}(\mu)$ is conjugate to the full orthogonal group $\mathscr{O} = \mathscr{O}(\mathbb{R}^d)$ so that $\mathbf{A}(\mu) = W^{-1}\mathscr{O}W$ for some positive-definite and symmetric W.

Proposition 3.9.6. *A measure μ on $\mathbb{R}^d$ is elliptically symmetric with $\mathscr{T}(\mathbf{D}(\mu)) \setminus \mathscr{T}(\mathbf{A}(\mu)) \neq \emptyset$ if and only if there exists a probability* measure ρ on $(0, \infty)$ such that, for $y \in \mathbb{R}^d$,

$$\log \hat{\mu}(y) = i\langle y, x\rangle - c_1\|Wy\|^2$$
$$+ c_2\Gamma\left(\frac{d}{2}\right)\int_0^\infty \sum_{j=1}^\infty (-1)^j\left(j!2j\Gamma\left(j + \frac{d}{2}\right)\right)^{-1}$$
$$\times \frac{(\|Wy\|s/2)^{2j}}{\log(1 + s^2)}\rho(ds), \qquad (3.9.7)$$

where $c_i \geq 0$, $c_1 + c_2 > 0$, Γ is Euler's gamma function, $x \in \mathbb{R}^d$, and W is a positive-definite symmetric matrix.

Proof. Since $WA(\mu)W^{-1} = \mathcal{O}$, μ is full on $\mathbb{R}^d$ and, by Proposition 3.9.5, we see that μ is selfdecomposable. Since $-I \in A(\mu)$, we obtain that μ is a translation of a symmetric selfdecomposable measure and, by Corollary 3.9.4, we have

$$\log \hat{\mu}(y) = i\langle y, x\rangle - 4^{-1}\langle y, Ry\rangle$$
$$+ c_2\int_{\mathbb{R}^d\setminus\{0\}}\int_0^1 \frac{[\cos t\langle y, x\rangle - 1]t^{-1}\,dt\,m(dx)}{\log(1 + \|x\|^2)}, \qquad (3.9.8)$$

where m is a probability measure such that $c_2m(dx) := \log(1 + \|x\|^2)M(dx)$ is a finite measure on $\mathbb{R}^d \setminus \{0\}$ and $c_2 \geq 0$. Note that M integrates $\|x\|^2$ on the unit ball and $\log(1 + s^2) \leq s^2$ for $s \geq 0$.

Assume $A(\mu) = \mathcal{O}$. Uniqueness of the Gaussian and Poissonian parts in the Lévy–Khintchine formula for an infinitely divisible measure implies that

$$ARA^{-1} = R \quad \text{and} \quad Am(\cdot) = m(\cdot) \quad \text{for all } A \in \mathcal{O}.$$

From Schur's lemma, we obtain $R = c_1I$ for some $c_1 \geq 0$. To solve the above measure equation, let us define measures H_d and ρ on the unit sphere S^{d-1} and the positive half-line as images of m under the mappings $x \mapsto x/\|x\|$ and $x \to \|x\|$, respectively. Then H_d is the Haar measure on the homogeneous space $S^{d-1} = \mathcal{O}(\mathbb{R}^d)/\mathcal{O}(\mathbb{R}^{d-1})$ and its Fourier transform is given by

$$\hat{H}_d(y) = \Gamma\left(\frac{d}{2}\right)\sum_{j=0}^\infty (-1)^j\left(j!\Gamma\left(j + \frac{d}{2}\right)\right)^{-1}\left(\frac{\|y\|}{2}\right)^{2j}.$$

Note that $\hat{H}_1(y) = \cos|y|$ and, for $d \geq 2$, $\hat{H}_d$ can be expressed in terms of Bessel functions. Finally, $\mathbb{R}^d \setminus \{0\}$ is isomorphic to $S^{d-1} \times \mathbb{R}^+$ and $(u, t) \to ut$ is a product of two independent random variables under m which gives

$$m(F) = \int_0^\infty H_d(t^{-1}F)\rho(dt) \quad \text{for all Borel sets } F. \quad (3.9.9)$$

Putting this into the formula for $\hat{\mu}$ and integrating, we obtain Proposition 3.9.6 with $W = I$.

The general case, $\mathbf{A}(\mu) = W^{-1}\mathscr{O}W$, is reduced to the previous one by the following observations:

$$\mathbf{A}(W\mu) = \mathscr{O}, \quad \mathbf{D}(W\mu) = W\mathbf{D}(\mu)W^{-1},$$
$$\mathscr{T}(\mathbf{D}(W\mu)) = W\mathscr{T}(\mathbf{D}(\mu))W^{-1},$$

which completes the necessity.

The sufficiency is a consequence of the observation that (3.9.7) can be written as (3.9.8) with m given by (3.9.9), thus μ is selfdecomposable; and then referring to Proposition 3.9.5. Q.E.D.

3.10 BIBLIOGRAPHIC COMMENTS

The results in Sections 3.1 and 3.2 are due to Urbanik (1972), where $\mathbb{R}^d$-valued random variables are discussed. However, the proofs of all the auxiliary lemmas in Section 3.3 which lead to the fundamental characterization of operator-selfdecomposable measures given in Theorem 3.3.5 are taken from Urbanik (1978), and they are valid in a Banach space. The notion of U-exponents and their properties discussed in the second part of Section 3.3 are from Jurek (1991).

The solution of the measure inequality given in Theorems 3.4.4 and 3.4.5 is from Jurek (1982b), but the polar coordinates system and the new norm are from Jurek (1984). Theorem 3.4.4 was also proved by Wolfe (1980). Theorem 3.4.6 was first proved in Urbanik (1972), but the proof was based on Choquet's theorem on extreme points of compact convex sets. Lemma 3.4.7 is also from Urbanik (1972). Properties of the support of Lévy spectral measures and Gaussian covariance operators of operator-selfdecomposable measures follow Jurek (1989a).

Generators for $L_0(Q)$ in Section 3.5 may be found in Jurek (1982b). Section 3.6 on the random integral representation is based on Jurek

(1982c) which was a follow up of the basic paper by Jurek and Vervaat (1983). In Wolfe (1982a) and (1982b), similar results may be found, but with different proofs. The continuity of the random integral mapping in Theorem 3.6.13 is proved in Sato and Yamazato (1984b) for $\mathbb{R}^d$. A different proof which is valid in the Banach space case is given in Jurek and Rosinski (1988). The material in Section 3.7.1 on one-parameter semigroups of operators on $C_0(\mathbb{R}^d)$ is standard and appears in Freedman (1971) and Williams (1979). The infinitesimal generators of Ornstein–Uhlenbeck type processes associated with $\exp(-tQ)$-decomposable measures in Theorem 3.7.8 may be found in Sato and Yamazato (1984a). Section 3.8 on the absolute continuity of $\exp(-tQ)$-decomposable measures is from Yamazato (1983), which is a generalization of Sato (1985). Theorem 3.8.1 is due to Sato. In our presentation of Theorem 3.8.1, we use the random integral presentation and Lemma 3.8.4 from Jurek (1988b). Theorem 3.8.10 is from Sato and Yamazato (1984a). The characterization of selfdecomposable measures on Banach spaces in Section 3.9 is due to Kumar and Schreiber (1975). The random integral representation for those measures given in Theorem 3.9.3 is due to Jurek and Vervaat (1983). Finally, elliptically symmetric operator-selfdecomposable measures are described in Jurek (1991).

CHAPTER 4

Operator-Stable Measures

Certain measures $\mu \in \text{ID}(\mathbb{R}^d)$ have the property that, for some $B \in \text{Aut}(\mathbb{R}^d)$, $\mu^t = t^B \mu * \delta(b(t))$ for all $t > 0$, with $b(t) \in \mathbb{R}^d$. These measures are called *operator-stable* and B is called an *exponent*, and they arise in a natural way (cf. Section 4.2). Section 4.4 deals with showing that the fixed points of the mapping $\mathcal{T}_B$ [cf. (3.6.17) and Proposition 3.6.10] are the operator-stable measures with exponent B. In Section 4.6, a decomposition of a full operator-stable measure into its Gaussian component and its compound Poisson component is explicitly obtained in terms of the decomposition of $\mathbb{R}^d$ by "eigenspaces" of the exponent B. Also, in Section 4.6, the class of all exponents of a full operator-stable measure is described. Section 4.8 deals with the special case of symmetric measures. In Section 4.9, it is shown that all full operator-stable measures are absolutely continuous. This follows from the fact that full operator-selfdecomposable measures are absolutely continuous, but a simpler proof is presented. The question of moments is dealt with in Section 4.12. Domain of attraction problems are treated in Sections 4.13 and 4.14. Finally, some special cases are considered in Section 4.15.

4.1 STATEMENT OF THE PROBLEM

In the preceding chapter, we considered limit measures obtained from $A_n(\mu_1 * \cdots * \mu_n) * \delta(a_n)$, where each A_n is a linear operator on $\mathbb{R}^d$, each $a_n \in \mathbb{R}^d$, and $\{A_n \mu_k : 1 \leq k \leq n, 1 \leq n\}$ is an infinitesimal system. We now specialize to the important case when $\mu_n = \nu$ for all $n \geq 1$, that is, independent identically distributed summands. Most results require that the limit measure be full. With this assumption, we see that each A_n is invertible and ν is full (cf. Corollary 2.2.1).

Also, $\{A_n\nu: n \geq 1\}$ is an infinitesimal system. To see this, note that the assumption $A_n\nu^n * \delta(a_n) \Rightarrow \mu$ implies that $A_n(\nu^0)^n \Rightarrow \mu^0$. (The notation ν^n, where ν is a measure and n is a positive integer, means the n-fold convolution of ν with itself.) Since $(\mu^0)\hat{\,}(0) = 1$ and $(\mu^0)\hat{\,}$ is continuous, there is $a > 0$ such that $|(\mu^0)\hat{\,}(x)| > 0$ for all $\|x\| \leq a$. Hence, for $\|x\| \leq a$, $(A_n\nu^0)\hat{\,}(x) \to 1$. By Proposition 1.7.5 (e), this convergence holds for all $x \in \mathbb{R}^d$. Therefore, $A_n\nu^0 \Rightarrow \delta(0)$. Hence, $\{A_n\nu^0: n \geq 1\}$ is an infinitesimal array. Since μ^0 is full, by Proposition 3.2.1, $A_n \to 0$. Therefore, $\{A_n\nu: n \geq 1\}$ is also an infinitesimal array. Thus, the class of all operator-stable measures is contained in the class of all operator-selfdecomposable measures.

4.2 SAKOVIC–SHARPE CHARACTERIZATION

We call a measure $\mu \in \mathscr{P}$ *operator-stable* provided there are a measure $\nu \in \mathscr{P}$, a sequence $\{A_n\}$ of linear operators, and a sequence $\{a_n\}$ of vectors such that $A_n\nu^n * \delta(a_n) \Rightarrow \mu$. In terms of random vectors, ξ is called *operator-stable* if there is a sequence $\{\xi_n\}$ of independent identically distributed random vectors which converges in law to ξ when suitably normed and centered. Sharpe's basic characterization of such full μ is that $\mu^t = t^B\mu * \delta(b(t))$ for *all* $t \ll 0$, for some linear operator B and vectors $b(t)$. Before obtaining this characterization for all $t \ll 0$, we show that its analog is valid for positive integers.

Theorem 4.2.1. *Let μ be a full measure. Then μ is operator-stable if and only if, for every $n \geq 1$, there are linear operators B_n on $\mathbb{R}^d$ and vectors b_n in $\mathbb{R}^d$ such that $\mu^n = B_n\mu * \delta(b_n)$, that is, μ and μ^n are of the same type.*

Proof. Assume $\mu^n = B_n\mu * \delta(b_n)$ for every $n \geq 1$. Since μ is full, so are the μ^n for all $n \geq 1$. Hence, each B_n is invertible (cf. Corollary 2.1.3). Solving $\mu^n = B_n\mu * \delta(b_n)$ for μ, we obtain $\mu = B_n^{-1}\mu^n * \delta(-B_n^{-1}b_n)$. Thus, μ is operator-stable.

Now, assume μ is operator-stable. Then there are $\nu \in \mathscr{P}(\mathbb{R}^d)$, $A_k \in \text{End}$, $a_k \in \mathbb{R}^d$ such that $A_k\nu^k * \delta(a_k) \Rightarrow \mu$. Since μ is full, we may assume that A_k are invertible (cf. Corollary 2.2.1). For $n \geq 1$, $\mu^n = (\lim_{k \to \infty} A_k\nu^k * \delta(a_k))^n = \lim_{k \to \infty} A_k\nu^{kn} * \delta(na_k)$. Also, $\mu = \lim_{k \to \infty} A_{kn}\nu^{kn} * \delta(a_{kn})$. Setting $\mu_k := A_{kn}\nu^{kn} * \delta(a_{kn})$, we have $\mu_k \Rightarrow \mu$. But, with $C_k := A_k A_{kn}^{-1}$ and $c_k := na_k - A_k A_{kn}^{-1}a_{kn}$, we also

have $C_k\mu_k * \delta(c_k) = A_k\nu^{kn} * \delta(na_k) \Rightarrow \mu^n$. Since μ and μ^n are full, they are of the same type by the convergence of types theorem 2.2.10
Q.E.D.

Remark 4.2.2

(a) Each full operator-stable μ is infinitely divisible.
(b) The B_n in Theorem 4.2.1 is a limit point of the sequence $\{A_k A_{kn}^{-1}: k \geq 1\}$, where $\{A_k\}$ is a norming sequence of the measure μ.

This follows from the preceding theorem since, for every $n \geq 1$, $\mu = B_n^{-1}\mu^n * \delta(-B_n^{-1}b_n) = (B_n^{-1}\mu * \delta(-B_n^{-1}b_n/n))^n$. Also, operator-stable measures are defined as limit measures of infinitesimal systems so that they must be infinitely divisible.

Let $\mathbf{OS}(\mathbb{R}^d)$ denote the class of all operator-stable measures on $\mathbb{R}^d$. Occasionally, we simply write **OS**. Since the characteristic function $\hat{\nu}$ of an infinitely divisible measure ν has the property that $\hat{\nu}^t$, for $t > 0$, is again a characteristic function, we let ν^t denote the corresponding infinitely divisible measure. The desired characterization for full operator-stable measures is obtained through a series of lemmas. For $t > 0$, let $G_t := \{A \in \mathrm{Aut}(\mathbb{R}^d)$: for some $a \in \mathbb{R}^d$, $\mu^t = A\mu * \delta(a)\}$, and let $G := \bigcup_{t>0} G_t$. We show by the following lemmas that G is a closed subgroup of $\mathrm{Aut}(\mathbb{R}^d)$ and that $G_1 = \mathbf{A}(\mu)$, the symmetry group of μ, is a compact normal subgroup of G.

Lemma 4.2.3. *For all $t > 0$, $G_t \neq \varnothing$.*

Proof. By the previous theorem, $G_t \neq \varnothing$ when t is a positive integer. For now, assume t is the ratio of two positive integers k and j, so $t = k/j$. Since $G_j \neq \varnothing$ and $G_k \neq \varnothing$, there are A_j, A_k, a_j, a_k such that $\mu^j = A_j\mu * \delta(a_j)$ and $\mu^k = A_k\mu * \delta(a_k)$. Solving the first equation, we have $\mu = A_j^{-1}\mu^j * \delta(-A_j^{-1}a_j)$, and, substituting into the second equation, we obtain $\mu^k = A_k A_j^{-1}\mu^j * \delta(a_k - A_k A_j^{-1}a_j)$, that is, μ^j and μ^k are of the same type. Let $B := A_k A_j^{-1}$ and $b := a_k - A_k A_j^{-1}a_j$, so $\mu^k = B\mu^j * \delta(b)$. Thus, $\mu^k = (B\mu * \delta(b/j))^j$, so $B\mu * \delta(b/j)$ is the unique j^{th} root of the infinitely divisible measure μ^k. Therefore, $B\mu * \delta(b/j) = (\mu^k)^{1/j} = \mu^t$, so $G_t \neq \varnothing$ when t is a positive rational number. Now, for t a positive irrational number, select $0 < t_n \to t$ with each t_n rational. Then $\mu^{t_n} \Rightarrow \mu^t$. Since each $G_{t_n} \neq \varnothing$, there are B_n and b_n such that $\mu^{t_n} = B_n\mu * \delta(b_n)$. Hence,

$\mu = B_n^{-1} \mu^{t_n} * \delta(-B_n^{-1} b_n)$. By the convergence of types theorem 2.2.10, μ and μ^t are of the same type, so $G_t \neq \emptyset$. Q.E.D.

Lemma 4.2.4. *For all $s, t > 0$, $G_t^{-1} = G_{1/t}$ and $G_{st} = G_s G_t$.*

Proof. Let $t > 0$ be fixed. Then $B \in G_{1/t}$ if and only if $\mu^{1/t} = B\mu * \delta(b)$ for some vector b if and only if $\mu^t = B^{-1}\mu * \delta(-tB^{-1}b)$ if and only if $B^{-1} \in G_t$. Therefore, $G_{1/t} = G_t^{-1}$ for all $t > 0$.

Let $B \in G_s G_t$. Then $B = AC$ for some $A \in G_s$ and some $C \in G_t$. Hence, $\mu^{ts} = (A\mu * \delta(a))^t = A\mu^t * \delta(ta) = AC\mu * \delta(Ac + ta)$, for some a and c in $\mathbb{R}^d$. Thus, $B = AC \in G_{ts}$, so $G_s G_t \subseteq G_{ts}$ for all $s, t > 0$. To obtain the reverse inclusion, set $s = 1/u$ and $t = uv$ to obtain $G_{1/u} G_{uv} \subseteq G_v$. Then $G_{uv} \subseteq G_{1/u}^{-1} G_v = G_u G_v$ for all $u, v > 0$. Q.E.D.

Note that $G_t G_s = G_s G_t$ for all $t, s > 0$, but $A \in G_t$ and $B \in G_s$ need not commute.

Lemma 4.2.5. *When $s \neq t$, $G_s \cap G_t = \emptyset$.*

Proof. Suppose that for some $s \neq t$ there is $A \in G_s \cap G_t$. Hence $(\mu^0)^s = A\mu * A\overline{\mu} = A(\mu^0) = (\mu^0)^t$. Thus, $(\mu^0)^{\wedge s-t}(y) = 1$ for all $y \in \mathbb{R}^d$. Since $s \neq t$, this implies that μ^0 is not full. This contradiction establishes the lemma. Q.E.D.

Lemma 4.2.6. *If $A_n \to A \in \mathbf{Aut}(\mathbb{R}^d)$ with $A_n \in G_{t_n}$, then there is a unique $t > 0$ such that $A \in G_t$ and $t_n \to t$.*

Proof. First, we show that $0 < \inf_{n \geq 1} t_n \leq \sup_{n \geq 1} t_n < \infty$. Suppose $\inf_{n \geq 1} t_n = 0$. For convenience, we suppose $t_n \to 0$ instead of some subsequence of $\{t_n\}$. Then $(\mu^0)^{t_n} \Rightarrow \delta(0)$. But, $(\mu^0)^{t_n} = A_n(\mu^0) \Rightarrow A(\mu^0)$. Since A is invertible, μ^0 is not full. Therefore, $0 < \inf_{n \geq 1} t_n$. Suppose $\sup_{n \geq 1} t_n = \infty$. Again, we suppose $t_n \to \infty$, so $1/t_n \to 0$. Let $B_n = A_n^{-1}$, so $B_n \in G_{1/t_n}$ and $B_n \to A^{-1} \in \mathbf{Aut}(\mathbb{R}^d)$. The preceding argument again leads to a contradiction, so $\sup_{n \geq 1} t_n < \infty$.

Next, let t be any cluster point of $\{t_n : n \geq 1\}$. Then t is positive and finite. Let $t_{n_k} \to t$, so that $A_{n_k} \mu \Rightarrow A\mu$ and $\mu^{t_{n_k}} \Rightarrow \mu^t$. Since $A_{n_k} \mu = \mu^{t_{n_k}} * \delta(a_k)$ for some a_k, we have $a_k \to$ (some) a. Hence, $A\mu = \lim_{k \to \infty} A_{n_k} \mu = \lim_k \mu^{t_{n_k}} * \delta(a_k) = \mu^t * \delta(a)$, so $A \in G_t$. Since $G_t \cap G_s = \emptyset$ for $t \neq s$, we have that $\{t_n : n \geq 1\}$ has exactly one cluster point. Therefore, the sequence $\{t_n\}$ converges to some $t > 0$ and $A \in G_t$. Q.E.D.

Lemma 4.2.7. *The set $G := \bigcup_{t>0} G_t$ is a closed subgroup of* $\mathbf{A}ut(\mathbb{R}^d)$.

Proof. By Lemma 4.2.4, G is closed with respect to products and contains its inverses, so G is a subgroup of $\mathbf{A}ut(\mathbb{R}^d)$. By Lemma 4.2.6, G is closed. Q.E.D.

Let l denote the mapping from G to the positive reals given by $l(A) = t$ for $A \in G_t$. By Lemma 4.2.5, l is a well-defined mapping.

Lemma 4.2.8. *The mapping l is a continuous homeomorphism from the group G to the group $\mathbb{R}^+$ and the kernel of l, G_1, is a compact normal subgroup of G.*

Proof. For $A \in G_t$ and $B \in G_s$, $AB \in G_{ts}$, so $l(AB) = ts = l(A)l(B)$. Thus, l is a homeomorphism. By Lemma 4.2.6, l is continuous. The kernel of l is simply the symmetry group $\mathbf{A}(\mu)$, which is a compact subgroup of G by Corollary 2.3.2. To see that G_1 is normal, note that, for fixed $g \in G_t$,

$$gG_1 g^{-1} \subseteq G_t G_1 G_{1/t} = G_1^2 = G_1. \qquad \text{Q.E.D.}$$

Now, define $L: \mathcal{T}(G) \to \mathbb{R}$ by

$$L(A) := \log(l(\exp A)),$$

where $\mathcal{T}(G)$ is the tangent space of the group G (cf. Section 1.5).

Lemma 4.2.9. *The function L defined above is a continuous linear functional on $\mathcal{T}(G)$.*

Proof. First, we show that for all $t \in \mathbb{R}$ and for all $A \in \mathcal{T}(G)$, $L(tA) = t \cdot L(A)$. Fix $A \in \mathcal{T}(G)$ and let $h: \mathbb{R} \to \mathbb{R}$ be given by

$$h(t) := L(tA) \quad \text{for } t \in \mathbb{R}.$$

Since sA and tA commute,

$$h(s+t) = \log l(\exp(sA) \cdot \exp(tA)) = \log(l(\exp(sA)) \cdot l(\exp(tA)))$$
$$= \log l(\exp(sA)) + \log l(\exp(tA)) = h(s) + h(t).$$

Hence, h is a continuous homeomorphism from $\mathbb{R}$ to $\mathbb{R}$. Therefore, there is $a \in \mathbb{R}$ such that

$$h(t) = ta \quad \text{for } t \in \mathbb{R}.$$

Since $L(A) = h(1) = a$, we have

$$L(tA) = h(t) = ta = tL(A) \quad \text{for all } t \in \mathbb{R}, \text{ all } A \in \mathcal{T}(G).$$

Second, we show that for all $A, B \in \mathcal{T}(G)$, $L(A + B) = L(A) + L(B)$. Fix $A, B \in \mathcal{T}(G)$. By Lemma 1.5.5, the function

$$f(t) := \exp^{-1}(\exp(tA) \cdot \exp(tB))$$

is well defined for t sufficiently near zero, has values in $\mathcal{T}(G)$, is analytic at zero, and $f'(0) = A + B$. Also,

$$L(f(t)) = \log l(\exp(tA) \cdot \exp(tB)) = L(tA) + L(tB),$$

since h is a homeomorphism. Therefore,

$$\begin{aligned}
L(A + B) = L(f'(0)) &= L\left(\lim_{t \to 0} \frac{f(t)}{t}\right) = \lim_{t \to 0} L\left(\frac{f(t)}{t}\right) \\
&= \lim_{t \to 0} \frac{1}{t} L(f(t)) = \lim_{t \to 0} \frac{1}{t}(L(tA) + L(tB)) \\
&= \lim_{t \to 0} \frac{1}{t}(tL(A) + tL(B)) = L(A) + L(B).
\end{aligned}$$

Therefore, L is a continuous linear functional. Q.E.D.

Lemma 4.2.10. *The kernel of L is $\mathcal{T}(G_1)$.*

Proof. Since $G_1 = \mathbf{A}(\mu)$ is a compact subgroup of G, its tangent space is well defined. Let $A \in \mathcal{T}(G_1)$. Then $\exp(A) \in G_1$, so $l(\exp(A)) = 1$. Thus, $L(A) = 0$ and $A \in \ker L$. For $A \in \ker L$, $L(A) = 0$, so $L(tA) = 0$ for all $t \in \mathbb{R}$. Thus, $\exp(tA) \in G_1$ for all $t \in \mathbb{R}$. Since $A = \lim_{t \to 0} t^{-1}(\exp(tA) - I)$, we have $A \in \mathcal{T}(G_1)$. Q.E.D.

Lemma 4.2.11. *The group G_1 does not contain an open neighborhood of I in G.*

Proof. Let $1 \neq t_n \to 1$. Select $A_n \in G_{t_n}$ for all $n \geq 1$. Then there are $a_n \in \mathbb{R}^d$ such that $\mu^{t_n} = A_n\mu * \delta(a_n)$. Since $\mu^{t_n} \Rightarrow \mu$, the sequence $\{A_n\}$ is conditionally compact in $\mathrm{Aut}(\mathbb{R}^d)$ because μ is full (cf. Lemma 2.2.3). Let $\{A_{n_k}\}$ be a subsequence of $\{A_n\}$ such that $A_{n_k} \to$ (some) $A \in \mathrm{Aut}(\mathbb{R}^d)$. Since $t_{n_k} \to 1$, by Lemma 4.2.6, we have that $A \in G_1$. Set $C_k := A_{n_k}A^{-1}$ for each $k \geq 1$. Then $C_k \in G_{t_{n_k}} \neq G_1$ by Lemma 4.2.4. Clearly, $C_k \to I$. Q.E.D.

Summarizing the above, we have the following diagram:

$$\begin{array}{ccc} G & \xrightarrow{l} & (\mathbb{R}^+, \cdot) \\ {\scriptstyle \exp}\uparrow & & \downarrow{\scriptstyle \log} \\ \mathcal{T}(G) & \xrightarrow{L} & (\mathbb{R}, +) \end{array}$$

where $(\mathbb{R}^+, \cdot)$ is the multiplicative group, $(\mathbb{R}, +)$ is the additive group, and all the mappings $\exp, l, \log$, and L are continuous homeomorphisms between the appropriate algebraic structures (groups or linear spaces). Moreover, $\ker L = \mathcal{T}(G_1)$ and G_1 does not contain any open neighborhood of the identity. Now, we are ready to present the main characterization of the class $\mathbf{OS}(\mathbb{R}^d)$.

Theorem 4.2.12. *If μ is full and operator-stable on $\mathbb{R}^d$, then there are a linear operator B and a function $b: (0, \infty) \to \mathbb{R}^d$ such that, for all $t > 0$,*

$$\mu^t = t^B\mu * \delta(b(t)). \tag{4.2.1}$$

Conversely, any measure satisfying (4.2.1) is operator-stable.

Proof. By Theorem 1.5.6, $I \in G$ belongs to the interior of $\exp(\mathcal{T}(G))$. Thus, by Lemma 4.2.11,

$$\exp(\mathcal{T}(G)) \not\subseteq G_1.$$

Let $C \in \mathcal{T}(G)$ be such that $\exp(C) \notin G_1$. Since $h(t) = \log l(\exp(tC))$ is a continuous homeomorphism from $\mathbb{R}$ to $\mathbb{R}$, there is $a \in \mathbb{R}$ such that, for all $t \in \mathbb{R}$, $\log l(\exp(tC)) = ta$. Since $a = 0$ implies $l(\exp(tC)) = 1$ for all $t \in \mathbb{R}$ which in turn implies $\exp(C) \in G_1$, a contradiction, we see that $a \neq 0$. Let $B := (1/a)C$. Then, for all $t \in \mathbb{R}$,

$$\log l(\exp(tB)) = \left(\frac{t}{a}\right)a = t.$$

Thus, for all $t > 0$,

$$\log l(t^B) = \log t,$$

that is,

$$l(t^B) = t.$$

Thus, $t^B \in G_t$ for all $t > 0$ and there is $b(t) \in \mathbb{R}^d$ such that

$$\mu^t = t^B \mu * \delta(b(t)).$$

For the converse, let $\mu_n := \mu$, $A_n := n^{-B}$, and $a_n := -n^{-B}b(n)$ for $n \geq 1$. Then $A_n(\mu_1 * \cdots * \mu_n) * \delta(a_n) = n^{-B}\mu^n * \delta(-n^{-B}b(n)) = \mu$ for all n. Thus, μ is operator-stable. Q.E.D.

Corollary 4.2.13. *Let μ be full operator-stable and let B be a linear operator such that $\mu^t = t^B \mu * \delta(b(t))$ for all $t > 0$. Then*

$$t^B \to 0 \quad as \ t \to 0;$$

equivalently, the eigenvalues of B all have positive real parts.

Proof. Since $\mu^t \Rightarrow \delta(0)$, as $t \to 0$, $t^B \mu * \delta(b(t))$ converges as $t \to 0$. Since μ is full, $\{\langle t^B; b(t)\rangle\}_{t>0}$ is conditionally compact in the space of all affine transformations. Let α be a limit point of $\{\langle t^B; b(t)\rangle\}_{t>0}$ as t goes to zero on some sequence. Then $\alpha\mu = \delta(0)$. Since μ is full, $\alpha = 0$, so $t^B \to 0$ as $t \to 0$. Q.E.D.

Corollary 4.2.14. *Let μ be full and operator-stable on $\mathbb{R}^d$. Then $|\hat{\mu}(y)| < 1$ for all $y \in \mathbb{R}^d \setminus \{0\}$.*

Proof. Let $H := \{y \in \mathbb{R}^d : |\hat{\mu}(y)|^2 = 1\}$. As in Section 2.1, H is a closed subgroup of $\mathbb{R}^d$. By Proposition 2.1.1, since μ is full, H does not contain any one-dimensional subspace of $\mathbb{R}^d$. Hence, H must be either $\{0\}$ or a lattice. But, since $\mu^t = t^B \mu * \delta(b(t))$ for all $t > 0$, we have $|\hat{\mu}(y)|^t = |\hat{\mu}(t^{B^*}y)|$ and $|\hat{\mu}(t^{-B^*}y)|^t = |\hat{\mu}(y)|$. Therefore, $t^{B^*}H = H$ for all $t > 0$. By Corollary 4.2.13, H cannot be a lattice. Therefore, $H = \{0\}$. Q.E.D.

A linear operator B satisfying Theorem 4.2.12 is called an *exponent* of the measure μ and the function $b(\cdot)$ is called a *centering function* of μ. From the proof of the theorem, we see that the exponent B

need not be unique. Later, in Section 4.6, we characterize the class of all exponents for a given μ.

4.3 A CLASS OF t^B-STABLE MEASURES

For a fixed linear operator B such that $t^B \to 0$ (cf. Corollary 4.2.13), we introduce the class $\mathscr{S}(B)$ of all t^B-*stable* measures, that is, $\mu \in \mathscr{S}(B)$ provided, for each $t > 0$, $\mu^t = t^B \mu * \delta(b(t))$ for some vector $b(t)$. We have $\mathscr{S}(B) \subseteq L_0(B)$, where $L_0(B)$ is given in Section 3.4, because, for each $t \geq 1$,

$$\mu = t^{-B}\big[\mu^t * \delta(-b(t))\big] = e^{-(\log t)B}\mu * t^{-B}\big[\mu^{t-1} * \delta(-b(t))\big]$$

or $\mu = e^{-sB}\mu * \nu_s$ for all $s \geq 0$, where $\nu_s := e^{-sB}[\mu^{e^s-1} * \delta(-b(e^s))]$. The algebraic and topological structures of the class $\mathscr{S}(B)$ are given in the following proposition.

Proposition 4.3.1

(a) *The class $\mathscr{S}(B)$ is a closed subsemigroup of* ID.
(b) *If the linear operator A commutes with B, then $A\mathscr{S}(B) \subseteq \mathscr{S}(B)$; if, in addition, A is invertible, then $A\mathscr{S}(B) = \mathscr{S}(B)$.*
(c) *For each $x \in \mathbb{R}^d$ and $c > 0$, $T_c\mathscr{S}(B) * \delta(x) = \mathscr{S}(B)$, where $T_c y := cy$ for $y \in \mathbb{R}^d$.*

Proof

(a) It is easy to see that $\mathscr{S}(B)$ is a semigroup. Now, let $\{\mu_n\}$ be a sequence in $\mathscr{S}(B)$ and $\mu_n \Rightarrow$ (some) μ. Then μ is infinitely divisible, $\mu_n^t \Rightarrow \mu^t$ and $t^B\mu_n \Rightarrow t^B\mu$ as $n \to \infty$ for all $t > 0$. From the equality

$$\mu_n^t = t^B\mu_n * \delta(b_n(t)) \quad \text{for } n \geq 1 \text{ and } t > 0,$$

we obtain that $b_n(t) \to$ (some) $b(t) \in \mathbb{R}^d$ as $n \to \infty$ for all $t > 0$ (cf. Proposition 1.7.6, and Theorem 1.7.1). Therefore, $\mu^t = t^B\mu * \delta(b(t))$ for all $t > 0$, that is, $\mu \in \mathscr{S}(B)$.
(b) Assume A commutes with B and let $\mu \in \mathscr{S}(B)$. Then $(A\mu)^t = A(t^B\mu * \delta(b(t))) = t^B(A\mu) * \delta(Ab(t))$ for all $t > 0$, so

$A\mu \in \mathscr{S}(B)$. In addition, when A^{-1} exists, we have $\mathscr{S}(B) \subseteq A^{-1}\mathscr{S}(B) \subseteq \mathscr{S}(B)$, so $A\mathscr{S}(B) = \mathscr{S}(B)$.

(c) This follows from (b). Q.E.D.

For a measure $\mu = [a, R, M]$ to be t^B-stable imposes many relationships of B with R and with M. In this section, some of these are used to obtain a representation of the characteristic function of any t^B-stable measure. We begin with the most basic of these relationships. Before that, we mention that the notation $t \cdot N$ is different from tN, where $t \in \mathbb{R}^1 \setminus \{0\}$ and N is a measure. The latter means

$$tN(F) := t^I N(F) = N(t^{-I}(F)) = N(\{x \in \mathbb{R}^d : tx \in F\}),$$

whereas the former, $t \cdot N(F)$, means the product of the two numbers t and $N(F)$.

Proposition 4.3.2. *Let $\mu = [a, R, M]$ and let B be a linear operator. Then μ is t^B-stable if and only if, for all $t > 0$,*

$$tR = t^B R t^{B^*} \quad \text{and} \quad t \cdot M = t^B M.$$

Proof. Since for all $t > 0$ and for all $y \in \mathbb{R}^d$, we have

$$\left(t^B \mu * \delta(b(t))\right)^{\wedge}(y)$$

$$= \exp\left\{i\langle b(t) + t^B a, y\rangle - \frac{1}{2}\langle t^B R t^{B^*} y, y\rangle\right.$$

$$\left. + \int \left[\exp(i\langle x, t^{B^*} y\rangle) - 1 - \frac{i\langle x, t^{B^*} y\rangle}{1 + \langle x, x\rangle}\right] M(dx)\right\}$$

$$= \left[a_1(t), t^B R t^{B^*}, t^B M\right]^{\wedge}(t),$$

where

$$a_1(t) := b(t) + t^B a + \int \left[\frac{x}{1 + \|x\|^2} - \frac{x}{1 + \|t^{-B}x\|^2}\right] t^B M(dx),$$

and since

$$\mu^t = [ta, tR, t \cdot M],$$

so by the uniqueness of the Lévy–Khintchine representation we obtain the equalities. Q.E.D.

Proposition 4.3.3. *Let B and R be linear operators. Then, for all $t > 0$, $tR = t^B R t^{B^*}$ if and only if $R = BR + RB^*$.*

Proof. Assume $tR = t^B R t^{B^*}$ for all $t > 0$. Differentiating with respect to t, we obtain $R = t^{-1} B t^B R t^{B^*} + t^{-1} t^B R t^{B^*} B^*$. Evaluating this at $t = 1$, we have $R = BR + RB^*$.

Conversely, assume $R = BR + RB^*$. For fixed x and y, define

$$g(x, y, t) := \langle t^B R t^{B^*} x, y \rangle \quad \text{for } t > 0.$$

Then

$$g(x, y, t) = \langle t^B B R t^{B^*} x, y \rangle + \langle t^B R B^* t^{B^*} x, y \rangle = t \frac{\partial g}{\partial t}(x, y, t).$$

Hence, $g(x, y, t) = c(x, y) t$ for some $c(x, y) \in \mathbb{R}^1$. Clearly, $c(x, y) = g(x, y, 1) = \langle Rx, y \rangle$. Hence, $g(x, y, t) = \langle tRx, y \rangle$, that is, $\langle t^B R t^{B^*} x, y \rangle = \langle tRx, y \rangle$ for all $x, y \in \mathbb{R}^d$ and all $t > 0$. Q.E.D.

To solve the equation $t \cdot M = t^B M$ for all $t > 0$, we use the new norm on $\mathbb{R}^d$ introduced in (3.4.3), that is,

$$\|x\|_B := \int_0^\infty \|e^{-tB} x\| \, dt = \int_0^1 \|s^B x\| s^{-1} \, ds.$$

Since $t^{-B} \to 0$ as $t \to \infty$, $\|\cdot\|_B$ is well defined and gives a norm. Let

$$S_B := \{x \in \mathbb{R}^d : \|x\|_B = 1\},$$

that is, S_B is the unit sphere in $\mathbb{R}^d$ with respect to this norm. As before, define

$$\Phi: S_B \times \mathbb{R}^+ \to \mathbb{R}^d \setminus \{0\}$$

by

$$\Phi(x, t) := t^B x.$$

As usual, we suppress the dependency of Φ on B since B is fixed. Proposition 3.4.3 states that Φ is a homeomorphism and for x fixed, the function $t \to \|t^B x\|_B$, $t > 0$, is strictly increasing. Also, recall that for $F \subseteq S_B$ and $G \subseteq (0, \infty)$, $[F; G] := \{t^B x : x \in F, t \in G\}$. Hence, for $s > 0$, $s^B[F; G] = \{r^B x : x \in F, r \in sG\} = [F; sG]$.

Proposition 4.3.4. *Let G be a measure on $\mathcal{B}(\mathbb{R}^d \setminus \{0\})$ which is finite on any set bounded away from zero. Then, for all $t > 0$, $t \cdot G = t^B G$ if and only if*

$$G(A) = \int_{S_B} \int_0^\infty I_A(s^B x) s^{-2} \, ds \, m(dx) \qquad (4.3.1)$$

for all $A \in \mathcal{B}(\mathbb{R}^d \setminus \{0\})$, for some finite Borel measure m on S_B. In particular, for $F \in \mathcal{B}(S_B)$,

$$m(F) := G([F; [1, \infty)]).$$

Proof. Assume $t \cdot G = t^B G$ for all $t > 0$. For fixed $F \in \mathcal{B}(S_B)$, set $f(s) := G([F; [s, \infty)])$ for $s > 0$. Then, for all $t, s > 0$, we have

$$tf(s) = (t^B G)([F; [s, \infty)]) = G\left(\left[F; \left[\frac{s}{t}, \infty\right)\right]\right) = f\left(\frac{s}{t}\right).$$

Thus, $sf(s) = f(1)$ for all $s > 0$, so $G([F; [s, \infty)]) = m(F)/s$, where $m(F) := G([F; [1, \infty)])$. Clearly, m is a finite measure on $\mathcal{B}(S_B)$. For $0 < s < t$,

$$G([F; [s, t)]) = m(F)\left(\frac{1}{s} - \frac{1}{t}\right) = m(F)\int_s^t u^{-2} \, du$$

$$= \int_{S_B} \int_0^\infty I_{[F;[s,t)]}(u^B x) u^{-2} \, du \, m(dx).$$

Consequently, (4.3.1) holds for all $A \in \mathcal{B}(\mathbb{R}^d \setminus \{0\})$.

Conversely, assume (4.3.1) holds. Then, for $t > 0$,

$$(t^B G)(A) = G(t^{-B} A) = \int_{S_B} \int_0^\infty I_A\big((st)^B\big) s^{-2} \, ds \, m(dx)$$

$$= t \int_{S_B} \int_0^\infty I_A(r^B x) r^{-2} \, dr \, m(dx) = t \cdot G(A). \quad \text{Q.E.D.}$$

Corollary 4.3.5. *Let G be a measure on $\mathscr{B}(\mathbb{R}^d \setminus \{0\})$ satisfying (4.3.1) for the finite measure m given by $m(F) := G([F;[1,\infty)])$ for $F \in \mathscr{B}(S_B)$. Then*

$$\operatorname{supp} G = \bigcup_{t>0} t^B(\operatorname{supp} m) \cup \{0\}$$

and $\operatorname{lin}(\operatorname{supp} G) = \operatorname{lin}_B(\operatorname{supp} m)$. [Recall that $\operatorname{lin}(C)$ denotes the linear subspace generated by the set C of vectors, whereas $\operatorname{lin}_B(C)$ denotes the smallest B-invariant subspace generated by C.]

Proof. For product measures, it is easy to see that $\operatorname{supp}(\nu_1 \times \nu_2) = (\operatorname{supp} \nu_1) \times (\operatorname{supp} \nu_2)$. Let λ be the measure on $\mathscr{B}((0,\infty))$ given by $\lambda(A) := \int_A u^{-2}\, du$. Then (4.3.1) implies that $G = \Phi(m \times \lambda)$, so $\operatorname{supp} G = \Phi(\operatorname{supp} m \times [0,\infty))$.

The proof that $\operatorname{lin}(\operatorname{supp} G) = \operatorname{lin}_B(\operatorname{supp} m)$ is similar to the proof of Corollary 3.4.10. Q.E.D.

The next result concerning M is easily obtained due to the fact that t^B-stable implies $\exp(-sB)$-decomposable. Hence, from Corollary 3.4.9, we have the following corollary.

Corollary 4.3.6. *Let M be the Lévy measure of a t^B-stable measure.*

(a) *If V is a B-invariant subspace of $\mathbb{R}^d$, then $M(V)$ is either zero or infinite. In particular, $M(\mathbb{R}^d)$ is either zero or infinite.*
(b) *If C is a countable subset of $(0,\infty)$, then $M([S_B;C]) = 0$.*
(c) *In particular, M has no atoms.*

The results of this section yield the following representation of the characteristic function. Its proof is immediate.

Theorem 4.3.7. *The measure $\mu = [a, R, M]$ is t^B-stable if and only if its characteristic function is of the form*

$$\hat{\mu}(y) = \exp\left\{i\langle a, y\rangle - \frac{1}{2}\langle Ry, y\rangle \right. \\ \left. + \int_{S_B}\int_0^\infty \left[e^{i\langle y, t^B x\rangle} - 1 - \frac{i\langle y, t^B x\rangle}{1 + \|t^B x\|^2}\right] t^{-2}\, dt\, m(dx)\right\}$$

and R satisfies the relation $R = BR + RB^*$, where m is the finite measure on $\mathscr{B}(S_B)$ given by $m(F) := M([F;[1,\infty)])$ such that $\operatorname{lin}(\operatorname{supp} M) = \operatorname{lin}_B(\operatorname{supp} m)$.

Remark 4.3.8. From Theorem 4.2.12, we see that Corollaries 4.3.5 and 4.3.6, Theorem 4.3.7, and Propositions 4.3.3 and 4.3.4 also give information on full operator-stable measures, since in that case t^B always converges to zero as $t \to 0$. Actually, the complete description of the spectrum of any exponent B is given in Section 4.6.

4.4 THE CLASS $\mathscr{S}(B)$ AND FIX POINTS OF THE MAPPING $\mathscr{T}_B$

In Section 3.6, there was defined a random integral mapping $\mathscr{T}_B$: $\mathrm{ID}_{\log} \to L_0(B)$ given by

$$\mathscr{T}_B(\mu) := \mathscr{L}\left(\int_0^\infty e^{-tB}\, dY(t)\right),$$

where Y is a $D(\mathbb{R}^d, [0,\infty))$-valued random variable with stationary independent increments, $Y(0) = 0$ a.s., and $\mathscr{L}(Y(1)) = \mu \in \mathrm{ID}_{\log}$. Proposition 3.6.10 and Theorem 3.6.13 gave some of the basic algebraic and topological properties of this mapping; in particular, $\mathscr{T}_B$ is an isomorphism between the semigroups $\mathrm{ID}_{\log}$ and $L_0(B)$. Since $\mathscr{S}(B) \subseteq L_0(B)$ and $L_0(B) = \mathscr{T}_B(\mathrm{ID}_{\log})$, the obvious problem then is to characterize $\mathscr{S}(B)$ in terms of the mapping $\mathscr{T}_B$.

Proposition 4.4.1. *Each t^B-stable measure μ belongs to $\mathrm{ID}_{\log}$.*

Proof. From Propositions 4.3.4 and 3.4.3 and Remark 4.3.8, we have

$$\int_{\|x\|_B > 1} \log\|x\|_B M(dx) = \int_{S_B}\int_1^\infty (\log\|t^B u\|_B) t^{-2}\, dt\, m(du)$$

$$\leq m(S_B) \int_1^\infty t^{-2} \|B\|_B \log t\, dt < \infty.$$

Since the norms $\|\cdot\|$ and $\|\cdot\|_B$ are equivalent, from Proposition 1.8.13 we obtain $\mathscr{S}(B) \subseteq \mathrm{ID}_{\log}$. Q.E.D.

Theorem 4.4.2. *For any t^B-stable measure μ, there exists a vector x such that $\mathcal{T}_B(\mu) = \mu * \delta(x)$.*

Conversely, if $\mu \in \mathrm{ID}_{\log}$ is such that $\mathcal{T}_B(\mu) = \mu^c * \delta(z)$ for some $c > 0$ and $z \in \mathbb{R}^d$, then μ is t^{cB}-stable

Proof. Let $\mu = [a, R, M]$ be t^B-stable. Then R and M satisfy Propositions 4.3.2 and 4.3.3. Since $\mathcal{T}_B(\mu) = [a^\infty, R^\infty, M^\infty]$, from (3.6.9) to (3.6.11) we have

$$R^\infty = \int_0^\infty e^{-tB}(BR + RB^*)e^{-tB^*}\,dt$$

$$= \int_0^\infty (e^{-tB})' R(-e^{-tB^*})\,dt + \int_0^\infty e^{-tB}RB^* e^{-tB^*}\,dt = R,$$

where the last equality is by the integration by parts formula and the fact that $e^{-tB} \to 0$ as $t \to \infty$; also, for $F \in \mathcal{B}(\mathbb{R}^d \setminus \{0\})$,

$$M^\infty(F) = \int_0^\infty \int_{S_B} \int_0^\infty I_F\!\left((e^{-st})^B x\right) t^{-2}\,dt\,m(dx)\,ds$$

$$= \int_0^\infty \int_{S_B} \int_0^\infty I_F(r^B x) r^{-2} e^{-s}\,dr\,m(dx)\,ds = M(F);$$

and, finally, $a^\infty = B^{-1}a$, so $x = (B^{-1} - I)a$.

Conversely, if $\mu = [b, T, N] \in \mathrm{ID}_{\log}$, then $\mathcal{T}_B(\mu) = \mu^c * \delta(z)$ for some $c > 0$ and vector z is equivalent to

$$\begin{aligned} b^\infty &= cb + z, \\ c \cdot T &= \int_0^\infty e^{-tB} T e^{-tB^*}\,dt, \\ c \cdot N(F) &= \int_0^\infty N(e^{tB}F)\,dt \end{aligned} \qquad (4.4.1)$$

for all $F \in \mathcal{B}(\mathbb{R}^d \setminus \{0\})$. By integration by parts, we obtain

$$(cB)T + T(cB^*) = \int_0^\infty (-e^{-tB})' T e^{-tB^*}\,dt$$

$$+ \int_0^\infty (-e^{-tB})(T e^{-tB^*})'\,dt = T;$$

so, from Proposition 4.3.3 and Remark 4.3.8, we see that T is the covariance operator of a t^{cB}-stable measure. The last equality in (4.4.1) expressed in terms of the B-Lévy spectral function (cf. Section 3.4) becomes

$$cL_N^B(A;t) = \int_t^\infty L_N^B(A;s)s^{-1}\,ds$$

for $t > 0$ and $A \in \mathscr{B}(S_B)$. Hence,

$$c\frac{dL_N^B}{dt}(A;t) = -t^{-1}L_N^B(A;t)$$

except for countably many values of t (cf. Proposition 3.4.1 and Theorem 3.4.4). Therefore,

$$L_N^B(A;t) = -c\gamma(A)t^{-c^{-1}}$$

for some positive constant $\gamma(A)$. Consequently, γ is a finite Borel measure on S_B and

$$N([A,[t,s)]) = \gamma(A)\int_t^s r^{-c^{-1}-1}\,dr$$

$$= \int_{S_B}\int_0^\infty I_{[A,[t,s)]}(r^B u)r^{-c^{-1}-1}\,dr\,\gamma(du).$$

Hence, for $F \in \mathscr{B}(\mathbb{R}^d \setminus \{0\})$ and $\tilde{\gamma}(A) := c\gamma(c^{-1}A)$ for $A \in \mathscr{B}(S_B)$ (cf. the proof of Remark 3.5.6), we have

$$N(F) = \int_{S_B}\int_0^\infty I_F(r^B u)r^{-c^{-1}-1}\,dr\,\gamma(du)$$

$$= c\int_{S_B}\int_0^\infty I_F(s^{cB} u)s^{-2}\,ds\,\gamma(du)$$

$$= \int_{S_{cB}}\int_0^\infty I_{cF}(s^{cB} x)s^{-2}\,ds\,\tilde{\gamma}(dx),$$

since $cS_B = S_{cB}$. By Proposition 4.3.4, this implies that $T_{c^{-1}}N$ is the Lévy measure of some t^{cB}-stable measure. Finally, Proposition 4.3.1 shows that N is the Lévy measure of some t^{cB}-stable measure.

Q.E.D.

Corollary 4.4.3. *A measure $[a, R, M]$ is t^B-stable if and only if there is $c > 0$ such that, for all $F \in \mathscr{B}(\mathbb{R}^d \setminus \{0\})$,*

$$M(F) = c \int_0^\infty M(e^{tB} F) \, dt$$

and

$$R = c \int_0^\infty e^{-tB} R e^{-tB^*} \, dt.$$

Corollary 4.4.4. *A measure $\mu \in \mathrm{ID}_{\log}$ is t^B-stable if and only if $\mathscr{T}_B(\mu) = \mu * \delta(z)$ for some $z \in \mathbb{R}^d$. Consequently, we have $\mathscr{S}(B) = \mathscr{T}_B(\mathscr{S}(B))$, that is, the class $\mathscr{S}(B)$ is invariant under the mapping $\mathscr{T}_B$.*

4.5 NORMING SEQUENCES

A norming sequence $\{A_n\}$ for an operator-stable measure μ is defined analogously to the case of an operator-selfdecomposable measure, that is,

$$A_n \nu^n * \delta(a_n) \Rightarrow \mu \tag{4.5.1}$$

for some measure ν and some sequence $\{a_n\}$ of vectors. Therefore, all of the results in Section 3.2 apply to norming sequences $\{A_n\}$ of an operator-stable measure; in particular, (a) $A_n \to 0$ as $n \to \infty$, and (b) there is a norming sequence $\{B_n\}$ such that $B_{n+1} B_n^{-1} \to I$. In this section, we establish useful bounds on $\|A_n\|$ and $\|A_n^{-1}\|$. Throughout this section, μ is a full operator-stable measure with a norming sequence $\{A_n\}$ and μ is t^B-stable.

Proposition 4.5.1. *Let k and r be integers with $0 \leq r < k$. Then for any norming sequence $\{A_n\}$ of a full operator-stable measure μ, the sequences $\{A_{nk+r} A_n^{-1} k^B\}_{n \geq 1}$ and $\{k^{-B} A_n A_{nk+r}^{-1}\}_{n \geq 1}$ are conditionally compact and their limit points belong to $\mathbf{A}(\mu)$, the symmetry group of μ, where B is an exponent of μ.*

Proof. Let ν and a_n be as in (4.5.1). By Proposition 3.2.1, $A_n \to 0$ as $n \to \infty$, so $A_{nk+r} \nu^r \Rightarrow \delta(0)$ as $n \to \infty$. Consequently,

$$A_{nk+r} \nu^{nk} * \delta(a_{nk+r}) \Rightarrow \mu \quad \text{as } n \to \infty. \tag{4.5.2}$$

Also, $A_n \nu^{nk} * \delta(ka_n) \Rightarrow \mu^k$ as $n \to \infty$. But, $\mu^k = k^B \mu * \delta(b(k))$ for some vector $b(k)$. Thus,

$$k^{-B} A_n \nu^{nk} * \delta(k^{-B-1} a_n - k^{-B} b(k)) \Rightarrow \mu \qquad (4.5.3)$$

as $n \to \infty$. The convergences in (4.5.2) and (4.5.3), by Theorem 2.2.10, imply that $k^{-B} A_n = B_n C_n A_{nk+r}$ with $B_n \to I$ as $n \to \infty$ and each $C_n \in \mathbf{A}(\mu)$. Thus, $A_{nk+r} A_n^{-1} k^B = (B_n C_n)^{-1}$. Since $\mathbf{A}(\mu)$ is a compact group (cf. Corollary 2.3.2), we have the proposition for $\{A_{nk+r} A_n^{-1} k^B\}$. The remainder of the proof is obvious. Q.E.D.

Proposition 4.5.2. *Let k be a positive integer. Then for any norming sequence $\{A_n\}$ of the full operator-stable measure μ with an exponent B, we have*

$$\lim_{n \to \infty} \max_{0 \le r < k} \inf\{\|A_{nk+r} A_n^{-1} k^B - A\|: A \in \mathbf{A}(\mu)\} = 0$$

and

$$\lim_{n \to \infty} \max_{0 \le r < k} \inf\{\|k^{-B} A_n A_{nk+r}^{-1} - A\|: A \in \mathbf{A}(\mu)\} = 0.$$

Proof. Since the maximum is over a finite set, we may consider r as fixed. Then the limits are obviously zero from Proposition 4.5.1. Q.E.D.

Later in this chapter we show that for any exponent B of μ the real parts of its eigenvalues are greater than or equal to $1/2$. However, we already know that the real parts are positive (cf. Corollary 4.2.13). For now we establish some useful bounds for any norming sequence.

Theorem 4.5.3. *Let $\{A_n\}$ be an arbitrary norming sequence for the full operator-stable measure μ with an exponent B.*

(a) *For each positive a such that*

$$0 < a < \min\{\operatorname{Re}(x): x \text{ is an eigenvalue of } B\},$$

there is a constant $c_a > 0$ such that, for all $n \ge 1$,

$$\|A_n\| \le c_a n^{-a}.$$

(b) *For each positive b such that*

$$b > \max\{\operatorname{Re}(x): x \text{ is an eigenvalue of } B\},$$

there is a constant $c_b > 0$ such that, for all $n \geq 1$,

$$\|A_n^{-1}\| \leq c_b n^{-b}.$$

Proof

(a) Since $\mathbf{A}(\mu)$ is compact, $c_1 := \sup\{\|A\|: A \in \mathbf{A}(\mu)\}$ is finite. Select a positive integer k so large that $(c_1 + 1)\|k^{aI-B}\| < 1$ (cf. Corollary 4.2.13). Fix $n \geq 1$. Then n can be represented as $n = \sum_{i=0}^{s} p_i k^i$ for some integers p_i and s such that $0 \leq p_i < k$ for all i, $p_s > 0$, and $k^s \leq n < k^{s+1}$. Thus, there are nonnegative integers $m_1, \ldots, m_s, r_1, \ldots, r_s$ such that, for $1 \leq j < s$,

$$m_{j+1} = m_j k + r_j \quad \text{and} \quad n = m_s k + r_s$$

with $0 \leq r_j < k$. By Proposition 4.5.2, there is $m_0 > 0$ such that, for $m \geq m_0$,

$$\max_{0 \leq r < k} \|A_{mk+r} A_m^{-1} k^B\| \leq c_1 + 1.$$

Hence, for $m \geq m_0$ and $0 \leq r < k$, $\|A_{mk+r} A_m^{-1}\| \leq (c_1 + 1)\|k^{-B}\|$. Select n_0 so that $k^{n_0-1} \geq m_0$. Since $m_j \geq k^{j-1}$, for $j \geq n_0$, $\|A_{m_{j+1}} A_{m_j}^{-1}\| \leq (c_1 + 1)\|k^{-B}\|$. Therefore,

$$\|A_n\| \leq \|A_{m_s k + r_s} A_{m_s}^{-1}\| \|A_{m_1}\| \left(\prod_{j=2}^{s} \|A_{m_j} A_{m_{j-1}}^{-1}\|\right)$$

$$\leq \|A_{m_1}\| \left(\prod_{j=2}^{n_0} \|A_{m_j} A_{m_{j-1}}^{-1}\|\right) \|k^{-B}\|^{-n_0+1} (c_1 + 1)^s \|k^{-B}\|^s.$$

Since

$$\|A_{m_1}\| \left(\prod_{j=2}^{n_0} \|A_{m_j} A_{m_{j-1}}^{-1}\|\right) \|k^{-B}\|^{-n_0+1}$$

$$\leq \left(\max_{1 \leq j < k} \|A_j\|\right) \left(\max_{1 \leq j < l \leq k} \|A_l A_j^{-1}\|\right)^{n_0 - 1} \|k^{-B}\|^{-n_0+1},$$

letting c_3 denote this last quantity, we have

$$\|A_n\| \le c_3(c_1 + 1)^s \|k^{-B}\|^s = c_3 k^{-sa}(c_1 + 1)^s \|k^{aI-B}\|^s \le c_3 k^{-sa}.$$

The inequality $n < k^{s+1}$ implies that $k^{-sa} < k^a n^{-a}$. Therefore, $\|A_n\| \le c_3 k^a n^{-a}$. Set $c_a := c_3 k^a$ to obtain part (a).

(b) Now, assume $b > \max\{\mathrm{Re}(x): x$ is an eigenvalue of $B\}$. The constant c_1 is as before, but now k is a positive integer such that $(c_1 + 1)\|k^{B-bI}\| < 1$. Fix $n \ge 1$ and let the m_j's and r_j's be defined as in the proof of part (a). From Proposition 4.5.2, there is $m_0 > 0$ such that, for $m \ge m_0$ and for all $r \in [0, k)$,

$$\|A_m A_{mk+r}^{-1}\| \le k^b \|k^{B-bI}\|(c_1 + 1) < k^b.$$

Select n_0 so large that $k^{n_0-1} \ge m_0$. Since $m_j \ge k^{j-1}$, we have, for $j \ge n_0$, $\|A_{m_j} A_{m_{j+1}}^{-1}\| \le k^b$. The rest of the proof proceeds as in part (a). Q.E.D.

Corollary 4.5.4. *Let $\{A_n\}$ and b be as in Theorem 4.5.3. Then there are constants c'_b and N such that, for all $n \ge N$ and for all $m > 0$,*

$$\|A_n A_{nm}^{-1}\| \le c'_b m^b.$$

Proof. Since all norms on $\mathbb{R}^d$ are equivalent, in this proof $\|\cdot\|$ denotes the norm $\|\cdot\|_\mu$ defined in Section 4.7. Hence, for $A \in \mathbf{A}(\mu)$, $\|A\| = 1$. Select q so large that $2\|q^{B-bI}\| < 1$, let $c := [\log_q m]$, and write $A(n)$ for A_n. Then

$$A_n A_{nm}^{-1} = \left(\prod_{j=0}^{c-1} A(nq^j) A(nq^{j+1})^{-1}\right) A(nq^c) A(nm)^{-1}.$$

For $0 \le j \le c - 1$, in Proposition 4.5.2, replace n, k, r by $nq^j, q, 0$, respectively, to obtain that the limit points of $A(nq^j) A(nq^{j+1})^{-1}$ are in $q^B \mathbf{A}(\mu)$. Hence, for n sufficiently large and for all $j = 0, \ldots, c - 1$,

$$\|A(nq^j) A(nq^{j+1})^{-1}\| < 2\|q^B\|.$$

Similarly, the limit points of $A(nq^c) A(nm)^{-1}$ are in $\mathbf{A}(\mu)$, so for n sufficiently large

$$\|A(nq^c) A(nm)^{-1}\| < 2.$$

Hence, for n sufficiently large,

$$\|A_n A_{nm}^{-1}\| \le 2^{c+1}\|q^B\|^c \le 2m^b. \qquad \text{Q.E.D.}$$

4.6 STRUCTURAL CHARACTERIZATIONS OF OPERATOR-STABLE MEASURES AND THEIR EXPONENTS

The relationships of an exponent B with the covariance operator and with the Lévy measure of a t^B-stable measure suggest the possibility of obtaining the t^B-stable measure as a convolution of two measures, one Gaussian and the other with no Gaussian component. For full t^B-stable measures, this is the first theorem of this section. To obtain this decomposition, we factor the minimal polynomial of an exponent B into the product of two polynomials, $g_B(\cdot)h_B(\cdot)$, where the roots of g_B all have real parts equal to $1/2$ and those of h_B are unequal to $1/2$. (Later, we see that those unequal to $1/2$ must be greater than $1/2$; cf. Theorem 4.6.5.) Keep in mind that since $t^B \to 0$ as $t \to 0$, the roots of $g_B h_B$ all have positive real parts. When the exponent B is understood, we simply write g and h for g_B and h_B, respectively, By the primary decomposition theorem (cf. Theorem 1.2.3), the space $\mathbb{R}^d$ may be written as $\mathbb{R}^d = V_1 \oplus V_2$, where $V_1 = \ker g(B)$ and $V_2 = \ker h(B)$. For full t^B-stable measures, V_1 will support the Gaussian component and V_2 will support the non-Gaussian component.

Notation
$\text{Aut}_+(\mathbb{R}^d) := \{B \in \text{Aut}(\mathbb{R}^d): \text{all eigenvalues of } B \text{ have positive real parts}\}$. For $B \in \text{Aut}_+(\mathbb{R}^d)$, let $g \cdot h_1 \cdot h_2$ denote its minimal polynomial factored so that the roots of g have real parts equal to $1/2$, those of h_1 are greater than $1/2$, and those of h_2 are less than $1/2$, whenever B has such eigenvalues.

Lemma 4.6.1. *Let $B \in \text{Aut}_+(\mathbb{R}^d)$. Then the class of all Lévy measures M on $\mathbb{R}^d$ such that $t \cdot M = t^B M$ for all $t > 0$ is generated, under weak convergence, by linear combinations of those Lévy measures N such that $t \cdot N = t^B N$ and $\operatorname{supp} N = \{t^B x_N: t > 0\} \cup \{0\}$ for some $x_N \in \mathbb{R}^d$.*

Proof. Let $\Lambda := \{$Lévy measures M on $\mathbb{R}^d: t \cdot M = t^B M$ for all $t > 0\}$. For $M \in \Lambda$, set $W_M(dx) := \|x\|^2 (1 + \|x\|^2)^{-1} M(dx)$. Then $\Lambda_1 := \{W_M: M \in \Lambda\}$ is a convex set of finite Borel measures. Also, by

Proposition 4.3.4 and Corollary 4.3.5, the support of each $M \in \Lambda$ is the union of orbits of t^B, so the same is true for each $W \in \Lambda_1$. Now, let $\Lambda_2 := \{W \in \Lambda_1 : W(\mathbb{R}^d) \leq 1\}$. Then Λ_2 is convex and, by the Helly compactness theorem, it is also compact. We assert that the set of all extreme points of Λ_2 is equal to the set Λ_3 of those measures in Λ_2 whose support is exactly one orbit of t^B. To see this, assume ρ is an extreme point of Λ_2. Suppose the support of ρ properly contains $A := \{t^B y : t > 0\}$ for some $y \in \mathbb{R}^d$. For $E \in \mathscr{B}(\mathbb{R}^d)$, let $\rho_1(E) := \rho(E \cap A)\rho(\mathbb{R}^d)/\rho(A)$ and $\rho_2(E) := \rho(E \cap (\mathbb{R}^d \setminus A))\rho(\mathbb{R}^d)/\rho(\mathbb{R}^d \setminus A)$. Since the supports of ρ_1 and ρ_2 are t^B-invariant, ρ_1 and ρ_2 are in Λ_2. For $\alpha := \rho(A)/\rho(\mathbb{R}^d)$, we have $0 < \alpha < 1$ and $\rho = \alpha \rho_1 + (1 - \alpha)\rho_2$, which contradicts ρ being an extreme point of Λ_2. Therefore, the set of extreme points of Λ_2 is contained in Λ_3. The reverse inclusion is obvious.

Hence, the set of all convex combinations of measures in Λ_3 is dense in Λ_1. Since there is a one-to-one correspondence between Λ_1 and Λ, linear combinations of those $M \in \Lambda$ whose supports are exactly one orbit of t^B are dense in Λ.

Finally, since $t \cdot N = t^B N$ for all $t > 0$ implies that $N(\{t^B x : t > s\}) = s^{-1} N(\{t^B x : t > 1\})$ for all $s > 0$, we see that, for $N \in \Lambda$ with $W_N \in \Lambda_3$, $\operatorname{supp} N = \{t^B x_N : t > 0\} \cup \{0\}$ for some $x_N \in \mathbb{R}^d$. Q.E.D.

Lemma 4.6.2. *Let $B \in \operatorname{Aut}_+(\mathbb{R}^d)$ and let G be a measure on $\mathscr{B}(\mathbb{R}^d \setminus \{0\})$ such that $t \cdot G = t^B G$ for all $t > 0$ and $\operatorname{supp} G = \{t^B x_0 : t > 0\} \cup \{0\}$ for some $x_0 \in \mathbb{R}^d$. Then G is a Lévy measure, that is, $\int (\|x\|^2 \wedge 1) G(dx) < \infty$, if and only if $x_0 \in \ker h_1(B)$.*

Proof. Let $y \in S_B$ be such that $y = t_0^B x_0$ for some $t_0 > 0$. Then $\operatorname{supp} G = \{t^B y : t > 0\} \cup \{0\}$. For the first part of the proof, assume G is a Lévy measure. Hence, by Proposition 4.3.4, $G(A) = c \int_0^\infty I_A(t^B y) t^{-2} dt$, where $c := m(\{y\}) = G(\{s^B y : s \geq 1\})$. Thus,

$$c \int_0^1 \|t^B y\|_B^2 t^{-2} \, dt = \int_{\|x\|_B \leq 1} \|x\|_B^2 G(dx) < \infty.$$

Let $X := \operatorname{lin}_B\{y\}$, that is, the smallest B-invariant subspace containing y. Since B is invertible, $BX = X$ so we may consider B' on X, where B' is the restriction of B to X. By Theorem 1.2.3, X may be decomposed into the direct sum of subspaces $X_1, \ldots, X_k$ with $B'X_j = X_j$ and with the minimal polynomial $q_j^{m_j}$ of B_j being a power of a polynomial which is irreducible over the reals, where B_j is the restriction of B' to X_j, $1 \leq j \leq k$. Write $y = \sum_{r=1}^{k} y_r$ with each $y_r \in X_r$. Then $y_r \neq 0$ since y generates X. Clearly, $t^B y = \sum_{r=1}^{k} t^{B_r} y_r$. The problem now is to compute $t^{B_r} y_r$ in terms of the eigenvalues of B, that is, of B_r. Let $T_p(a_1, \ldots, a_p)$ denote the upper triangular matrix with a_1 on the principal diagonal, a_2 on the super diagonal, and so on, and a_p as the $(1, p)$ entry.

First, assume that q_r is linear, that is, $q_r(t) = t - \alpha$ with $\alpha > 0$. Since $T_{m_r}(\alpha, 1, 0, \ldots, 0) = \alpha I + N$, where I is the $m_r \times m_r$ identity matrix and N is the $m_r \times m_r$ matrix with all entries being zero except for all ones on the super diagonal, and because $N^j \neq 0$ for $0 \leq j \leq m_r - 1$, $N^{m_r} = 0$, we see that $q_r^{m_r}(T_{m_r}(\alpha, 1, 0, \ldots, 0)) = 0$ and no polynomial of lower degree has this property. Hence, $q_r^{m_r}$ is the minimal polynomial of $T_{m_r}(\alpha, 1, 0, \ldots, 0)$; consequently, there exists a basis $\{z_1, \ldots, z_{m_r}\}$ for X_r such that $T_{m_r}(\alpha, 1, 0, \ldots, 0)$ is the matrix representation of B_r with respect to this basis. Also,

$$t^{B_r} = t^\alpha t^N = t^\alpha \sum_{j=0}^{m_r - 1} (\log t)^j (j!)^{-1} N^j$$

$$= T_{m_r}\left(t^\alpha, t^\alpha \log t, \ldots, t^\alpha (\log t)^{m_r - 1}/(m_r - 1)!\right).$$

Define a norm $\|\cdot\|_r$ on X_r by $\|\sum_j a_j z_j\|_r := \sum_j |a_j|$. Therefore, $\|t^{B_r} y_r\|_r^2$ is equal to a sum of terms of the form $c_1 t^{2\alpha} (\log t)^{c_2}$ with constants $c_1 > 0$ and $0 \leq c_2 \leq 2m_r - 2$. Now, we may define a norm $\|\cdot\|_X$ on X by $\|\sum_{r=1}^{k} x_r\|_X := \sum_{r=1}^{k} \|x\|_r$. Then $\|\sum_{r=1}^{k} x_r\|_X^2 \geq \|x_r\|_r^2$ for $1 \leq r \leq k$. Since $\int_0^1 \|t^B y\|_B^2 t^{-2} \, dt < \infty$, we have $\int_0^1 \|t^{B_r} y_r\|_r^2 t^{-2} \, dt < \infty$ for each $1 \leq r \leq k$. However, the last integral is equal to the sum of terms of the form $c \int_0^1 t^{2\alpha - 2} (\log t)^b \, dt$ with $c > 0$ and $0 \leq b \leq 2m_r - 2$. Hence, each of these integrals is finite which implies that $2\alpha - 2 > -1$, that is, $\alpha > 1/2$. Therefore, $X \subseteq \ker h_1(B)$. Consequently, $y \in \ker h_1(B)$, so we have $x_0 \in \ker h_1(B)$.

Second, assume q_r is quadratic, that is, $q_r(t) = t^2 - 2\alpha t + \alpha^2 + \beta^2$ with $\alpha > 0$ and $\beta > 0$. In this case, we must extend X_r and B_r to their complexifications, that is, $X_r^c := X_r + iX_r$ and $B_r^c(x + ix) := B_r x + iB_r x$ for $x \in X_r$. Proceeding as before, we find there exists a complex basis $\{z_1, \ldots, z_{2m_r}\}$ of X_r^c such that the matrix representation of B_r^c with respect to this basis is

$$\begin{pmatrix} T_{m_r}(\alpha + i\beta, 1, 0, \ldots, 0) & 0 \\ 0 & T_{m_r}(\alpha - i\beta, 1, 0, \ldots, 0) \end{pmatrix}.$$

Ideas similar to those when q_r was linear lead to obtaining

$$\int_0^1 t^{2\alpha-2}(\log t)^b \, dt < \infty \quad \text{with } b \geq 0.$$

Therefore, as before, $\alpha > 1/2$ which implies that $x_0 \in \ker h_1(B)$.

To obtain the converse, assume $x_0 \in \ker h_1(B)$. Using the norms constructed in the necessary part of this proof, we have

$$\int_0^1 \|t^B y_r\|_r^2 t^{-2} \, dt < \infty,$$

each $1 \leq r \leq k$. Hence,

$$\int_0^1 \|t^B y\|_X^2 t^{-2} \, dt < \infty.$$

Consequently,

$$\int_0^1 \|t^B y\|_B^2 t^{-2} \, dt < \infty,$$

which implies $\int (\|x\|^2 \wedge 1) G(dx) < \infty$, that is, G is a Lévy measure.
Q.E.D.

Lemma 4.6.3. *Let $B \in \text{Aut}_+(\mathbb{R}^d)$ and let G be a Lévy measure on $\mathbb{R}^d$ such that $t \cdot G = t^B G$ for all $t > 0$. Then $\text{supp } G \subseteq \ker h_1(B)$.*

Proof. By Lemma 4.6.1, there is a sequence $\{N_n\}$ of Lévy measures such that, for all $n \geq 1$, $t \cdot N_n = t^B N_n$ for all $t > 0$, $\text{supp } N_n = \bigcup_{j=1}^{k_n} \{t^B x_{j,n} : t > 0\} \cup \{0\}$ and $N_n \Rightarrow G$. By Lemma 4.6.2, $x_n \in \ker h_1(B)$ for all $n \geq 1$, so $\text{supp } N_n \subseteq \ker h_1(B)$. Since $\ker h_1(B)$ is closed, $\overline{(\bigcup_{n=1}^\infty \text{supp } N_n)} \subseteq \ker h_1(B)$. Let $x \notin \overline{(\bigcup_{n=1}^\infty \text{supp } N_n)}$. Select an open neighborhood U of x sufficiently small so that $\overline{U} \cap (\bigcup_{n=1}^\infty \text{supp } N_n) = \emptyset$. Also, select U so that $G(\partial U) = 0$. Then $G(U) = \lim N_n(U) = 0$. Therefore, $x \notin \text{supp } G$. Consequently, we have $\text{supp } G \subseteq \overline{(\bigcup_{n=1}^\infty \text{supp } N_n)} \subseteq \ker h_1(B)$. Q.E.D.

Lemma 4.6.4. *Let $B \in \text{Aut}_+(\mathbb{R}^d)$, $\mu = [a, R, M]$ be t^B-stable, not necessarily full, and $\gamma := [0, R, 0]$ be the zero-mean Gaussian compo-*

nent of μ. Then

(a) supp γ is a B-invariant subspace;
(b) supp $\gamma \subseteq \ker g(B)$.

Proof. For (a), let $\eta := [a, 0, M]$, so $\mu = \gamma * \eta$. Then $\mu^t = t^B\mu * \delta(b(t))$ for all $t > 0$ implies that

$$\gamma^t * \eta^t = t^B\gamma * t^B\eta * \delta(b(t))$$

for all $t > 0$. Since γ^t and $t^B\gamma$ are zero-mean Gaussian measures and the other factors have no Gaussian components, we see that $\gamma^t = t^B\gamma$ for all $t > 0$, that is, γ is t^B-stable. From Theorem 1.8.20, since γ has mean zero and covariance operator R, its support is given by $(\ker R)^\perp$, so it is a subspace of $\mathbb{R}^d$. All this implies that

$$\operatorname{supp} \gamma = \operatorname{supp} \gamma^t = \operatorname{supp} t^B\gamma = t^B(\operatorname{supp} \gamma).$$

Let $v \in \operatorname{supp} \gamma$. Then $(t-1)^{-1}(t^B - I)v \in \operatorname{supp} \gamma$ for all $t > 0$. Since these vectors converge to Bv as $t \to 1$, we obtain $Bv \in \operatorname{supp} \gamma$, so supp γ is B-invariant.

For (b), let $V := \operatorname{supp} \gamma$, $B_1 := B|V$, and I_1 be the identity on V. Since, for $t > 0$, $\hat{\gamma}^t(y) = \hat{\gamma}(t^{1/2}y)$, $t^{I_1/2}$ is an exponent for γ. Hence, $t^{B_1}\gamma = \gamma^t = t^{I_1/2}\gamma$ implies $t^{B_1 - I_1/2} \in \mathbf{A}_V(\gamma) := \{A \in \operatorname{Aut}(V): A\gamma = \gamma\}$. Since γ is full Gaussian on V, $\mathbf{A}_V(\gamma) = W^{-1}\mathcal{O}_V W$ for some $W \in \operatorname{Aut}(V)$, where $\mathcal{O}_V$ is the group of all orthogonal transformations on V (cf. Proposition 2.3.11). Hence,

$$W(t^{B_1 - I_1/2})W^{-1} = t^{W(B_1 - I_1/2)W^{-1}} \in \mathcal{O}_V$$

for all $t > 0$. Hence, $t^{W(B_1 - I_1/2)W^{-1}}$ commutes with its adjoint and

$$t^{W(B_1 - I_1/2)W^{-1} + [W(B_1 - I_1/2)W^{-1}]^*} = I$$

for all $t > 0$. Differentiating with respect to t and evaluating at $t = 1$, we see that $W(B_1 - I_1/2)W^{-1}$ is skew-symmetric. Since the eigenvalues of any skew-symmetric operator are always simple and have real parts equal to zero, the same holds for $B_1 - I_1/2$. Therefore, the eigenvalues of B_1 are simple and have real parts equal to $1/2$, so this is also true for B. Let g_1 be the minimal polynomial of B_1. Then the roots of g_1 are simple and have real parts equal to $1/2$. Thus, g_1 divides g. (Recall that g is that factor of the minimal polynomial of B

containing all roots whose real parts are equal to $1/2$.) Hence, for $x \in \operatorname{supp} \gamma$, $g_1(B)(x) = g_1(B_1)(x) = 0$ since $g_1(B_1)$ is identically equal to zero. Thus, $g(B)(x) = g_1(B)(x) = 0$, so $x \in \ker g(B)$. Therefore, $\operatorname{supp} \gamma \subseteq \ker g(B)$. Q.E.D.

Theorem 4.6.5 (Structural Theorem). *Let μ be a full operator-stable measure with exponent B. Then the minimal polynomial of B may be factored into $g \cdot h$, where the roots of g have real parts equal to $1/2$ and the roots of h have real parts strictly greater than $1/2$. Furthermore, $\mu = \mu_1 * \mu_2$, where μ_1 is Gaussian and μ_2 has no Gaussian component, $\operatorname{supp} \mu_1 = \ker g(B)$, $\operatorname{lin}(\operatorname{supp} \mu_2) = \ker h(B)$, and $\mathbb{R}^d = \operatorname{supp} \mu_1 \oplus \operatorname{lin}(\operatorname{supp} \mu_2)$. Finally, for $i = 1, 2$, letting B_i denote the restriction of B to $\operatorname{lin}(\operatorname{supp} \mu_i)$, we have that each μ_i is a full t^{B_i}-stable measure on the space $\operatorname{lin}(\operatorname{supp} \mu_i)$. (Later, we show that the roots of g are simple; cf. Theorem 4.6.12.)*

Proof. Let $\mu = [a, R, M]$ and set $\gamma \coloneqq [0, R, 0]$, $\eta \coloneqq [0, 0, M]$, so that $\mu = \gamma * \eta * \delta(a)$. By Theorem 1.2.3,

$$\mathbb{R}^d = \ker g(B) \oplus \ker h_1(B) \oplus \ker h_2(B),$$

where g, h_1, and h_2 are as in Lemmas 4.6.1 and 4.6.2. By Lemma 4.6.3, $\operatorname{lin}(\operatorname{supp} \eta) \subseteq \operatorname{lin}(\operatorname{supp} M) \subseteq \ker h_1(B)$. By Lemma 4.6.4, $\operatorname{supp} \gamma \subseteq \ker g(B)$. Since μ is full, we have

$$\mathbb{R} = \operatorname{supp}(\mu * \delta(-a)) = \operatorname{supp} \gamma * \eta = \overline{\operatorname{supp} \gamma + \operatorname{supp} \eta}$$
$$\subseteq \ker g(B) \oplus \ker h_1(B) \subseteq \mathbb{R}^d.$$

Therefore, $\mathbb{R}^d = \ker g(B) \oplus \ker h_1(B)$, $\operatorname{supp} \gamma = \ker g(B)$, $\operatorname{lin}(\operatorname{supp} \eta) = \ker h_1(B)$, and $h_2 = 0$.

Now, represent a as $a_1 + a_2$ with $a_1 \in \ker g(B)$ and $a_2 \in \ker h_1(B)$. Set $\mu_1 \coloneqq \gamma * \delta(a_1)$ and $\mu_2 \coloneqq \eta * \delta(a_2)$, thereby establishing the second statement in the theorem.

Because μ is t^B-stable, $\mu_1^t * \mu_2^t = t^B \mu_1 * t^B \mu_2 * \delta(b(t))$ for some vectors $b(t)$ and for all $t > 0$. Thus, μ_1^t and $t^B \mu_1$ are the Gaussian components of the same infinitely divisible measure, so $\mu_1^t = t^B \mu_1 * \delta(c_1(t))$ for some vectors $c_1(t) \in \mathbb{R}^d$ and for all $t > 0$. Hence, $\gamma^t = t^B \gamma * \delta(-ta_1 + t^B a_1 + c_1(t))$ for all $t > 0$. Also, $\operatorname{supp} \gamma^t = \operatorname{supp} \gamma = \ker g(B)$ and $\operatorname{supp} t^B \gamma = t^B(\operatorname{supp} \gamma) = t^B(\ker g(B)) = \ker g(B)$ since $\ker g(B)$ is B-invariant. Hence, $-ta_1 + t^B a_1 + c_1(t) \in \ker g(B)$. Since $a_1 \in \ker g(B)$, $c_1(t) \in \ker g(B)$ for all $t > 0$. Therefore, μ_1 is full and t^{B_1}-stable on $\ker g(B) = \operatorname{supp} \mu_1$. Similarly, $\eta' = t^B \eta * \delta(-ta_2 + t^B a_2 + c_2(t))$ for some vectors $c_2(t) \in \mathbb{R}^d$ and for all $t > 0$. By Proposition 3.4.10, $\operatorname{lin}(\operatorname{supp} \eta') = \operatorname{lin}(\operatorname{supp} t \cdot M) = \operatorname{lin}(\operatorname{supp} M) = \ker h_1(B)$. Also, since t^B is linear and invertible, $\operatorname{lin}(\operatorname{supp} t^B \eta) = t^B(\operatorname{lin}(\operatorname{supp} \eta)) = t^B(\ker h_1(B)) = \ker h_1(B)$. Thus, $c_2(t) \in \ker h_1(B)$ for all $t > 0$, and so μ_2 is full and t^{B_2}-stable on $\ker h_1(B) = \operatorname{lin}(\operatorname{supp} \mu_2)$. Q.E.D.

The next goal of this section is to consider, for a fixed full operator-stable measure μ, a characterization of the class of all possible exponents B. We denote this class by $\mathscr{E}_s(\mu)$, that is,

$$\mathscr{E}_s(\mu) := \{B \in \operatorname{Aut}(\mathbb{R}^d) \colon \mu \in \mathscr{S}(B)\},$$

where $\mathscr{S}(B)$ is the class of all t^B-stable measures. We first describe some basic properties of $\mathscr{E}_s(\mu)$, recalling that **OS** denotes the class of all t^B-stable measures for some B.

Proposition 4.6.6

(a) *For $A \in \operatorname{Aut}$, $A\mathscr{S}(B) = \mathscr{S}(ABA^{-1})$.*
(b) *For $A \in \operatorname{Aut}$ and $\mu \in$ **OS**, $A\mathscr{E}_s(\mu)A^{-1} = \mathscr{E}_s(A\mu)$.*
(c) *For $A \in \mathbf{A}(\mu)$, $\mu \in$ **OS**, and $B \in \mathscr{E}_s(\mu)$, $ABA^{-1} \in \mathscr{E}_s(\mu)$.*
(d) *For any vector a, $\mathscr{E}_s(\mu) = \mathscr{E}_s(\mu * \delta(a))$.*
(e) *The set $\mathscr{E}_s(\mu)$ is closed.*

Proof. For (a) and (b), since $\mu^t = t^B \mu * \delta(b(t))$ for all $t > 0$, $t^{ABA^{-1}}(A\mu) * \delta(Ab(t)) = A(t^B \mu * \delta(b(t))) = (A\mu)^t$. Therefore, $\mu \in \mathscr{S}(B)$ and $A \in \operatorname{Aut}$ implies that $A\mu \in \mathscr{S}(ABA^{-1})$ and $ABA^{-1} \in \mathscr{E}_s(A\mu)$. Hence, $A\mathscr{S}(B) \subseteq \mathscr{S}(ABA^{-1})$ and $A\mathscr{E}_s(\mu)A^{-1} \subseteq \mathscr{E}_s(A\mu)$. Similarly, $A^{-1}\mathscr{S}(ABA^{-1}) \subseteq \mathscr{S}(B)$ and $A^{-1}\mathscr{E}_s(A\mu)A \subseteq \mathscr{E}_s(\mu)$.

For (c), since $\mu = A\mu * \delta(a)$ for some vector a, we have

$$A\mu^t = At^B \mu * \delta(Ab(t)) = t^{ABA^{-1}}(A\mu) * \delta(Ab(t))$$
$$= t^{ABA^{-1}} \mu * \delta(Ab(t) - t^{ABA^{-1}} a).$$

Hence,

$$\mu^t = (A\mu * \delta(a))^t = A\mu^t * \delta(ta) = t^{ABA^{-1}} \mu * \delta(ta + Ab(t) - t^{ABA^{-1}} a).$$

For (d), let $B \in \mathscr{E}_s(\mu)$. Then $\mu^t = t^B\mu * \delta(b(t))$ implies $(\mu * \delta(a))^t = t^B(\mu * \delta(a)) * \delta(b(t) + ta - t^Ba)$, so $B \in \mathscr{E}_s(\mu * \delta(a))$. Thus, $\mathscr{E}_s(\mu) \subseteq \mathscr{E}_s(\mu * \delta(a))$. The reverse inclusion is just as easy.

For (e), let $B_n \in \mathscr{E}_s(\mu)$ be such that $B_n \to B \in \text{End}$. Then, for all $n \geq 1$ and all $t > 0$, $\mu^t = t^{B_n}\mu * \delta(b_n(t))$ for some vectors $b_n(t)$. Since $t^{B_n}\mu \Rightarrow t^B\mu$, we see that for each $t > 0$ there is a vector $b(t)$ such that $b_n(t) \to b(t)$. Therefore, $\mu^t = t^B\mu * \delta(b(t))$, so $B \in \mathscr{E}_s(\mu)$. Q.E.D.

Before stating the theorem which characterizes the class $\mathscr{E}_s(\mu)$, we recall some definitions and notation. In Section 2.3, we showed that the symmetry semigroup $\mathbf{A}(\mu) := \{A \in \text{Aut}: \mu = A\mu * \delta(a) \text{ for some vector } a\}$ of a full measure μ is a compact subgroup of Aut (cf. Corollary 2.3.2). In Section 1.5, the tangent space of any closed subgroup G of Aut was defined and denoted by $\mathscr{T}(G)$, that is, $\mathscr{T}(G) := \{A \in \text{End}: A = \lim_{n \to \infty} b_n^{-1}(g_n - I) \text{ for some sequence } \{g_n\}_{n \geq 1} \text{ in } G \text{ with } 0 < b_n \to 0 \text{ as } n \to \infty\}$. Many of the results on Lie algebras collected in Chapter 1 come to play in proving the following theorem.

Theorem 4.6.7 (Structure of the Exponents). *Let μ be full and operator-stable on $\mathbb{R}^d$ with an exponent B. Then*

$$\mathscr{E}_s(\mu) = B + \mathscr{T}(\mathbf{A}(\mu)).$$

Proof. We first recall many of the results used to prove Sharpe's representation theorem 4.2.12. For $t > 0$, let $G_t := \{A \in \text{Aut}: \mu^t = A\mu * \delta(a) \text{ for some vector } a\}$ and set $G := \bigcup_{t > 0} G_t$. Then $G_1 = \mathbf{A}(\mu)$ is a compact subgroup of the closed group G (cf. Lemmas 4.2.7 and 4.2.8). Also, the function $L: \mathscr{T}(G) \to \mathbb{R}^1$ given by $L(A) := \log(l(\exp A))$, where $l(C) = t$ for $C \in G_t$, is a continuous linear functional (cf. Lemma 4.2.9).

Since $A \in \mathscr{E}_s(\mu)$ implies that $t^A \in G$ for all $t > 0$ and A is the derivative of t^A at $t = 1$, we see that $\mathscr{E}_s(\mu) \subseteq \mathscr{T}(G)$. We claim that $\mathscr{E}_s(\mu) = \{A \in \mathscr{T}(G): L(A) = 1\}$. For $A \in \mathscr{E}_s(\mu)$, in particular, we have $\mu^e = e^A\mu * \delta(b(e))$, that is, $e^A \in G_e$. Hence, $l(e^A) = e$ so $L(A) = 1$. Conversely, let $A \in \mathscr{T}(G)$ with $L(A) = 1$. By the linearity of L, $L((\log t)A) = \log t$ for all $t > 0$. Thus, $l(t^A) = t$, so $t^A \in G_t$ for all $t > 0$, that is, $A \in \mathscr{E}_s(\mu)$. Therefore, $\mathscr{E}_s(\mu) = \{A \in \mathscr{T}(G): L(A) = 1\}$. By Lemma 4.2.10, $\mathscr{T}(\mathbf{A}(\mu)) = \ker L$.

Now, let B be any exponent of μ. Then $B \in \mathscr{T}(G)$ and $L(B) = 1$. For any other exponent C of μ, we have $L(B - C) = 0$, so $B - C \in \ker L = \mathscr{T}(\mathbf{A}(\mu))$. Hence, $\mathscr{E}_s(\mu) \subseteq B + \mathscr{T}(\mathbf{A}(\mu))$. Finally, let $D \in$

$\mathcal{T}(\mathbf{A}(\mu))$. Then $B + D \in \mathcal{T}(G)$ since $\mathcal{T}(G)$ is a vector space, and $L(B + D) = 1$. Thus, $B + D \in \mathcal{E}_s(\mu)$, so $\mathcal{E}_s(\mu) \supseteq B + \mathcal{T}(\mathbf{A}(\mu))$.
Q.E.D.

Corollary 4.6.8. *Let μ be full and operator-stable on $\mathbb{R}^d$. Then μ has exactly one exponent if and only if $\mathbf{A}(\mu)$ is finite.*

Proof. From Theorem 4.6.6, we have that μ has exactly one exponent if and only if $\mathcal{T}(\mathbf{A}(\mu)) = \{0\}$. From Section 1.5, the image of $\mathcal{T}(\mathbf{A}(\mu))$ under the exponential mapping contains an open neighborhood U_I of the identity I in $\mathbf{A}(\mu)$. Hence, $\mathcal{T}(\mathbf{A}(\mu)) = \{0\}$ if and only if the one-point set $\{I\}$ is open in $\mathbf{A}(\mu)$. If $\mathcal{T}(\mathbf{A}(\mu)) = \{0\}$, then $\exp(\mathcal{T}(\mathbf{A}(\mu))) = \{I\} = U_I$ is an open set in $\mathbf{A}(\mu)$. Because $\mathbf{A}(\mu)$ is a group, each singleton set is open in $\mathbf{A}(\mu)$, and the compactness of $\mathbf{A}(\mu)$ implies it is a finite group. Conversely, if $\mathbf{A}(\mu)$ is finite, then $\{I\}$ is an open neighborhood of the identity, so $\mathcal{T}(\mathbf{A}(\mu)) = \{0\}$. Q.E.D.

Corollary 4.6.9. *Let μ be full and operator-stable on $\mathbb{R}^d$. Then the set $\mathcal{E}_s(\mu)$ is affine, that is, for all $\alpha \in \mathbb{R}^1$ and $B_1, B_2 \in \mathcal{E}_s(\mu)$, we have $\alpha B_1 + (1 - \alpha) B_2 \in \mathcal{E}_s(\mu)$.*

Proof. Fix $B \in \mathcal{E}_s(\mu)$ so that $\mathcal{E}_s(\mu) = B + \mathcal{T}(\mathbf{A}(\mu))$. Then, for $B_1, B_2 \in \mathcal{E}_s(\mu)$, there are $C_1, C_2 \in \mathcal{T}(\mathbf{A}(\mu))$ such that $B_i = B + C_i$, $i = 1, 2$. For any $\alpha \in \mathbb{R}^1$, $\alpha B_1 + (1 - \alpha) B_2 = B + \alpha C_1 + (1 - \alpha) C_2$. Since $\mathcal{T}(\mathbf{A}(\mu))$ is a vector space, we have $\alpha C_1 + (1 - \alpha) C_2 \in \mathcal{T}(\mathbf{A}(\mu))$, so $\alpha B_1 + (1 - \alpha) B_2 \in \mathcal{E}_s(\mu)$ Q.E.D.

For a Gaussian measure μ, the class $\mathcal{E}_s(\mu)$ of its exponents has a simple description given by the following theorem.

Theorem 4.6.10. *Let μ be a full Gaussian measure on $\mathbb{R}^d$ with covariance operator R. Then*

$$\mathcal{E}_s(\mu) = \frac{I}{2} + S^{-1} \mathcal{K} S,$$

where S is the unique positive-definite self-adjoint square root of R^{-1} and $\mathcal{K}$ is the class of all skew-symmetric linear operators on $\mathbb{R}^d$. Conversely, if $\mathcal{E}_s(\mu) = I/2 + T^{-1} \mathcal{K} T$ for some positive-definite self-adjoint linear operator T, then $R = \lambda T^{-2}$ for some $\lambda > 0$.

Proof. By Proposition 4.6.6(d), we may assume that μ has mean zero. Consider the case when $R = I$. We first show that $\mathcal{E}_s(\mu) \supseteq I/2 + \mathcal{K}$, so let $K \in \mathcal{K}$. Clearly, $I/2 \in \mathcal{E}_s(\mu)$; consequently, $t^{I/2+K}\mu = t^K\mu^t$ for all $t > 0$. However,

$$(t^K\mu^t)\hat{\,}(y) = \exp\{-t\langle t^{K^*}y, t^{K^*}y\rangle/2\} = \exp\{-t\langle y, y\rangle/2\}$$

since $K + K^* = 0$. Thus $t^{I/2+K}\mu = \mu^t$ for all $t > 0$, so $I/2 + K \in \mathcal{E}_s(\mu)$, that is, $\mathcal{E}_s(\mu) \supseteq I/2 + \mathcal{K}$. Now, let $B \in \mathcal{E}_s(\mu)$. By Proposition 4.3.2, $t \cdot I = t^B t^{B^*}$ for all $t > 0$, so $t^{B-I/2}t^{B^*-I/2} = I$. Taking the derivative at $t = 1$, we obtain $(B - I/2) + (B^* - I/2) = 0$, that is, $B - I/2 \in \mathcal{K}$, so $\mathcal{E}_s(\mu) \subseteq I/2 + \mathcal{K}$. Consequently, $\mathcal{E}_s(\mu) = I/2 + \mathcal{K}$ when $R = I$.

For general R,

$$(S\mu)\hat{\,}(y) = \exp\{-\langle RSy, Sy\rangle/2\} = \exp\{-\langle y, y\rangle/2\},$$

that is, the covariance operator of $S\mu$ is the identity, I. Hence, $\mathcal{E}_s(S\mu) = I/2 + \mathcal{K}$. By Proposition 4.6.6(b), $\mathcal{E}_s(\mu) = I/2 + S^{-1}\mathcal{K}S$.

For the converse, note that the covariance operator R of μ and the covariance operator U of $T\mu$ are related by $U = TRT$, since T is self-adjoint. Also, by Proposition 4.6.6(b), $\mathcal{E}_s(T\mu) = T\mathcal{E}_s(\mu)T^{-1} = I/2 + \mathcal{K}$. Hence, for all $K \in \mathcal{K}$ and for all $t > 0$,

$$t \cdot U = t^{I/2+K}Ut^{I/2+K^*} \quad \text{or} \quad Ut^K = t^K U.$$

Thus, U commutes with everything in $\exp(\mathcal{K})$. Since $\exp(\mathcal{K})$ contains all rotations on $\mathbb{R}^d$ (cf. Section 1.5), we see that U commutes with every rotation. Consequently, we have that $U = \lambda I$ for some real λ (cf. page 180). Since U is positive-definite, λ must be positive. Therefore, $TRT = \lambda I$ or $R = \lambda T^{-2}$. Q.E.D.

An easy corollary to the theorem is the following.

Corollary 4.6.11. *Let μ be a full Gaussian measure on $\mathbb{R}^d$ with covariance operator R and let B be an invertible linear operator on $\mathbb{R}^d$. Then B is an exponent for μ if and only if $S(B - I/2)S^{-1}$ is skew-symmetric, where S is the positive-definite self-adjoint square root of R^{-1}.*

Earlier (cf. Corollary 4.2.13), we showed that if B is an exponent for a full measure, then all of the eigenvalues of B have positive real parts. Then in Theorem 4.6.5 we established that those eigenvalues must have real parts greater than or equal to $1/2$. This naturally raises the question of whether or not every $B \in \text{Aut}(\mathbb{R}^d)$, whose eigenvalues have real parts greater than or equal to $1/2$, is an exponent for some full measure? The answer is in the affirmative with the additional requirement that the roots of the minimal polynomial of B, with real parts equal to $1/2$, are simple.

Theorem 4.6.12 (Spectral Characterization of Exponents). *A necessary and sufficient condition that a linear operator B on $\mathbb{R}^d$ be an exponent for some full operator-stable measure is that all of the roots of the minimal polynomial of B have real parts greater than or equal to $1/2$ and those with real parts equal to $1/2$ are simple roots.*

Proof. Necessity. Assume $B \in \mathscr{E}_s(\mu)$ for some full μ. As before, factor the minimal polynomial of B into the product $g \cdot h$, where the roots of g all have real parts equal to $1/2$ and those of h have real parts strictly greater than $1/2$ (cf. Theorem 4.6.3). It remains only to show that the roots of g are simple. Let μ_1 be the Gaussian component of μ and let B_1 be the restriction of B to $\ker g(B)$. Apply Corollary 4.6.11 to μ_1 and B_1 on the space $\ker g(B)$. Then $S(B_1 - I_1/2)S^{-1}$ is skew-symmetric for some invertible linear operator S on $\ker g(B)$, where I_1 is the identity on $\ker g(B)$. Hence, the minimal polynomial g_1 of B_1 has only simple roots with real parts equal to $1/2$ (cf. page 209). But, $g_1 = g$.

Sufficiency. Let g and h be as usual. By Theorem 1.2.3, decompose $\mathbb{R}^d$ into the direct sum of subspaces $V_1, \ldots, V_r$ such that $BV_j = V_j$ and the minimal polynomial of the restriction of B to V_j is a power of a real-irreducible polynomial, for $j = 1, \ldots, r$. Number the subspaces V_j so that the eigenvalues in $V_1, \ldots, V_k$ have real parts greater $1/2$ and those in $V_{k+1}, \ldots, V_r$ have real parts equal to $1/2$. Set $V_0 := V_{k+1} \oplus \cdots \oplus V_r$. Let $S_j := S_B \cap V_j$ for $j = 1, \ldots, k$, where S_B is the unit sphere under the norm $\|\cdot\|_B$. Let m_j be a Borel measure on S_B with $\text{supp}\, m_j = S_j$, $1 \le j \le k$. Hence, $\text{lin}_B(\text{supp}\, m_j) \subseteq V_j$, since V_j is B-invariant. Let $x \in V_j$. Then $x/\|x\|_B \in S_B \cap V_j = \text{supp}\, m_j$, which implies that $x \in \text{lin}_B(\text{supp}\, m_j)$. Therefore, $\text{lin}_B(\text{supp}\, m_j) = V_j$ for $j = 1, \ldots, k$. Define M_j on the Borel subsets of $\mathbb{R}^d \setminus \{0\}$ by $M_j(A) := \int_{S_B} \int_0^\infty I_A(s^B x) s^{-2}\, ds\, m_j(dx)$. Then M_j is a Lévy measure and $t \cdot M_j = t^B M_j$ for all $t > 0$ (cf. Proposition 4.3.4 and Lemma 4.6.1). Let $\nu_j := [0, 0, M_j]$. From Corollary 4.3.5, we see that $\text{lin}(\text{supp}\, \nu_j) =$

$\operatorname{lin}(\operatorname{supp} M_j) = \operatorname{lin}_B(\operatorname{supp} M_j) = V_j$, so ν_j is full on V_j. Therefore, $\nu := \nu_1 * \cdots * \nu_k = [0, 0, M]$, with $M := \sum_{j=1}^{k} M_j$, is t^B-stable and full on $V_1 \oplus \cdots \oplus V_k$.

Now, to construct a Gaussian measure γ which is full on V_0 and t^B-stable, let B_0 denote the restriction of B to V_0. Since the minimal polynomial of B_0 has only simple roots with real parts equal to $1/2$, there is some basis for V_0 such that the matrix representation of B_0 with respect to this basis is block-diagonal, with the blocks being either 1×1 matrices with entry $1/2$ or 2×2 matrices of the form

$$\begin{pmatrix} \frac{1}{2} & \beta \\ -\beta & \frac{1}{2} \end{pmatrix},$$

with $\beta \neq 0$. Let I_0 be the identity on V_0. Then $B_0 - I_0/2$ is skew-symmetric, so

$$t^{B_0 - I_0/2} t^{B_0^* - I_0/2} = I_0 \quad \text{or} \quad t \cdot I_0 = t^{B_0} I_0 t^{B_0^*}$$

By Proposition 4.3.2 and Theorem 1.8.20, $\gamma := [0, I_0, 0]$ is t^B-stable, Gaussian with zero mean, and full on V_0 with covariance operator I_0. Set $\mu = \gamma * \nu$. Then μ is full on $\mathbb{R}^d$ and t^B-stable. Q.E.D.

We conclude this section with a necessary and sufficient condition for the compatibility of a Lévy measure M and an exponent B.

Theorem 4.6.13. *Let M be a Lévy measure on $\mathbb{R}^d$ and $B \in \operatorname{Aut}$ with its minimal polynomial of the form $g \cdot h$, where the roots of g have real parts equal to $1/2$ and are simple, unless g is identically one, and the roots of h have real parts greater than $1/2$. Then there exists a full t^B-stable measure μ on $\mathbb{R}^d$ whose Lévy measure is M if and only if both $t \cdot M = t^B M$ for all $t > 0$ and $\operatorname{lin}(\operatorname{supp} M) = \ker h(B)$.*

Proof. Assume M is the Lévy measure of the full t^B-stable measure μ. By Proposition 4.3.2, $t \cdot M = t^B M$ for all $t > 0$. By Theorem 4.6.5, there is a measure μ_2 having no Gaussian component such that $\mu = \mu_1 * \mu_2$, $\operatorname{lin}(\operatorname{supp} \mu_2) = \ker h(B)$ and M is the Lévy measure of μ_2. Therefore, $\ker h(B) = \operatorname{lin}(\operatorname{supp} \mu_2) = \operatorname{lin}(\operatorname{supp} M)$.

Conversely, let $\mu_2 := [0, 0, M]$. Then μ_2 is full and operator-stable on $\ker h(B)$ with exponent $B_2 := B|\ker h(B)$. As in the proof of

Theorem 4.6.12, construct a Gaussian measure μ_1 on $\mathbb{R}^d$ such that μ_1 is full on $\ker g(B)$, $B_1 := B|\ker g(B)$ is an exponent for μ_1, and $\mu := \mu_1 * \mu_2$ is full on $\mathbb{R}^d$, t^B-stable, and with Lévy measure M.
Q.E.D.

In Theorem 4.6.5, we saw that a full t^B-stable measure μ has a t^{B_1}-stable Gaussian component μ_1 and a t^{B_2}-stable Poisson component μ_2, where $B_i = B|V_i$, $i = 1, 2$, even though μ_1 and μ_2 are not full. We will need a condition on a projection P on $\mathbb{R}^d$ so that when μ is t^B-stable, not necessarily full on $\mathbb{R}^d$, $P\mu$ is t^{BP}-stable.

Proposition 4.6.14. *Let μ be t^B-stable on $\mathbb{R}^d$ and let P be a projection on $\mathbb{R}^d$ which commutes with B. Then $P\mu$ is t^{BP}-stable on $\mathbb{R}^d$.*

Proof. Using the fact that P and B commute, we see that

$$(P\mu)^t = P(t^B \mu * \delta(b(t)))$$
$$= (t^B P)\mu * \delta(P(b(t))) = t^{BP}(P\mu) * \delta(P(b(t))). \quad \text{Q.E.D.}$$

Proposition 4.6.15. *Let μ be full and t^B-stable on $\mathbb{R}^d$ and let P be either the projection onto V_2 with $\ker P = V_1$ or the projection onto V_1 with $\ker P = V_2$. Then P and B commute.*

Proof. From Theorem 4.6.5, we know that both V_1 and V_2 are B-invariant subspaces. Let P be the projection onto V_2 with $\ker P = V_1$. Since $B(I - P)y \in V_1$, we see that $PBy = PBPy + PB(I - P)y = PBPy = BPy$ for all y. Hence, $PB = BP$. The other case is as simple.
Q.E.D.

Finally, we will need a partial converse to Proposition 4.6.14.

Proposition 4.6.16. *Let V be a proper subspace of $\mathbb{R}^d$ and let ν be a full measure on V. Let $\mu(F) := \nu(F \cap V)$ for $F \in \mathscr{B}(\mathbb{R}^d)$. If ν is t^C-stable on V for some $C \in \text{Aut}(V)$, then $\mu^t = t^D \mu * \delta(c(t))$ for all $t > 0$, where D is any linear operator on $\mathbb{R}^d$ whose restriction to V is C, and $c(t) \in V$.*

Proof. Let P be the orthogonal projection of $\mathbb{R}^d$ onto V. Then, for any $t > 0$ and $y \in \mathbb{R}^d$,

$$\hat{\mu}^t(y) = \left(\int_V e^{i\langle Py, x\rangle} \nu(dx)\right)^t = \hat{\nu}^t(Py),$$

$$(t^C \nu)\hat{\ }(Py) = \int_V e^{i\langle Py, t^C x\rangle} \nu(dx)$$

$$= \int_{\mathbb{R}^d} e^{i\langle y, t^D x\rangle} \mu(dx) = (t^D \mu)\hat{\ }(y).$$

Therefore,

$$\hat{\mu}^t(y) = \hat{\nu}^t(Py) = (t^C \nu)\hat{\ }(Py) \exp(i\langle Py, c(t)\rangle)$$
$$= (t^D \mu)\hat{\ }(y) \exp(i\langle y, c(t)\rangle). \qquad \text{Q.E.D.}$$

Remark 4.6.17. From the decomposition of a full operator-stable measure $\mu = [a, R, M]$ into $\mu_1 * \mu_2$ with $\mu_1 = [a_1, R, 0]$ and $\mu_2 = [a_2, 0, M]$, a natural supposition is that $Ry = 0$ for all $y \in \operatorname{supp} \mu_2$. The following example shows that this is false in general. It is true if and only if the kernels of $g(B)$ and $h(B)$ are orthogonal, since $\operatorname{supp} \mu_1 = \ker g(B)$ and $\operatorname{supp} \mu_2 \subseteq \ker h(B)$ (cf. Theorem 4.6.5).

Let $d = 2$ and let R be the orthogonal projection onto the x-axis. Then R is positive-semidefinite and self-adjoint. Let

$$B = \begin{pmatrix} \frac{1}{2} & \frac{1}{2} \\ 0 & 1 \end{pmatrix}.$$

Since $(1,0)^*$ and $(1,1)^*$ are eigenvectors of B corresponding to the eigenvalues $1/2$ and 1, respectively, we have

$$t^B = \begin{pmatrix} t^{1/2} & t - t^{1/2} \\ 0 & t \end{pmatrix}.$$

The minimal polynomial of B is $(x - 1/2)(x - 1)$, so $V_1 := \ker g(B) = x$-axis and $V_2 := \ker h(B) = \{(x, x)^* : x \in \mathbb{R}^1\}$. Define M for $E \in \mathscr{B}(\mathbb{R}^2 \setminus \{0\})$ by

$$M(E) := \int_0^\infty I_E\left(t^B \begin{pmatrix} 1 \\ 1 \end{pmatrix}\right) t^{-2} \, dt.$$

Then lin(supp M) = V_2. Since $t^B R t^{B^*} = t \cdot R$ and $t^B M = t \cdot M$ for all $t > 0$, by Proposition 4.3.2, $\mu := [0, R, M]$ is t^B-stable. Clearly, μ is full on $\mathbb{R}^2$, supp $\mu_1 = V_1$, and lin(supp μ_2) = V_2. However, $R(x, x)^* = x$.

4.7 COMMUTING EXPONENTS AND OTHER NORMS

In many problems, it would be an advantage to know that whenever $B \in \mathcal{E}_s(\mu)$ and $A \in \mathbf{A}(\mu)$ we have $t^B A = A t^B$ for all $t > 0$, or, equivalently, $BA = AB$. That this is not necessarily true is seen by the following example.

Example. Let μ be the symmetric Cauchy measure on $\mathbb{R}^2$, that is, $\hat{\mu}(y) = e^{-\|y\|}$ for $y \in \mathbb{R}^2$. Clearly, $\mathbf{A}(\mu)$ is the full orthogonal group. Also, $(t^I \mu)\hat{\,}(y) = e^{-t\|y\|} = \hat{\mu}^t(y)$ for all $t > 0$ and for all $y \in \mathbb{R}^2$. Consequently, $I \in \mathcal{E}_s(\mu)$. Therefore, by Theorem 4.6.7, $\mathcal{E}_s(\mu) = I + \mathcal{K}$, where $\mathcal{K}$ is the class of all skew-symmetric linear operators on $\mathbb{R}^2$, since $\mathcal{T}(\mathcal{O}) = \mathcal{K}$ (cf. Section 1.5). Hence,

$$B := \begin{pmatrix} 1 & 1 \\ -1 & 1 \end{pmatrix} \in \mathcal{E}_s(\mu).$$

But, for

$$A := \begin{pmatrix} 1 & 0 \\ 0 & -1 \end{pmatrix} \in \mathbf{A}(\mu),$$

we have $AB \ne BA$.

Note that in this example there does exist an exponent, namely the identity, which commutes with $\mathbf{A}(\mu)$. This raises the question: when does such an exponent exist? We show that the answer to this question is: always! This prompts other problems.

1. Describe the class of all exponents of μ which commute with $\mathbf{A}(\mu)$.
2. Find a necessary and sufficient condition so that *all* exponents of μ commute with $\mathbf{A}(\mu)$.
3. Knowing that such "commuting" exponents always exist, what can one say about the measure m_B on S_B given in the represen-

tation of Theorem 4.3.7 (cf. Proposition 4.3.4), that is, if $\mu = [a, R, M]$ and $B \in \mathscr{E}_s(\mu)$, then

$$M(A) = \int_{S_B} \int_0^\infty 1_A(s^B x) s^{-2} \, ds \, m_B(dx) \quad \text{for } A \in \mathscr{B}(\mathbb{R}^d \setminus \{0\}).$$

We now proceed to solve these problems. Let μ be operator-stable on $\mathbb{R}^d$ and denote the class of "commuting" exponents by $\mathscr{E}_{cs}(\mu)$, that is,

$$\mathscr{E}_{cs}(\mu) := \{B \in \mathscr{E}_s(\mu): BA = AB \text{ for all } A \in \mathbf{A}(\mu)\}.$$

We call each $B \in \mathscr{E}_{cs}(\mu)$ a *commuting* exponent of μ.

Theorem 4.7.1. *Let μ be a full and operator-stable measure on $\mathbb{R}^d$. Then $\mathscr{E}_{cs}(\mu)$ is nonempty.*

Proof. Since $\mathscr{E}_s(\mu) \neq \varnothing$ by Theorem 4.2.12, we may select $B \in \mathscr{E}_s(\mu)$. By Corollary 2.3.2, $\mathbf{A}(\mu)$ is a compact group, so let H denote the normalized Harr measure on $\mathbf{A}(\mu)$. Define

$$B_c := \int_{\mathbf{A}(\mu)} ABA^{-1} H(dA).$$

Note that, for all $A \in \mathbf{A}(\mu)$, $ABA^{-1} \in \mathscr{E}_s(\mu)$ by Proposition 4.6.6(c). Also, $\mathscr{E}_s(\mu)$ is closed and affine [cf. Proposition 4.6.6(e) and Corollary 4.6.9]. Thus, $B_c \in \mathscr{E}_s(\mu)$. Using the invariance property of the Harr measure, we obtain, for $D \in \mathbf{A}(\mu)$,

$$DB_c D^{-1} = \int_{\mathbf{A}(\mu)} (DA) B (DA)^{-1} H(dA) = B_c.$$

Therefore, $DB_c = B_c D$, which implies $B_c \in \mathscr{E}_{cs}(\mu)$. Q.E.D.

Corollary 4.7.2. *When $\mathbf{A}(\mu)$ is finite, the unique exponent B of the full operator-stable μ is commuting.*

This corollary follows from Corollary 4.6.8 and Theorem 4.7.1.

Lemma 4.7.3. *Let μ be a full and operator-stable measure and let $W \in \mathbf{Aut}$. Then $\mathscr{E}_{cs}(W\mu) = W\mathscr{E}_{cs}(\mu)W^{-1}$.*

Proof. Since

$$\mathscr{E}_{cs}(\mu) = \mathscr{E}_s(\mu) \cap \{B: BA = AB \text{ for all } A \in \mathbf{A}(\mu)\},$$

from Proposition 4.6.6(b) and Lemma 2.3.3, we have

$$W\mathscr{E}_{cs}(\mu)W^{-1} = \mathscr{E}_s(W\mu) \cap \{WBW^{-1}: BA = AB \text{ for all } A \in \mathbf{A}(\mu)\}$$
$$= \mathscr{E}_s(W\mu) \cap \{C: CD = DC \text{ for all } D \in \mathbf{A}(W\mu)\}$$
$$= \mathscr{E}_{cs}(W\mu). \qquad \text{Q.E.D.}$$

Later, we shall often use the following simple observation.

Lemma 4.7.4. *Let μ be a full and operator-stable measure. If $B \in \mathscr{E}_s(\mu)$ and $B_c \in \mathscr{E}_{cs}(\mu)$, then B and B_c commute and $\exp s(B - B_c) \in \mathbf{A}(\mu)$ for every $s \in \mathbb{R}$.*

Proof. Since $B - B_c \in \mathscr{T}(\mathbf{A}(\mu))$ by Theorem 4.6.7, $B - B_c$ is the limit of $(D_n - I)/d_n$ for some $D_n \in \mathbf{A}(\mu)$ and $0 < d_n \to 0$. Since B_c commutes with each D_n, B_c must commute with $B - B_c$, which is equivalent to B_c commutes with B. We have that $\exp s(B - B_c) \in \mathbf{A}(\mu)$ for every $s \in \mathbb{R}^1$. Q.E.D.

Thus, a commuting exponent is an exponent which commutes both with the symmetry group and with the class of all exponents of μ.

Now that we know that $\mathscr{E}_{cs}(\mu)$ is nonempty, we wish to characterize it in a manner similar to the characterization of $\mathscr{E}_s(\mu)$ in Theorem 4.6.7. Since $\mathscr{E}_{cs}(\mu)$ concerns operators which commute with $\mathbf{A}(\mu)$, it is natural to expect that its characterization involves those operators in $\mathbf{A}(\mu)$ which commute with all of $\mathbf{A}(\mu)$, the so-called *center* of the group $\mathbf{A}(\mu)$. We denote the center of $\mathbf{A}(\mu)$ by $\text{Cent}(\mathbf{A}(\mu))$, that is, $\text{Cent}(\mathbf{A}(\mu)) := \{C \in \mathbf{A}(\mu): CA = AC \text{ for all } A \in \mathbf{A}(\mu)\}$. When μ is full, it is easy to see that $\text{Cent}(\mathbf{A}(\mu))$ is a compact subgroup of the compact group $\mathbf{A}(\mu)$. Hence, the tangent space $\mathscr{T}(\text{Cent}(\mathbf{A}(\mu)))$ exists and is a subspace of $\mathscr{T}(\mathbf{A}(\mu))$.

Theorem 4.7.5. *Let μ be a full and operator-stable measure on $\mathbb{R}^d$. Then, for any $B \in \mathscr{E}_{cs}(\mu)$,*

$$\mathscr{E}_{cs}(\mu) = B + \mathscr{T}(\text{Cent}(\mathbf{A}(\mu))).$$

Proof. For any $B_c \in \mathcal{E}_{cs}(\mu)$, $\exp s(B_c - B) \in \mathbf{A}(\mu)$ for all $s \in \mathbb{R}^1$ (cf. Lemma 4.7.4). Also, since $B_c - B$ commutes with $\mathbf{A}(\mu)$, for $A \in \mathbf{A}(\mu)$ and $s \in \mathbb{R}^1$, we have $A(\exp s(B_c - B))A^{-1} = \exp sA(B_c - B)A^{-1} = \exp s(B_c - B)$. Consequently, $\exp s(B_c - B) \in \text{Cent}(\mathbf{A}(\mu))$ for all s. Thus, $B_c - B \in \mathcal{T}(\text{Cent}(\mathbf{A}(\mu)))$. Therefore, $\mathcal{E}_{cs}(\mu) \subseteq B + \mathcal{T}(\text{Cent}(\mathbf{A}(\mu)))$.

To obtain the reverse inclusion, let $B' := B + D$ for some $D \in \mathcal{T}(\text{Cent}(\mathbf{A}(\mu)))$. Since $\mathcal{T}(\text{Cent}(\mathbf{A}(\mu))) \subseteq \mathcal{T}(\mathbf{A}(\mu))$, we have $B' \in \mathcal{E}_s(\mu)$ (cf. Theorem 4.6.7). Also, for all s, $\exp s(B' - B) \in \text{Cent}(\mathbf{A}(\mu))$. Hence, $\exp s(B' - B)$ commutes with $\mathbf{A}(\mu)$, and consequently $B' - B$ commutes with $\mathbf{A}(\mu)$. Since B commutes with $\mathbf{A}(\mu)$ by definition, we have that B' also commutes with $\mathbf{A}(\mu)$. Therefore, $B' \in \mathcal{E}_{cs}(\mu)$.

Q.E.D.

The following corollary provides an answer to: when is every exponent commuting?

Corollary 4.7.6. *Let μ be a full and operator-stable measure. Then every exponent of μ is commuting if and only if $\mathcal{T}(\mathbf{A}(\mu)) = \mathcal{T}(\text{Cent}(\mathbf{A}(\mu)))$.*

In Section 3.4, a very useful norm was defined by

$$\|x\|_B := \int_0^\infty \|e^{-tB}x\| \, dt = \int_0^1 \|s^B x\| s^{-1} \, ds,$$

where all of B's eigenvalues have positive real parts. This norm was used to solve the equation $t \cdot M = t^B M$ for all $t > 0$. The solution was that M must be of the form

$$M(A) = \int_{S_B} \int_0^\infty \|t^B x\|_B t^{-2} \, dt \, m_B(dx),$$

where $S_B := \{x \in \mathbb{R}^d : \|x\|_B = 1\}$ and m_B is a Borel measure on S_B. A disadvantage of this representation is that the sphere S_B and the measure m_B depend on the exponent B and may not be invariant under the action of the symmetry group of $[a, R, M]$. The knowledge that there is always a commuting exponent, when $[a, R, M]$ is full, allows us to define another norm such that the unit sphere under this new norm does not depend on the exponent and is invariant under the action of the symmetry group. A unit sphere in this norm can be used to solve $t \cdot M = t^B M$, $t > 0$, obtaining a measure which also does not depend on the exponent and is invariant under the action of the symmetry group.

Let $B \in \mathscr{E}_s(\mu)$ with μ full. For $x \in \mathbb{R}^d$, define

$$\|x\|_{\mu, B} := \int_{\mathbf{A}(\mu)} \int_0^1 \|At^B x\| t^{-1} \, dt \, H(dA), \qquad (4.7.1)$$

where H is the normalized Haar measure on the compact group $\mathbf{A}(\mu)$. Since $\|x\|_{\mu, B} < \infty$ for all $x \in \mathbb{R}^d$, it is obviously a norm on $\mathbb{R}^d$ (cf. Section 3.4). For μ full, the norm $\|\cdot\|_{\mu, B}$ does not depend on the exponent B of μ. Simply note that for $B \in \mathscr{E}_{cs}(\mu)$ and $B' \in \mathscr{E}_s(\mu)$, B and $B' - B$ commute and $\exp(s(B' - B)) \in \mathbf{A}(\mu)$ for all $s \in \mathbb{R}^1$ (cf. Lemma 4.7.4). Therefore, by the invariance of the Haar measure,

$$\|x\|_{\mu, B'} = \int_0^1 \int_{\mathbf{A}(\mu)} \|At^{B'-B} t^B x\| t^{-1} H(dA) \, dt = \|x\|_{\mu, B}.$$

Hence, we may define the norm $\|\cdot\|_\mu$ for μ full and operator-stable by

$$\|x\|_\mu := \|x\|_{\mu, B} \qquad (4.7.2)$$

for some $B \in \mathscr{E}_s(\mu)$. Letting $B \in \mathscr{E}_{cs}(\mu)$, we see that the norm $\|\cdot\|_\mu$ is invariant under $\mathbf{A}(\mu)$, that is, for $A \in \mathbf{A}(\mu)$, $\|Ax\|_\mu = \|x\|_\mu$ for all x. Let S_μ denote the unit sphere with respect to this norm, that is, $S_\mu := \{x \in \mathbb{R}^d : \|x\|_\mu = 1\}$.

In Section 3.4, we defined the norm $\|\cdot\|_Q$ for any Q such that $\exp(-tQ) \to 0$ as $t \to \infty$ by

$$\|x\|_Q = \int_0^1 \|t^Q x\| t^{-1} \, dt.$$

Using $B \in \mathscr{E}_{cs}(\mu)$ for Q, we see that

$$\|x\|_\mu = \int_{\mathbf{A}(\mu)} \|Ax\|_B H(dA). \qquad (4.7.3)$$

Similar to the function Φ_Q defined in Section 3.4, for μ full and operator-stable with $B \in \mathscr{E}_s(\mu)$, we define $\Psi_B : S_\mu \times (0, \infty) \to \mathbb{R}^d \setminus \{0\}$ by

$$\Psi_B(u, t) := t^B u. \qquad (4.7.4)$$

Proposition 4.7.7. Let μ be a full operator-stable measure and let $B \in \mathcal{E}_s(\mu)$.

(a) Ψ_B is a homeomorphism between $S_\mu \times (0, \infty)$ and $\mathbb{R}^d \setminus \{0\}$.
(b) For each nonzero vector x, the function $0 < s \to \|s^B x\|_\mu$ is strictly increasing.
(c) For $E \in \mathcal{B}(S_\mu)$, $G \in \mathcal{B}((0, \infty))$, and $s > 0$,
$$s^B \Psi_B(E \times G) = \Psi_B(E \times sG).$$
(d) For $E \in \mathcal{B}(S_\mu)$ and $G \in \mathcal{B}((0, \infty))$, we have, for any $A \in \mathbf{A}(\mu)$ which commutes with B,
$$A\Psi_B(E \times G) = \Psi_B(AE \times G).$$
(e) For $B_c \in \mathcal{E}_{cs}(\mu)$, $u \in S_\mu$, and $t > 0$, we have
$$\Psi_B(u, t) = t^{B-B_c} \Psi_{B_c}(u, t).$$

Proof. The proofs of (a), (b), and (c) are similar to the proof of Proposition 3.4.3, so they are omitted here.

For (d), since S_μ is $\mathbf{A}(\mu)$-invariant,
$$A\Psi_B(E \times G) = \{t^B(Au): u \in E, t \in G\} = \Psi_B(AE \times G).$$

For (e), by Lemma 4.7.4, B and $B - B_c$ commute so
$$\Psi_B(u, t) = t^{B-B_c} t^{B_c} u = t^{B-B_c} \Psi_{B_c}(u, t). \qquad \text{Q.E.D.}$$

Define functions $u_B: \mathbb{R}^d \setminus \{0\} \to S_\mu$ and $\tau_B: \mathbb{R}^d \setminus \{0\} \to (0, \infty)$ by
$$(u_B(x), \tau_B(x)) := \Psi_B^{-1}(x),$$
that is, $u_B(x)$ is that unique point on the sphere S_μ which is on the orbit $t^B x$ and $\tau_B(x)$ is the "time" to reach x from the origin traveling on the orbit $t^B x$. This gives two families of functions indexed by the class of exponents of μ. However, the family of "times" actually consists of exactly one function. Also, the u_B's are closely related.

Proposition 4.7.8. Let μ be a full operator-stable measure. The functions τ_B for $B \in \mathcal{E}_s(\mu)$ are all the same and may be simply denoted by τ. The functions u_B satisfy

$$u_{B_c}(x) = (\tau(x))^{B-B_c} u_B(x)$$

for all $B \in \mathscr{E}_s(\mu)$ and for all $B_c \in \mathscr{E}_{cs}(\mu)$.

Proof. For each $x \in \mathbb{R}^d \setminus \{0\}$, we have the unique representations $x = r^B w$ and $x = s^{B_c} v$ for some $r, s > 0$ and $w, v, \in S_\mu$. By Lemma 4.7.4, B_c and $B - B_c$ commute and $t^{B-B_c} \in \mathbf{A}(\mu)$ for all $t > 0$. Hence, $r^B w = r^{B_c}(r^{B-B_c} w) = s^{B_c} v$ and $r^{B-B_c} w \in S_\mu$, since S_μ is $\mathbf{A}(\mu)$-invariant. Because Ψ_{B_c} is a homeomorphism, $s = r$ and $v = r^{B-B_c} w$. Q.E.D.

Proposition 4.7.7 gives the properties of Ψ_B which are analogous to those of Φ_Q in Section 3.4. Those were the properties used in Section 4.3 to solve $t \cdot M = t^B M$ in terms of the norm $\|\cdot\|_B$ and Φ_B defined on S_B. Now, we may solve the equality in terms of the new norm and new sphere S_μ. The advantage is that the sphere used in the integral representation is $\mathbf{A}(\mu)$-invariant and does not depend on the exponent. It would appear that the mixing measure does depend on the exponent; however, we will show in Proposition 4.7.10 that it does not.

Proposition 4.7.9. *Let $\mu = [a, R, M]$ be full and operator-stable with an exponent B. Then, for all $E \in \mathscr{B}(\mathbb{R}^d \setminus \{0\})$,*

$$M(E) = \int_{S_\mu} \int_0^\infty I_E(s^B x) s^{-2} \, ds \, \tilde{m}_B(dx)$$

for the finite measure $\tilde{m}_B$ given by

$$\tilde{m}_B(F) := M(\Psi_B(F \times [1, \infty)))$$

for $F \in \mathscr{B}(S_\mu)$, and

$$\operatorname{supp} M = \bigcup_{t > 0} \left(t^B (\operatorname{supp} \tilde{m}_B) \right) \cup \{0\}.$$

Also, for M satisfying the above, we have $t \cdot M = t^B M$ for all $t > 0$.

The proof of this proposition is similar to the proofs of Proposition 4.3.4 and Corollary 4.3.5.

Proposition 4.7.10. *The mixing measure $\tilde{m}_B$ in Proposition 4.7.9 does not depend on the choice of the exponent B.*

Proof

Step 1. For $A \in \mathbf{A}(\mu)$ commuting with $B \in \mathscr{E}_s(\mu)$, we have $A\tilde{m}_B = \tilde{m}_B$. Let $F \in \mathscr{B}(S_\mu)$. Then, noting that $AM = M$ and $A^{-1}(F) \subseteq S_\mu$,

$$(A\tilde{m}_B)(F) = M(\{t^B A^{-1} u : t \geq 1, u \in F\})$$
$$= (AM)(\{t^B u : t \geq 1, u \in F\}) = \tilde{m}_B(F).$$

Step 2. For $B \in \mathscr{E}_s(\mu)$ and $B_c \in \mathscr{E}_{cs}(\mu)$, we have $\tilde{m}_B = \tilde{m}_{B_c}$. Note that, for any exponent B,

$$M(E) = \int_0^\infty \int_{S_\mu} I_E(t^B x) t^{-2} \tilde{m}_B(dx)\, dt$$
$$= \int_0^\infty \tilde{m}_B\big((t^{-B} E) \cap S_\mu\big) t^{-2}\, dt.$$

Also, for the sets of the form $T = \{s^B x : s \geq 1, x \in F\} = \Psi_B(F \times [1, \infty))$ for some $F \in \mathscr{B}(S_\mu)$, we have

$$(t^{-B} T) \cap S_\epsilon = \begin{cases} \varnothing & \text{if } t < 1, \\ F & \text{if } t \geq 1. \end{cases}$$

Furthermore, by Proposition 4.7.7, Lemma 4.7.4, and Step 1, we obtain

$$\tilde{m}_{B_c}\big((t^{-B_c} T) \cap S_\mu\big) = \tilde{m}_{B_c}\big(t^{B - B_c}((t^{-B} T) \cap S_\mu)\big) = \tilde{m}_{B_c}\big((t^{-B} T) \cap S_\mu\big)$$

for $B_c \in \mathscr{E}_{cs}(\mu)$. Hence,

$$\tilde{m}_B(F) = M(T) = \int_0^\infty \tilde{m}_{B_c}\big((t^{-B_c} T) \cap S_\mu\big) t^{-2}\, dt$$
$$= \int_0^\infty \tilde{m}_{B_c}\big((t^{-B} T) \cap S_\mu\big) t^{-2}\, dt$$
$$= \int_1^\infty \tilde{m}_{B_c}(F) t^{-2}\, dt = \tilde{m}_{B_c}(F)$$

for all $F \in \mathscr{B}(S_\mu)$. Q.E.D.

COMMUTING EXPONENTS AND OTHER NORMS

Corollary 4.7.11. *Let μ be a full operator-stable measure with the mixing measure $\tilde{m}$. Then, for all $A \in \mathbf{A}(\mu)$, we have $A\tilde{m} = \tilde{m}$ (cf. Step 1 of the proof of Proposition 4.7.10).*

Remark 4.7.12. In Theorem 4.3.7, the sphere S_B and the measure m, which may depend on B, can be replaced by S_μ and $\tilde{m}$, respectively.

A partial converse of this remark is given in the following proposition.

Proposition 4.7.13. *Let $\mu = [a, R, M]$ be a full operator-stable measure with mixing measure $\tilde{m}$ on S_μ. Assume $A \in \mathbf{Aut}$ is such that $(A\tilde{m})(E) = \tilde{m}(E)$ for all Borel $E \subseteq \operatorname{supp} \tilde{m}$. If A commutes with some exponent B of μ, and if either $R = 0$ or $R = ARA^*$, then $A \in \mathbf{A}(\mu)$.*

Proof. Note that $A\tilde{m} = \tilde{m}$ on $\operatorname{supp} \tilde{m}$ implies $A^{-1}(\operatorname{supp} \tilde{m}) \supseteq \operatorname{supp} \tilde{m}$ and $A(\operatorname{supp} \tilde{m}) \supseteq \operatorname{supp} \tilde{m}$. Hence, $A(\operatorname{supp} \tilde{m}) = \operatorname{supp} \tilde{m}$. For $F \in \mathscr{B}(\mathbb{R}^d \setminus \{0\})$,

$$(AM)(F) = \int_{\operatorname{supp} \tilde{m}} \int_0^\infty I_{A^{-1}F}(t^B x) t^{-2} \, dt \, \tilde{m}(dx)$$

$$= \int_{\operatorname{supp} \tilde{m}} \int_0^\infty I_F(t^B A x) t^{-2} \, dt \, \tilde{m}(dx)$$

$$= \int_{A(\operatorname{supp} \tilde{m})} \int_0^\infty I_F(t^B x) t^{-2} \, dt \, (A\tilde{m})(dx)$$

$$= M(F). \qquad \text{Q.E.D.}$$

Example 4.7.14. In Proposition 4.7.13, it is not possible to remove the assumption that A commutes with some exponent of μ. Consider the following example.

Let $d = 2$ and $B = \operatorname{diag}(1, 2)$. For $F \in \mathscr{B}(\mathbb{R}^2 \setminus \{0\})$, set

$$M(F) := \int_0^\infty I_F\left(t^B \begin{pmatrix} 1 \\ 1 \end{pmatrix}\right) t^{-2} \, dt.$$

Then $\operatorname{supp} M = \left\{ \begin{pmatrix} s \\ s^2 \end{pmatrix} : s \geq 0 \right\}$ is a half-parabola and, for $s > 0$, $s \cdot M = s^B M$. Let $\mu := [0, 0, M]$. Since $\operatorname{lin}(\operatorname{supp} M) = \mathbb{R}^2$, μ is full. Hence, μ is a full operator-stable measure with an exponent B, since $\operatorname{lin}(\operatorname{supp} \mu) = \operatorname{lin}(\operatorname{supp} M) = \mathbb{R}^2$.

To find $\mathbf{A}(\mu)$, note that $A \in \mathbf{A}(\mu)$ if and only if $AM = M$. However, $AM = M$ implies that $A(\mathrm{supp}\, M) = \mathrm{supp}\, M$. The only $A \in \mathrm{Aut}$ taking a half-parabola onto itself is the identity. Therefore, $\mathbf{A}(\mu) = \{I\}$, so $B = \mathrm{diag}(1, 2)$ is the unique exponent of μ. Hence, for $x = \begin{pmatrix} x_1 \\ x_2 \end{pmatrix}$,

$$\|x\|_\mu = \|x\|_B$$

$$= \int_0^1 \begin{pmatrix} t & 0 \\ 0 & t^2 \end{pmatrix} \begin{pmatrix} x_1 \\ x_2 \end{pmatrix} t^{-1}\, dt$$

$$= \int_0^1 (x_1^2 + x_2^2 t^2)^{1/2}\, dt$$

$$= \begin{cases} \|x\| & \text{if } x_2 = 0, \\ \dfrac{\|x\|}{2} & \text{if } x_1 = 0, \\ \dfrac{\|x\|}{2} + \dfrac{x_1^2}{2|x_2|} \log\left(\dfrac{|x_2| + \|x\|}{|x_1|}\right) & \text{if } x_1 \ne 0,\, x_2 \ne 0. \end{cases}$$

The intersection of $\mathrm{supp}\, M$ and S_μ is the singleton $u := \left\{\begin{pmatrix} 0.893 \\ 0.797 \end{pmatrix}\right\}$. Define $\tilde{m}$ on $\mathscr{B}(S_\mu)$ by

$$\tilde{m}(E) = \begin{cases} \dfrac{1}{0.893} & \text{if } u \in E, \\ 0 & \text{if } u \notin E. \end{cases}$$

Then

$$\int_{S_\mu} \int_0^\infty I_F(t^B x) t^{-2}\, dt\, \tilde{m}(dx)$$

$$= \tilde{m}(\{u\}) \int_0^\infty I_F(t^B u) t^{-2}\, dt$$

$$= \int_0^\infty I_F\left(t^B \begin{pmatrix} 1 \\ 1 \end{pmatrix}\right) t^{-2}\, dt = M(F),$$

that is, $\tilde{m}$ is the mixing measure for μ. Hence, all $A \in \mathrm{Aut}(\mathbb{R}^2)$ with u for a fixed point satisfies $(A\tilde{m})(E) = \tilde{m}(E)$ for all $E \subseteq \mathrm{supp}\, \tilde{m} = \{u\}$. But, the only $A \in \mathbf{A}(\mu)$ is $A = I$.

Remark 4.7.15. Instead of the norm $\|\cdot\|_\mu$ given in (4.7.1), one can use the norm induced by the inner product

$$\langle x, y \rangle_1 := \int_{S_\mu} \int_0^1 \langle At^B x, At^B y \rangle t^{-1} \, dt \, H(dA),$$

where $\langle \cdot, \cdot \rangle$ is the usual inner product on $\mathbb{R}^d$.

Another possible norm uses

$$\|x\|_{p,B} := \left(\int_0^1 \|t^B x\|^p t^{-1} \, dt \right)^{1/p}, \qquad p \geq 1.$$

In particular, when $p = 2$, this norm is easy to compute in the previous example.

4.8 ELLIPTICALLY SYMMETRIC OPERATOR-STABLE MEASURES

For a full Gaussian measure μ on $\mathbb{R}^2$ with covariance operator the identity, the level curves of its density are circles and its symmetry group $\mathbf{A}(\mu)$ is the group $\mathcal{O}$ of all orthogonal operators. For a general invertible covariance operator R, the level curves become ellipses and the symmetry group becomes $S^{-1} \mathcal{O} S$, where S is the positive-definite self-adjoint square root of R^{-1} (cf. Proposition 2.3.11 and Theorem 4.6.10). This suggests the following definition. A measure μ on $\mathbb{R}^d$ is said to be *elliptically symmetric* provided its symmetry group $\mathbf{A}(\mu)$ is W-conjugate to $\mathcal{O}$ for some $W \in \mathbf{Aut}(\mathbb{R}^d)$, that is, $\mathbf{A}(\mu) = W^{-1} \mathcal{O} W$.

The characteristic function of a full operator-stable elliptically symmetric measure μ has a particularly simple form which shows that μ is either purely Gaussian or has no Gaussian component. This can also be obtained from the special form of the class of its exponents, $\mathscr{E}_s(\mu)$.

Theorem 4.8.1. *Let μ be full and operator-stable on $\mathbb{R}^d$ and elliptically symmetric, that is, $\mathbf{A}(\mu) = W^{-1} \mathcal{O} W$ for some $W \in \mathbf{Aut}$. Then there is $c \geq 1/2$ such that*

$$\mathscr{E}_s(\mu) = cI + W^{-1} \mathcal{K} W,$$

where $\mathcal{K}$ is the class of all skew-symmetric operators. Furthermore, μ is purely Gaussian if and only if $c = 1/2$; and μ has no Gaussian component if and only if $c > 1/2$.

Proof. Assume $W = I$. Select $B \in \mathscr{E}_{cs}(\mu)$. Then B commutes with all of $\mathbf{A}(\mu) = \mathscr{O}$. Hence, $A^*B^* = B^*A^*$ for all $A \in \mathscr{O}$, so B^* also commutes with all of $\mathscr{O}$. Consequently, $AB^*B = B^*BA$ for all $A \in \mathscr{O}$, that is, B^*B commutes with all of $\mathscr{O}$. By Schur's lemma, $B^*B = c_1 I$ for some constant c_1. Hence, B and B^* commute and both B and B^* commute with all of $\mathscr{O}$. By a corollary of Schur's lemma, $B = cI$ for some constant c. By Theorem 4.6.12, $c \geq 1/2$. Since $\mathscr{T}(\mathscr{O}) = \mathscr{K}$ (cf. Section 1.5), by Theorem 4.6.7, $\mathscr{E}_s(\mu) = cI + \mathscr{K}$ when $W = I$.

For general W, $\mathbf{A}(\mu) = W^{-1}\mathscr{O}W$ implies that $\mathbf{A}(W\mu) = \mathscr{O}$, so by Proposition 4.6.6(b) we have that $\mathscr{E}_s(W\mu) = cI + \mathscr{K}$ which implies $\mathscr{E}_s(\mu) = cI + W^{-1}\mathscr{K}W$.

The last statement of the theorem follows from Theorem 4.6.5.
Q.E.D.

Now, we proceed to the characterization of elliptically symmetric operator-stable measures in terms of characteristic functions. We begin with two auxiliary lemmas.

Lemma 4.8.2. *If $\mathbf{A}(\mu) = \mathscr{O}$, then*

(a) $\|x\| = c\|x\|_\mu$ *for some $c \geq 1/2$;*
(b) $S_\mu = S_c := \{x \in \mathbb{R}^d : \|x\| = c\}$;
(c) *when $c > 1/2$ and $\mu = [a, 0, M]$, there is a constant $k > 0$ such that, for all M-integrable functions f,*

$$\int_{\mathbb{R}^d \setminus \{0\}} f(x) M(dx) = k \int_{S_1} \int_0^\infty f(tx) t^{-(1+1/c)} \, dt \, m_1(dx),$$

where S_1 is the usual unit sphere in $\mathbb{R}^d$ and m_1 is a normalized $(d-1)$-dimensional Hausdorff measure on $\mathscr{S}_1$ [cf. Billingsley (1986), Section 19].

Proof

(a) By Theorem 4.8.1, $cI \in \mathscr{E}_s(\mu)$ for some $c \geq 1/2$. Since $\|x\|_{cI} = \int_0^1 \|s^c x\| s^{-1} \, ds = c^{-1}\|x\|$, we have

$$\|x\|_\mu = \int \|Ax\|_{cI} H(dA) = c^{-1}\|x\|.$$

(b) This follows immediately from (a).

(c) Let $\tilde{m}$ be the measure on $S_\mu = S_c$ given in Proposition 4.7.9. By Corollary 4.7.11, $A\tilde{m} = \tilde{m}$ for all $A \in \mathbf{A}(\mu) = \mathcal{O}$. Therefore, $\tilde{m} = k_1 \cdot m_c$ for some constant $k_1 > 0$, where m_c is a normalized Haar measure on S_c. Hence, using $m_c(cA) = c^{d-1} m_1(A)$ for $A \subseteq S_1$,

$$\int_{\mathbb{R}^d \setminus \{0\}} f(x) M(dx) = k_1 \int_{S_c} \int_0^\infty f(s^c x) s^{-2} \, ds \, m_c(ds)$$

$$= k_1 c^{d-1} \int_{S_1} \int_0^\infty f(s^c c x) s^{-2} \, ds \, m_1(dx)$$

$$= k_1 c^{d-2+1/c} \int_{S_1} \int_0^\infty f(tx) t^{-(1+1/c)} \, dt \, m_1(dx).$$

Q.E.D.

Lemma 4.8.3. *For $c > 1/2$ and $y \in \mathbb{R}^d$, we have*

$$\int_{S_1} \int_0^\infty \left(e^{it\langle y, x \rangle} - 1 - \frac{it\langle y, x \rangle}{1+t^2} \right) t^{-(1+1/c)} \, dt \, M_1(dx)$$
$$= -\beta \|y\|^{1/c},$$

where $\beta > 0$ depends on c.

Proof. Let $T(y)$ denote the term on the left and consider the three cases: $c > 1$, $c < 1$, and $c = 1$. First, $c > 1$. Since $\int_{S_1} \langle y, x \rangle m_1(dx) = 0$, we have

$$T(y) = \int_{S_1} \int_0^\infty (e^{it\langle y, x \rangle} - 1) t^{-(1+1/c)} \, dt \, m_1(dx).$$

Let $k_1 := \int_0^\infty (e^{it} - 1) t^{-(1+1/c)} \, dt$. Since $c > 1$, we have $|k_1| < \infty$, and by the Cauchy theorem, $k_1 = e^{-i\pi/2c} \int_0^\infty (e^{-t} - 1) t^{-(1+1/c)} \, dt$, so $\operatorname{Re}(k_1) < 0$. Hence,

$$\int_0^\infty (e^{it\langle y, x \rangle} - 1) t^{-(1+1/c)} \, dt = \begin{cases} k_1(\langle y, x \rangle)^{1/c} & \text{if } \langle y, x \rangle \geq 0, \\ \overline{k_1} |\langle y, x \rangle|^{1/c} & \text{if } \langle y, x \rangle < 0, \end{cases}$$

where $\bar{k}_1$ is the complex conjugate of k_1. Consequently,

$$T(y) = k_1 \int_{S_1^+} (\langle y, x \rangle)^{1/c} m_1(dx) + \bar{k}_1 \int_{S_1^-} |\langle y, x \rangle|^{1/c} m_1(dx),$$

where $S_1^+ := \{x \in S_1: \langle y, x \rangle > 0\}$ and $S_1^- := \{x \in S_1: \langle y, x \rangle < 0\}$. Letting $y' := y/\|y\|$, we have

$$T(y) = \|y\|^{1/c} \left(k_1 \int_{S_1^+} (\langle y', x \rangle)^{1/c} m_1(dx) + \bar{k}_1 \int_{S_1^-} |\langle y', x \rangle|^{1/c} m_1(dx) \right).$$

But, by the $\mathcal{O}$-invariance of m_1, these two integrals are the same with a common positive value independent of $y' \in S_1$. Therefore, $T(y) = -\beta \|y\|^{1/c}$, where $0 < \beta := (\text{Re}(k_1)) \int_{S_1} |\langle u, x \rangle|^{1/c} m_1(dx)$ is constant for any $u \in S_1$.

Second, $c < 1$. Since $1/2 \le c < 1$, the integral

$$\int_0^\infty (1 + t^2)^{-1} t^{2-1/c} \, dt < \infty.$$

Hence, $T(y)$ may be written as

$$T(y) = \int_{S_1} \int_0^\infty \left(e^{it\langle y, x \rangle} - 1 - it\langle y, x \rangle \right) t^{-(1+1/c)} \, dt \, m_1(dx)$$

$$+ i \int_{S_1} \int_0^\infty \langle y, x \rangle (1 + t^2)^{-1} t^{2-1/c} \, dt \, m_1(dx).$$

By the Fubini theorem and the $\mathcal{O}$-invariance of m_1, the last integral is equal to zero. Note that, since $1/2 \le c < 1$,

$$k_2 := \int_0^\infty (e^{it} - 1 - it) t^{-(1+1/c)} \, dt$$

$$= e^{-i\pi/2c} \int_0^\infty (e^{-t} - 1 + t) t^{-(1+1/c)} \, dt,$$

so $|k_2| < \infty$; moreover, $\text{Re}(k_2) < 0$. Hence,

$$\int_0^\infty \left(e^{it\langle y, x \rangle} - 1 - it\langle y, x \rangle \right) t^{-(1+1/c)} \, dt$$

$$= \begin{cases} k_2 (\langle y, x \rangle)^{1/c} & \text{if } \langle y, x \rangle \ge 0, \\ \bar{k}_2 |\langle y, x \rangle|^{1/c} & \text{if } \langle y, x \rangle < 0. \end{cases}$$

Similar to the argument when $c > 1$, this implies that $T(y) = -\beta \|y\|^{1/c}$, where $0 < \beta := -(\operatorname{Re}(k_2)) \int_{S_1} |\langle u, x \rangle|^{1/c} m_1(dx)$ is constant for any $u \in S_1$.

Finally, $c = 1$. Rewrite the inner integral in $T(y)$ in terms of its real and imaginary parts

$$\int_0^\infty (\cos(t\langle y, x \rangle) - 1) t^{-2} \, dt + i \int_0^\infty \left(\sin(t\langle y, x \rangle) - \frac{t\langle y, x \rangle}{1 + t^2} \right) t^{-2} \, dt.$$

Since $\int_0^\infty (\cos t - 1) t^{-2} \, dt = -\pi/2$, the first integral is equal to $-|\langle y, x \rangle| \pi/2$. For the second integral, consider the case when $\langle y, x \rangle > 0$. Then it is equal to

$$\lim_{\varepsilon \downarrow 0} \left(\langle y, x \rangle \int_{\varepsilon \langle y, x \rangle}^\infty \frac{\sin t}{t^2} \, dt - \langle y, x \rangle \int_\varepsilon^\infty \frac{1}{t(1 + t^2)} \, dt \right)$$

$$= \langle y, x \rangle \lim_{\varepsilon \downarrow 0} \left(-\int_\varepsilon^{\varepsilon \langle y, x \rangle} \frac{\sin t}{t^2} \, dt + \int_\varepsilon^\infty \left(\frac{\sin t}{t^2} - \frac{1}{t(1 + t^2)} \right) dt \right)$$

$$= \langle y, x \rangle (-\log \langle y, x \rangle + k_3),$$

since

$$\lim_{\varepsilon \downarrow 0} \int_\varepsilon^{\varepsilon a} \frac{\sin t}{t^2} \, dt = \log|a|$$

and

$$k_3 := \int_0^\infty \left(\frac{\sin t}{t^2} - \frac{1}{t(1 + t^2)} \right) dt$$

is a finite real constant. Similarly, when $\langle y, x \rangle < 0$, the second integral is equal to $\langle y, x \rangle (-\log|\langle y, x \rangle| + k_3)$. Hence,

$$T(y) = \int_{S_1} \left(-\frac{\pi}{2} |\langle y, x \rangle| - i \langle y, x \rangle \log|\langle y, x \rangle| + i k_3 \langle y, x \rangle \right) m_1(dx).$$

The last two integrals are both equal to zero, so $T(y) = -\beta \|y\|$, where $0 < \beta := (\pi/2) \int_{S_1} |\langle u, x \rangle| m_1(dx)$ is constant for any $u \in S_1$.

Q.E.D.

Theorem 4.8.4. *A measure μ on $\mathbb{R}^d$ is full, elliptically symmetric, and operator-stable if and only if*

$$\hat{\mu}(y) = \exp\{i\langle a, y \rangle - \beta \|W^{*-1}y\|^\alpha\}$$

for all $y \in \mathbb{R}^d$, for some $W \in \text{Aut}$, $a \in \mathbb{R}^d$, $\beta > 0$, and $\alpha \in (0, 2]$; in which case, $\mathbf{A}(\mu) = W\mathcal{O}W^{-1}$.

Proof. By Theorem 4.8.1, $cI \in \mathcal{E}_s(\mu)$; and μ is purely Gaussian when $c = 1/2$, and μ has no Gaussian component when $c > 1/2$.

Assume $c = 1/2$. Then μ is Gaussian and $\mathbf{A}(W\mu) = \mathcal{O}$. Hence, $(W\mu)\hat{\ }(y) = \exp\{i\langle b, y\rangle - (1/2)\beta'\|y\|^2\}$ for some $\beta' > 0$, because the covariance operator of $W\mu$ commutes with all of $\mathcal{O}$. Therefore, $\hat{\mu}(y) = \exp\{i\langle W^{-1}b, y\rangle - \beta\|W^{*-1}y\|^2\}$.

Assume $c > 1/2$ and $\mathbf{A}(\mu) = \mathcal{O}$. By Lemmas 4.8.2 and 4.8.3, $\hat{\mu}(y) = \exp\{i\langle a, y\rangle - \beta\|y\|^{1/c}\}$, for some $a \in \mathbb{R}^d$ and $\beta > 0$. Let $\alpha := 1/c$, so $0 < \alpha < 2$.

Assume $c > 1/2$ and $\mathbf{A}(\mu) = W^{-1}\mathcal{O}W$. Analogous to the case $c = 1/2$, we obtain the desired result.

For the converse part, fullness of μ follows from Proposition 2.1.1(c). Furthermore, μ is operator-stable with the exponent αI. Finally, $\mathcal{O} = \mathbf{A}(W^{-1}\mu)$. Q.E.D.

In Theorem 4.8.1, an exponent of the form cI, $c \geq 1/2$, was obtained by applying Schur's lemma and one of its corollaries via the orthogonal group. However, by using the full force of Schur's lemma, the assumption of elliptically symmetric may be slightly relaxed. First, let us note the following simple observation.

Remark 4.8.5. Let $A, B \in \text{Aut}$ be W-conjugate, that is, $A = W^{-1}BW$, with $W \in \text{Aut}$. Then a subspace V of $\mathbb{R}^d$ is A-invariant if and only if $W(V)$ is B-invariant. Furthermore, $\{0\}$ and $\mathbb{R}^d$ are the only A-invariant subspaces if and only if they are only B-invariant subspaces.

Proof. Note that $AV = V$ is equivalent to $B(W(V)) = W(V)$. Also, $W(\{0\}) = \{0\}$ and $W(\mathbb{R}^d) = \mathbb{R}^d$. Q.E.D.

Theorem 4.8.6. *Assume that μ is a full operator-stable measure and the only subspaces of $\mathbb{R}^d$ which are $\mathbf{A}(\mu)$-invariant are $\{0\}$ and $\mathbb{R}^d$. Then there are $c \geq 1/2$ and a positive-definite self-adjoint W such that, for* each $B \in \mathscr{E}_{cs}(\mu)$, one can find a skew-symmetric K_B such that

$$B = cI + W^{-1}K_B W.$$

Furthermore, either $K_B = 0$ or $K_B^2 = -\gamma_B I$ for some $\gamma_B > 0$.

Proof. Since $\mathbf{A}(\mu)$ is compact, $\mathbf{A}(\mu) = W\mathscr{O}_0 W^{-1}$ for some closed subgroup $\mathscr{O}_0$ of the orthogonal group and for some positive-definite self-adjoint W (cf. Corollary 2.4.2). By Remark 4.8.5, the only $\mathscr{O}_0$-invariant subspaces of $\mathbb{R}^d$ are $\{0\}$ and $\mathbb{R}^d$. Let $B \in \mathscr{E}_{cs}(\mu)$ and set $B_0 := W^{-1}BW$. Then $B_0 \in \mathscr{E}_{cs}(W^{-1}\mu)$ and $\mathbf{A}(W^{-1}\mu) = \mathscr{O}_0$ (cf. Lemmas 4.7.3 and 2.2.6, respectively). Let $B_1 := (B_0 + B_0^*)/2$ and $B_2 := (B_0 - B_0^*)/2$. Then $B_0 = B_1 + B_2$ with B_1 self-adjoint and B_2 skew-symmetric. Since B_0 commutes with the orthogonal subgroup $\mathscr{O}_0$, so does B_0^*, as $A^* = A^{-1}$ for $A \in \mathscr{O}_0$. Hence, both B_1 and B_2 commute with $\mathscr{O}_0$. Since B_1 is self-adjoint, by Schur's lemma $B_1 = cI$ for some $c \in \mathbb{R}^1$.

Since B_2 is skew-symmetric, its minimal polynomial is the product of $k \geq 1$ distinct irreducible polynomials, say $p_1, \ldots, p_k$. Furthermore, since it commutes with $\mathscr{O}_0$, $\ker p_i(B_2)$ is an $\mathscr{O}_0$-invariant subspace of $\mathscr{O}_0$, $1 \leq i \leq k$. Thus, $\ker p_i(B_2)$ is either $\{0\}$ or $\mathbb{R}^d$. Therefore, $k = 1$ and $p_1(x) = x$ or $p_1(x) = x^2 + \gamma$ for some $\gamma > 0$. Note that a skew-symmetric operator has purely imaginary eigenvalues. Consequently, $B_0 = cI + B_2$, where either $B_2 = 0$ or $B_2^2 = -\gamma I$. Finally, we conclude that $B = cI + W^{-1}K_B W$, where K_B is skew-symmetric and either $K_B = 0$ or $K_B^2 = -\gamma I$ for some $\gamma > 0$. From Theorem 4.6.5, we obtain $c \geq 1/2$. Q.E.D.

Corollary 4.8.7. *Additionally, if either d is odd or B is diagonalizable, then $B = cI$.*

Proof. Suppose d is odd. We know B_2 is skew-symmetric. Then $\det(B_2) = 0$. Hence, $\det(B_2^2) = 0$, so B_2^2 cannot be $-\gamma I$ for some $\gamma > 0$. Therefore, $B_2 = 0$ and $B = cI$.

Now, suppose B is diagonalizable. Let λ and v be an eigenvalue and corresponding eigenvector of B, respectively. Then, by Theorem 4.8.6, $WK_B W^{-1}v = (B - cI)v = (\lambda - c)v$. Hence, $\lambda - c$ is an eigenvalue of K_B. Since c and λ are real numbers and since the real part of any eigenvalue of a skew-symmetric operator is equal to zero, $c = \lambda$ for every eigenvalue λ of B, that is, $B = cI$. Q.E.D.

4.9 THE CENTERING FUNCTION $b(t)$

In the characterization of a full operator-stable measure μ given in Theorem 4.2.12, that is, $\mu^t = t^B\mu * \delta(b(t))$ for all $t > 0$, the centering function $b: (0, \infty) \to \mathbb{R}^d$ is unspecified. Its form can be described if 1 is not an eigenvalue of B. In this case, μ may be centered to eliminate b. This is analogous to the one-dimensional case since $B = 1$ gives the Cauchy distribution.

Lemma 4.9.1. *Let μ be t^B-stable with centering function $b: (0, \infty) \to \mathbb{R}^d$. Then, for all $t, s > 0$,*

$$b(st) = tb(s) + s^B b(t) \tag{4.9.1}$$

and $b(1) = 0$.

Proof. For $t, s > 0$,

$$\left(s^B\mu * \delta(b(s))\right)^t = s^B\mu^t * \delta(tb(s)) = \mu^{st}$$
$$= (st)^B\mu * \delta(b(st)) = s^B\mu^t * \delta(b(st) - s^B b(t)).$$

Therefore, $tb(s) = b(st) - s^B b(t)$. Clearly, $b(1) = 0$. Q.E.D.

Theorem 4.9.2. *Let μ be t^B-stable with centering function $b: (0, \infty) \to \mathbb{R}^d$ and assume that 1 is not an eigenvalue of B. Then for some $x_0 \in \mathbb{R}^d$ we have, for all $t > 0$,*

$$b(t) = (tI - t^B)x_0. \tag{4.9.2}$$

Proof. When $b(t) = 0$ for all $t > 0$, select $x_0 = 0$. Now, we only consider functions which nontrivially satisfy (4.9.1). Suppose $c: (0, \infty) \to \mathbb{R}^d$ satisfies (4.9.1) and $c(t_0) = 0$ for some $t_0 \neq 1$. Then $t_0^B c(t_0^{-1}) = c(t_0^{-1} t_0) = t_0 c(t_0^{-1})$, so t_0 is an eigenvalue of t_0^B. But, 1 is not an eigenvalue of B, so t_0 cannot be an eigenvalue of t_0^B. Hence, for any nontrivial solution c of (4.9.1), $c(t) \neq 0$ for all $t \neq 1$.

Now, assume c_1 and c_2 satisfy (4.9.1). Then $c := c_1 - c_2$ also satisfies (4.9.1). Therefore, if two functions satisfying (4.9.1) agree for some $t \neq 1$, then they agree for all $t > 0$. Clearly, the function $(tI - t^B)x$, for fixed $x \in \mathbb{R}^d$, satisfies (4.9.1). Given the centering function b, let $x_1 := b(2)$ and $(2I - 2^B)x_0 := x_1$. Since 1 is not an eigenvalue of B, x_0 is unique and always exists. Then b and the function $(t - t^B)x_0$ satisfy (4.9.1) and agree at $t = 2$. Therefore, $b(t) = (t - t^B)x_0$ for all $t > 0$. Q.E.D.

Corollary 4.9.3. *Let μ be as in Theorem 4.9.2 with centering function given by (4.9.2). Let $\nu := \mu * \delta(-x_0)$. Then $\nu^t = t^B \nu$ for all $t > 0$.*

The proof of this corollary is a simple calculation.

Remark 4.9.4. When 1 is an eigenvalue of B and μ is t^B-stable, it may not be possible to find an $x_0 \in \mathbb{R}^d$ so that $(\mu * \delta(-x_0))^t = t^B(\mu * \delta(-x_0))$ for all $t > 0$. Consider the following example. Let $d = 1$, $B = I$, and $\hat{\mu}(y) = \exp\{-|y| - i(2/\pi)y \log|y|\}$ for $y \in \mathbb{R}^1$. Then μ is an asymmetric Cauchy distribution. For $b(t) := (2/\pi)t \log t$ for $t > 0$, we have $\mu^t = t^I \mu * \delta(b(t))$ for all $t > 0$, so μ is t^I-stable with centering function b. It is easy to see that there does not exist an x_0 as in Corollary 4.9.3.

4.10 ABSOLUTE CONTINUITY OF FULL OPERATOR-STABLE MEASURES

In Theorem 3.8.9, it was shown that each full $\exp(-tQ)$-decomposable measure is absolutely continuous. It follows that each full operator-stable measure is also absolutely continuous. However, the proof in Section 3.8 was quite complicated, whereas a simpler proof for operator-stable measures is possible. This is done in this section. Moreover, it will be shown that, for each positive s and t, the functions $\|y\|^s |\hat{\mu}(y)|^t$ are Lebesgue integrable on $\mathbb{R}^d$. Thus, the probability density functions of μ^t, $t > 0$, are bounded and have partial derivatives of all orders [cf. Lukacs (1960)].

Lemma 4.10.1. *Let μ be full t^B-stable on $\mathbb{R}^d$ and let $\psi(y) := -\log|\hat{\mu}(y)|$ for $y \in \mathbb{R}^d$. Then there are positive a, b, and c such that for, $\|y\| \geq a$,*

$$\psi(y) \geq c\|y\|^b.$$

Proof. Since μ is infinitely divisible, ψ is finite everywhere. Also, μ being t^B-stable implies that $t\psi(y) = \psi(t^{B^*}y)$ for all $t > 0$ and $y \in \mathbb{R}^d$.

By Corollary 4.2.14, $\psi(y) > 0$ for all $y \in \mathbb{R}^d \setminus \{0\}$. Let $S := \{u \in \mathbb{R}^d: \|u\|_B = 1\}$ and $c := \min\{\psi(u): u \in S\}$, where $\|\cdot\|_B$ is the norm defined in Section 3.4. Since ψ is continuous and μ is full, we have $c > 0$. By Proposition 3.4.3, for each $y \neq 0$ there are unique $t_y > 0$ and $u_y \in S$ such that $y = t_y^{B^*} u_y$. If $\|y\|_B \geq 1$, then $t_y \geq 1$, so that $\|y\|_B \leq t_y^{\|B\|_B}$, and hence,

$$\psi(y) = \psi(t_y^{B^*} u_y) = t_y \psi(u_y) \geq c t_y \geq c \|y\|_B^{1/\|B\|_B}.$$

Since all norms on $\mathbb{R}^d$ are equivalent, the above inequality implies the desired conclusion. Q.E.D.

Theorem 4.10.2. *Let μ be full t^B-stable on $\mathbb{R}^d$. Then*

(a) *for all positive s and t,*

$$\int_{\mathbb{R}^d} \|y\|^s |\hat{\mu}(y)|^t \, dy < \infty;$$

(b) *for each $t > 0$, the measure μ^t has a bounded probability density function p_t which has partial derivatives of all orders;*

(c) *for all $t > 0$ and all $x \in \mathbb{R}^d$, we have*

$$p_t(x) = t^{-\beta} p_1(t^{-B}(x - b(t))),$$

where $b(\cdot)$ is the centering function for μ and $\beta := $ trace of B.

Proof

(a) Let a, b, and c be as given in Lemma 4.10.1 and set $\Delta := \{y \in \mathbb{R}^d: \|y\| > a\}$. Then we have

$$\int_\Delta \|y\|^s (\psi(y))^{-t} \, dy \leq c^{-t} \int_\Delta \|y\|^{-(tb-s)} \, dy$$

for all positive s and t. This last integral is finite for t sufficiently large. Therefore, for such t,

$$\int_\Delta \int_0^f r^{t-1} \mathbb{W} \, \mathbb{V} e^{-r\psi(y)} \, dr \, dy = \Gamma(t) \int_\Delta \mathbb{W} \, \mathbb{V} (\psi(y))^{-t} \, dy \, \mathbb{W} \, f \, ,$$

where $\Gamma(t)$ is Euler's gamma function. By the Fubini theorem, for a.a. $r > 0$,

$$\int_\Delta \|y\|^s e^{-r\psi(y)}\, dy < \infty.$$

But, this is a nonincreasing function at r, so the integral is finite for all $r > 0$. Clearly,

$$\int_{\mathbb{R}^d \setminus \Delta} \|y\|^s e^{-t\psi(y)}\, dy < \infty.$$

(b) For each $t > 0$, μ^t has a bounded probability density function p_t since its characteristic function is integrable. Also, p_t has partial derivatives of all orders since all "moments" of its characteristic function are finite.

(c) Simply calculate as follows:

$$\begin{aligned} p_t(x) &= \frac{1}{2\pi} \int_{\mathbb{R}^d} e^{-i\langle x, y\rangle} \hat{\mu}^t(y)\, dy \\ &= \frac{1}{2\pi} \int_{\mathbb{R}^d} e^{-i\langle x-b(t), y\rangle} \hat{\mu}(t^{B^*} y)\, dy \\ &= \frac{t^{-\beta}}{2\pi} \int_{\mathbb{R}^d} e^{-i\langle t^{-B}(x-b(t)), y\rangle} \hat{\mu}(y)\, dy \\ &= t^{-\beta} p_1(t^{-B}(x - b(t))). \end{aligned}$$ Q.E.D.

4.11 DOMAINS OF NORMAL ATTRACTION OF OPERATOR-STABLE MEASURES

In the definition of an operator-stable measure μ, it is assumed that

$$A_n \nu^n * \delta(a_n) \Rightarrow \mu, \qquad (4.11.1)$$

with $\langle A_n; a_n \rangle \in \mathscr{A}$ and $\nu \in \mathscr{P}$. The obvious problem is, given μ, identify those measures ν for which (4.11.1) holds. In this generality, we have the *generalized domain of attraction*, that is, $\nu \in \text{GDOA}(\mu)$ provided (4.11.1) holds for some sequence $\{\langle A_n; a_n \rangle\}$ in $\mathscr{A}$. This definition places no restriction on the form of A_n. Of special interest, suggested by the univariate case, is when $A_n = n^{-B}$ for some $B \in \mathcal{E}_s(\mu)$. This gives the *domain of normal attraction*, that is, $\nu \in \text{DONA}(\mu)$ provided, for some sequence $\{a_n\}$ in $\mathbb{R}^d$,

$$n^{-B}\nu^n * \delta(a_n) \Rightarrow \mu. \tag{4.11.2}$$

Hence, $\text{DONA}(\mu) \subset \text{GDOA}(\mu)$, and the containment is proper. Also, $\text{DONA}(\mu) \neq \emptyset$ when μ is full since μ itself is in $\text{DONA}(\mu)$ (cf. Theorem 4.2.12).

We use the following notation (cf. Theorem 4.6.5). Let μ be full and operator-stable on $\mathbb{R}^d$ with $B \in \mathcal{E}_s(\mu)$. Decompose $\mu = \mu_1 * \mu_2$ and $\mathbb{R}^d = V_1 \oplus V_2$, where μ_1 is Gaussian with $\text{supp}\,\mu_1 = V_1$ and μ_2 has no Gaussian component with $\text{lin}(\text{supp}\,\mu_2) = V_2$. Note that V_1 and V_2 need not be orthogonal. Let P_i, $i = 1, 2$, be the idempotents on V_i. Then $P_i(\mathbb{R}^d) = V_i$, $i = 1, 2$, and $P_1 P_2 = 0 = P_2 P_1$.

Proposition 4.11.1. *The exponent B commutes with each P_i, $i = 1, 2$. Thus, so does t^B, $t > 0$.*

Proof. For each $x \in \mathbb{R}^d$, $Bx = BP_1 x + BP_2 x$ and $Bx = P_1 Bx + P_2 Bx$. Since V_1 and V_2 are B-invariant subspaces, $BP_i x = P_i Bx$, $i = 1, 2$. Q.E.D.

Lemma 4.11.2. *Let μ be an operator-stable measure as described above. If ρ is an infinitely divisible measure on $\mathbb{R}^d$ such that $P_i \rho = \mu_i$, $i = 1, 2$, then $\rho = \mu$.*

Proof. Let $\mu = [a, R, M]$ and $\rho = [b, S, N]$. Recall that $A\rho = [b^*, ASA^*, AN]$ for $A \in \text{Aut}$, for some $b^* \in \mathbb{R}^d$ (cf. Section 3.4). Since $P_2 \rho = \mu_2$, we have

$$P_2 S P_2^* = 0 \quad \text{and} \quad P_2 N = M.$$

From $P_1 \rho = \mu_1$, we obtain

$$P_1 S P_1^* = R \quad \text{and} \quad P_1 N = 0.$$

Hence, $\text{supp}\, N \subset V_2$, so $N = P_2 N = M$.

Let $S^{1/2}$ denote the self-adjoint square root of S. Then $0 = P_2 S P_2^* = (P_2 S^{1/2})(P_2 S^{1/2})^*$, so $P_2 S^{1/2} = 0$. Hence, $P_2 S = 0 = SP_2^*$. Consequently, $S = (P_1 + P_2) S (P_1 + P_2)^* = P_1 S P_1^* = R$.

From the uniqueness of the representation $\mu = [a, R, M]$, it now follows that $a = b$. Therefore, $\rho = \mu$. Q.E.D.

We now restrict our attention to the domains of normal attraction. We find necessary and sufficient conditions for $\nu \in \mathscr{P}(\mathbb{R}^d)$ to be in DONA(μ), that is, ν satisfies (4.11.2). We also show that for two different full operator-stable distributions either their domains of normal attraction are disjoint or the distributions are of the same type, in which case their domains of normal attraction coincide.

It would appear that the DONA(μ) might depend on the choice of the exponent B of μ. This is not the case.

Proposition 4.11.3. *Assume that ν satisfies (4.11.2) for a particular exponent B of μ. Then ν satisfies (4.11.2) for every exponent of μ.*

Proof. Assume ν satisfies (4.11.2) for some $B \in \mathscr{E}_s(\mu)$ and $a_n \in \mathbb{R}^d$ and let $B' \in \mathscr{E}_s(\mu)$. Since both B and B' are in $\mathscr{E}_s(\mu)$, we have $n^B \mu * \delta(b_n) = \mu^n = n^{B'}\mu * \delta(b'_n)$ for every $n \geq 1$, for some $b_n, b'_n \in \mathbb{R}^d$. Hence, with $c_n := n^{-B}(b_n - b'_n)$, we have $\alpha_n := \langle n^{-B}n^{B'}; c_n \rangle \in \text{Inv}(\mu)$ for all $n \geq 1$ (cf. Section 2.2.5). Also, $(n^{-B}n^{B'})(n^{-B'}\nu^n) * \delta(a_n) \Rightarrow \mu$, that is, $\alpha'_n(n^{-B'}\nu^n) \Rightarrow \mu$, where $\alpha'_n := \langle n^{-B}n^{B'}; a_n \rangle$. Since $\alpha_n^{-1} \in \text{Inv}(\mu)$, we have, from Theorem 2.2.7, $\alpha_n^{-1}(\alpha'_n(n^{-B'}\nu^n)) \Rightarrow \mu$, that is, $n^{-B'}\nu^n * \delta(a'_n) \Rightarrow \mu$, where $a'_n := n^{-B'}(n^B a_n - b_n + b'_n)$. Therefore, ν satisfies (4.11.2) substituting B' and a'_n for B and a_n, respectively. Q.E.D.

Since μ can be decomposed into its Gaussian component μ_1 on V_1 and its Poisson component μ_2 on V_2, it is convenient to consider the projections of ν onto V_1 and V_2. For ν to be in DONA(μ), the projections $P_1 \nu$ and $P_2 \nu$ must behave properly.

Theorem 4.11.4. *We have $\nu \in \text{DONA}(\mu_1 * \mu_2)$ if and only if $P_i \nu \in \text{DONA}(\mu_i)$ for $i = 1, 2$.*

Proof. Assume $\nu \in \text{DONA}(\mu)$, that is, (4.11.2) holds. By Theorem 4.6.5, with $B_i := B|V_i$, μ_i is t^{B_i}-stable on V_i for $i = 1, 2$. By Proposition 4.11.1, n^{-B_i} commutes with P_i. Therefore, $n^{-B_i}(P_i \nu)^n * \delta(P_i a_n) = P_i(n^{-B}\nu^n * \delta(a_n)) \Rightarrow P_i \mu$, that is, $P_i \nu \in \text{DONA}(\mu_i)$ for $i = 1, 2$.

Now, assume $P_i \nu \in \text{DONA}(\mu_i)$ for $i = 1, 2$. Then

$$n^{-B_i}(P_i \nu)^n * \delta(b_{in}) \Rightarrow \mu_i$$

for some sequences $\{b_{1n}\}$ and $\{b_{2n}\}$ in V_1 and V_2, respectively, where

$B_i := B|V_i$ and $B \in \mathcal{E}_s(\mu)$. Consequently, $P_i(n^{-B}\nu^n * \delta(a_n)) \Rightarrow \mu_i$ for $i = 1, 2$, where $a_n := b_{1n} + b_{2n}$. Hence, the sequence $\{n^{-B}\nu^n * \delta(a_n)\}$ is tight. Let $\rho \in \mathcal{P}(\mathbb{R}^d)$ be the limit of some subsequence of $\{n^{-B}\nu^n * \delta(a_n)\}$. Then ρ is infinitely divisible and $P_i \rho = \mu_i$ for $i = 1, 2$. By Lemma 4.11.3, $\rho = \mu$. Therefore, $n^{-B}\nu^n * \delta(a_n) \Rightarrow \mu$. Q.E.D.

To state our necessary and sufficient condition for $\nu \in \mathrm{DONA}(\mu)$, recall that if $\mu = [a, R, M]$ is full and operator-stable with an exponent B, then, for every $E \in \mathcal{B}(\mathbb{R}^d \setminus \{0\})$,

$$M(E) = \int_{S_\mu} \int_0^\infty I_E(t^B x) t^{-2} \, dt \, \tilde{m}(dx), \qquad (4.11.3)$$

where $S_\mu = \{x \in \mathbb{R}^d : \|x\|_\mu = 1\}$ and $\tilde{m}$ is a finite measure on $\mathcal{B}(S_\mu)$ with $\mathrm{supp}\, \tilde{m} \subseteq V_2$ (cf. Propositions 4.7.9 and 4.3.4, Remark 4.7.12, and Theorem 4.3.7).

Remark. In what follows, one could substitute the norm $\|\cdot\|_B$ and the measure m_B, with a fixed $B \in \mathcal{E}_s(\mu)$.

From the central limit theorem (cf. Proposition 1.8.17), in order for $\nu \in \mathrm{DONA}(\mu)$ it is necessary and sufficient that for some $B \in \mathcal{E}_s(\mu)$ we have

$n \cdot n^{-B}\nu(G) \to M(G)$ for every Borel set G which is bounded away from the origin and $M(\partial G) = 0$,

$$(4.11.4)$$

and, for each $y \in \mathbb{R}^d$,

$$\lim_{\varepsilon \downarrow 0} \left\{ \begin{array}{c} \liminf_{n \to \infty} \\ \limsup_{n \to \infty} \end{array} \right\} n \left[\int_{\|x\| < \varepsilon} (\langle y, x \rangle)^2 (n^{-B}\nu)(dx) \right.$$

$$\left. - \left(\int_{\|x\| < \varepsilon} \langle y, x \rangle (n^{-B}\nu)(dx) \right)^2 \right] = \langle Ry, y \rangle. \quad (4.11.5)$$

From Theorem 4.11.4, we see that we may separately consider the cases when μ is Gaussian and when μ has no Gaussian component. We consider the "no Gaussian component" case first.

Theorem 4.11.5. *Let $\mu = [a, 0, M]$ be full and operator-stable on $\mathbb{R}^d$. In order for $\nu \in \mathrm{DONA}(\mu)$, it is necessary and sufficient that, for every $F \in \mathscr{B}(S_\mu)$ with $\tilde{m}(\partial' F) = 0$,*

$$\lim_{t \to \infty} t \cdot \nu(\{s^B x : x \in F, s > t\}) = \tilde{m}(F), \qquad (4.11.6)$$

where $\partial' F$ denotes the boundary of F relative to S_μ, for some $B \in \mathscr{E}_s(\mu)$; in which case, (4.11.6) holds for all exponents of μ.

Proof. We use the notation

$$[F; K] := \{t^B x : x \in F, t \in K\},$$

where $F \in \mathscr{B}(S_\mu)$, $K \in \mathscr{B}((0, \infty))$, and $B \in \mathscr{E}_s(\mu)$ is fixed. We know that, for each $s > 0$, $s^B[F; K] = [F; sK]$ and $\mathscr{B}(\mathbb{R}^d \setminus \{0\})$ is generated by the family of all $[F; K]$ (cf. Proposition 3.4.3). Note that $\partial[F; K] = [\partial' F; \overline{K}] \cup [\overline{F}; \partial'' K]$, where $\partial'' K$ is the boundary of K in $(0, \infty)$. In particular, if the Lebesgue measure of $\partial'' K$ is zero, then $M(\partial[F; K]) = 0$ if and only if $\tilde{m}(\partial' F) = 0$; simply note that [cf. (4.11.3)] for sets of the form $[F; K]$ we have

$$M([F; K]) = \tilde{m}(F) \int_K t^{-2}\, dt.$$

Now, assume that $\nu \in \mathrm{DONA}(\mu)$. Let $F \in \mathscr{B}(S_\mu)$ with $\tilde{m}(\partial' F) = 0$ and let K be a subinterval of $(0, \infty)$. Then $M(\partial[F; K]) = 0$ by the above. Consequently, from (4.11.3) and (4.11.4), we have for any $B \in \mathscr{E}_s(\mu)$ (cf. Proposition 4.11.3)

$$n \cdot \nu([F; nK]) = n \cdot n^{-B} \nu([F; K])$$
$$\to M([F; K]) = \tilde{m}(F) \int_K t^{-2}\, dt.$$

Hence, $n \cdot \nu([F; (n, \infty)]) \to \tilde{m}(F)$. Letting $[t]$ denote the greatest integer number, we see that

$$\tilde{m}(F) = \lim_{t \to \infty} [t] \cdot \nu([F; ([t], \infty)])$$
$$\geq \limsup_{t \to \infty} t \cdot \nu([F; (t, \infty)]) \geq \liminf_{t \to \infty} t \cdot \nu([F; (t, \infty)])$$
$$\geq \lim_{t \to \infty} ([t] + 1) \cdot \nu([F; ([t] + 1, \infty)]) = \tilde{m}(F).$$

Therefore, (4.11.6) holds for any $B \in \mathscr{E}_s(\mu)$.

Now, assume (4.11.6) holds for a particular $B \in \mathcal{E}_s(\mu)$. Let $F \in \mathcal{B}(S_\mu)$ with $\tilde{m}(\partial' F) = 0$. Then, for $s > 0$,

$$n \cdot (n^{-B}\nu)([F;(s,\infty)]) = n \cdot \nu([F;(ns,\infty)])$$
$$\to s^{-1}\tilde{m}(F) = \tilde{m}(F)\int_s^\infty t^{-2}\,dt.$$

Hence, for all subintervals K of $(0,\infty)$ we have

$$n \cdot n^{-B}\nu([F;K]) \to \tilde{m}(F)\int_K t^{-2}\,dt = M([F;K]).$$

Therefore, (4.11.4) holds because the sets $[F;K]$ form a convergence class. We next show that

$$\lim_{\varepsilon \downarrow 0} \limsup_{n\to\infty} n \int_{\|x\|_\mu < \varepsilon} \|x\|_\mu^2 (n^{-B}\nu)(dx) = 0, \quad (4.11.7)$$

from which (4.11.5) with $R = 0$ clearly follows. Let

$$A := \{\mathrm{Re}(z): z \text{ is an eigenvalue of } B\}.$$

Since μ has no Gaussian component, $\min A > 1/2$. Let α and β be such that $1/2 < \alpha < \min A \le \max A < \beta < \infty$. Then for some positive constants c_1 and c_2 we have

$$\|t^B\|_\mu \le \begin{cases} c_1 t^\alpha & \text{for } 0 \le t \le 1, \\ c_2 t^\beta & \text{for } 1 \le t < \infty. \end{cases} \quad (4.11.8)$$

For $t \ge 0$ and $n = 1, 2, \ldots$, let

$$F_n(t) := (n^{-B}\nu)\big(\{y \in \mathbb{R}^d : \|y\|_\mu > t\}\big).$$

Using (4.11.8), we obtain

$$F_n(t) = \nu\left(\left\{s^B x: 0 < s \le n,\, x \in S_\mu,\, \left\|\left(\frac{s}{n}\right)^B x\right\|_\mu > t\right\}\right)$$
$$+ \nu\left(\left\{s^B x: s > n,\, x \in S_\mu,\, \left\|\left(\frac{s}{n}\right)^B x\right\|_\mu > t\right\}\right)$$
$$\le \nu\big([S_\mu; (c_1^{-1/\alpha} t^{1/\alpha} n, \infty)]\big) + \nu\big([S_\mu; (c_2^{-1/\beta} t^{1/\beta} n, \infty)]\big).$$

Since (4.11.4) is already proved, we have

$$\lim_{n \to \infty} n \cdot \nu\left(\left[S_\mu; \left(c_1^{-1/\alpha} t^{1/\alpha} n, \infty\right)\right]\right) = M\left(\left[S_\mu; \left(c_1^{-1/\alpha} t^{1/\alpha}, \infty\right)\right]\right)$$
$$= \left(t^{-1/\alpha}\right)^B M\left(\left[S_\mu; \left(c_1^{-1/\alpha}, \infty\right)\right]\right)$$
$$= t^{-1/\alpha} \cdot M\left(\left[S_\mu; \left(c_1^{-1/\alpha}, \infty\right)\right]\right) < \infty. \tag{4.11.9}$$

A similar result holds for c_2 and β. In the following, the first equality is because $1 - F_n$ is the distribution function of the univariate $\|n^{-B}\xi\|_\mu$, where ξ has distribution ν, and the second equality is by integration by parts:

$$n \int_{\|x\|_\mu < \varepsilon} \|x\|_\mu^2 (n^{-B}\nu)(dx) = -n \int_0^\varepsilon t^2 F_n(dt)$$
$$= -n\left(\varepsilon^2 F_n(\varepsilon) - \int_0^\varepsilon 2t F_n(t)\, dt\right)$$
$$\leq 2n \int_0^\varepsilon t F_n(t)\, dt$$
$$\leq 2\int_0^\varepsilon tn \cdot \nu\left(\left[S_\mu; \left(c_1^{-1/\alpha} t^{1/\alpha} n, \infty\right)\right]\right) dt$$
$$+ 2\int_0^\varepsilon tn \cdot \nu\left(\left[S_\mu; \left(c_2^{-1/\beta} t^{1/\beta} n, \infty\right)\right]\right) dt.$$

By (4.11.9) and Fatou's lemma,

$$\limsup_{n \to \infty} n \int_{\|x\|_\mu < \varepsilon} \|x\|_\mu^2 (n^{-B}\nu)(dx)$$
$$\leq 2\int_0^\varepsilon t^{1-1/\alpha} \cdot M\left(\left[S_\mu; \left(c_1^{-1/\alpha}, \infty\right)\right]\right) dt$$
$$+ 2\int_0^\varepsilon t^{1-1/\beta} \cdot M\left(\left[S_\mu; \left(c_2^{-1/\beta}, \infty\right)\right]\right) dt,$$

which goes to zero as $\varepsilon \downarrow 0$. Hence, (4.11.7) holds. Therefore, $\nu \in$ DONA(μ). From the first part of this proof, we now see that (4.11.6) holds for every exponent of μ. Q.E.D.

We now consider the domain of normal attraction for a Gaussian distribution.

Theorem 4.11.6. *Let $\mu = [a, R, 0]$ be full Gaussian on $\mathbb{R}^d$. In order for $\nu \in \mathrm{DONA}(\mu)$, it is necessary and sufficient that the covariance operator, S, of ν exists and $S = R$.*

Proof. Assume $\nu \in \mathrm{DONA}(\mu)$. Since $I/2 \in \mathscr{E}_s(\mu)$, from (4.11.5) we have, for every $y \in \mathbb{R}^d$,

$$\langle Ry, y \rangle = \lim_{\varepsilon \downarrow 0} \left\{ \begin{array}{c} \liminf_{n \to \infty} \\ \limsup_{n \to \infty} \end{array} \right\} n \left[\int_{\|x\| < \varepsilon} (\langle y, x \rangle)^2 (n^{-1/2}\nu)(dx) \right.$$

$$\left. - \left(\int_{\|x\| < \varepsilon} \langle y, x \rangle (n^{-1/2}\nu)(dx) \right)^2 \right]$$

$$= \lim_{\varepsilon \downarrow 0} \left\{ \begin{array}{c} \liminf_{n \to \infty} \\ \limsup_{n \to \infty} \end{array} \right\} \left[\int_{\|x\| < \varepsilon\sqrt{n}} (\langle y, x \rangle)^2 \nu(dx) \right.$$

$$\left. - \left(\int_{\|x\| < \varepsilon\sqrt{n}} \langle y, x \rangle \nu(dx) \right)^2 \right] = \langle Sy, y \rangle,$$

which completes the necessity.

Now, assume that the covariance operator, S, of ν exists and that $S = R$. Reversing the above argument, we see that (4.11.5) holds with $B = I/2$. Next, to establish (4.11.4) with $M = 0$, observe that for $\varepsilon > 0$ we have

$$0 \leq n(n^{-1/2}\nu)(\{x : \|x\| \geq \varepsilon\}) = n\int_{\|x\| > \varepsilon\sqrt{n}} \nu(dx)$$

$$\leq \varepsilon^{-2} \int_{\|x\| > \varepsilon\sqrt{n}} \|x\|^2 \nu(dx) \to 0 \quad \text{as } n \to \infty.$$

Consequently, (4.11.4) holds with $B = I/2$ and $M = 0$. Q.E.D.

Combining these results, we have our main theorem concerning domains of normal attraction.

Theorem 4.11.7. *Let $\mu = [a, R, M]$ be full and operator-stable on $\mathbb{R}^d$ with decomposition $\mu = \mu_1 * \mu_2$, and let $\nu \in \mathscr{P}(\mathbb{R}^d)$. If for some $B \in \mathscr{E}_s(\mu)$ we have both*

(a) *the covariance operators of μ_1 and $P_1\nu$ are the same;*
(b) *for every Borel subset F of $S_\mu \cap V_2$ with $\tilde{m}(\partial' F) = 0$,*

$$\lim_{t \to \infty} t \cdot \nu(\{s^B x : x \in F, s > t\}) = \tilde{m}(F),$$

then $\nu \in \mathrm{DONA}(\mu)$ and (a) and (b) hold for all exponents of μ.

Conversely, if $\nu \in \mathrm{DONA}(\mu)$, then (a) and (b) hold for all exponents of μ.

The final question about domains of normal attraction is: are they disjoint or do they overlap in some manner?

Theorem 4.11.8. *Let μ and γ be full and operator-stable on $\mathbb{R}^d$. Then either $\mathrm{DONA}(\mu) = \mathrm{DONA}(\gamma)$ or they are disjoint. Furthermore, $\mathrm{DONA}(\mu) = \mathrm{DONA}(\gamma)$ if and only if the following three statements hold.*

(a) *μ and γ are of the same type, that is, for some $\langle A; a \rangle \in \mathscr{A}_1$, $\mu = \mathbf{A}\gamma * \delta(a)$.*
(b) *For any $B \in \mathscr{E}_s(\mu)$, there is a sequence $\{v_n\}$ in $\mathbb{R}^d$ such that the sequence $\{\langle n^{-B} A n^B; v_n \rangle\}$ is relative compact in $\mathscr{A}_1$, where A is from (a).*
(c) *Every limit $\langle D, v \rangle$ of the sequence $\{\langle n^{-B} A n^B; v_n \rangle\}$ in (b) belongs to $\mathrm{Inv}(\mu)$, that is, $\mu = \mathbf{D}\mu * \delta(v)$.*

Proof. First, we show that $\mathrm{DONA}(\mu) \cap \mathrm{DONA}(\gamma) \neq \varnothing$ implies (a), (b), and (c). Let $\nu \in \mathrm{DONA}(\mu) \cap \mathrm{DONA}(\gamma)$, $B \in \mathscr{E}_s(\mu)$, $C \in \mathscr{E}_s(\gamma)$, and $\{b_n\}$ and $\{c_n\}$ sequences in $\mathbb{R}^d$ such that

$$n^{-B}\nu^n * \delta(b_n) \Rightarrow \mu \quad \text{and} \quad n^{-C}\nu^n * \delta(c_n) \Rightarrow \gamma. \quad (4.11.10)$$

Then $n^{-B} n^C (n^{-C} \nu^n * \delta(c_n)) * \delta(b_n - n^{-B} n^C c_n) \Rightarrow \mu$. This convergence, (4.11.10), and Theorem 2.2.10 show that the set $\{\langle n^{-B} n^C; b_n - n^{-B} n^C c_n \rangle\}$ is relative compact in $\mathscr{A}_1$ and every limit $\langle A; a \rangle$ of this set satisfies $\mu = \mathbf{A}\gamma * \delta(a)$. Hence, (a) holds.

From Proposition 4.6.6(b), $A^{-1}BA \in \mathscr{E}_s(\gamma)$, so for some sequence $\{c_n'\}$ we have $n^{-A^{-1}BA}\nu^n * \delta(c_n') \Rightarrow \gamma$.

Consequently,

$$(n^{-B}An^B)(n^{-B}\nu^n * \delta(b_n)) * \delta(Ac_n' + a - n^{-B}An^B b_n)$$
$$= n^{-B}A\nu^n * \delta(Ac_n' + a)$$
$$= A(n^{-A^{-1}BA}\nu^n * \delta(c_n')) * \delta(a)$$
$$\Rightarrow A\gamma * \delta(a) = \mu.$$

This convergence with (4.11.10) and Theorem 2.2.7 shows that the set $\{\langle n^{-B}An^B; v_n\rangle\}$, with $v_n := Ac_n' + a - n^{-B}An^B b_n$, is relative compact in $\mathscr{A}_I$ and every limit of this set is in $\mathbf{Inv}(\mu)$. Therefore, (b) and (c) hold.

Second, assume (a), (b), and (c) hold. Let $\nu \in \mathrm{DONA}(\mu)$. Then $n^{-B}\nu^n * \delta(b_n) \Rightarrow \mu$ for some sequence $\{b_n\}$. From (b), we see that the sequence $\{n^{-B}An^B(n^{-B}\nu^n * \delta(b_n)) * \delta(v_n)\}$ is tight. From (c), every convergent subsequence of this sequence converges to μ. Consequently, $n^{-B}A\nu^n * \delta(n^{-B}An^B b_n + v_n) \Rightarrow \mu$. Hence, $n^{-A^{-1}BA}\nu^n * \delta(A^{-1}n^{-B}An^B b_n + A^{-1}v_n - A^{-1}a) \Rightarrow A^{-1}\mu * \delta(-A^{-1}a) = \gamma$. From (a) and Proposition 4.6.6, $A^{-1}BA \in \mathscr{E}_s(\gamma)$, so $\nu \in \mathrm{DONA}(\gamma)$. Therefore, $\mathrm{DONA}(\mu) \subseteq \mathrm{DONA}(\gamma)$.

The reverse inclusion is obtained in a similar manner, but there is a slight difference. Let $\nu \in \mathrm{DONA}(\gamma)$. Then $n^{-A^{-1}BA}\nu^n * \delta(d_n) \Rightarrow \gamma$ for some sequence $\{d_n\}$. Hence, $n^{-B}A\nu^n * \delta(Ad_n + a) \Rightarrow A\gamma * \delta(a) = \mu$. Consequently,

$$n^{-B}An^B[n^{-B}\nu^n * \delta(n^{-B}A^{-1}n^B(Ad_n + a - v_n))] * \delta(v_n) \Rightarrow \mu.$$

Since every limit of the set $\{\langle n^{-B}An^B; v_n\rangle\}$ is in $\mathbf{Inv}(\mu)$, which is a compact group (cf. Theorem 2.2.5), this implies that $n^{-B}\nu^n * \delta(n^{-B}A^{-1}n^B(Ad_n + a - v_n)) \Rightarrow \mu$, that is, $\nu \in \mathrm{DONA}(\mu)$. Therefore, $\mathrm{DONA}(\gamma) \subseteq \mathrm{DONA}(\mu)$. Q.E.D.

Remark 4.11.9. Theorem 4.11.4 suggests that any $\nu \in \mathrm{DONA}(\mu)$ might be decomposed into $\nu = \nu_1 * \nu_2$ with ν_i supported by V_i, $i = 1, 2$. However, this is false. Consider the following example.

Let $d = 2$ and $\hat{\mu}((x, y)) := \exp(-x^2/2 - |y|)$. Then μ is full operator-stable with $B := \mathrm{diag}(1/2, 1) \in \mathscr{E}_s(\mu)$ with $b(t) = 0$ for all $t > 0$. Note that V_1 is the x-axis, V_2 is the y-axis, μ_1 is the standard univariate normal distribution, and μ_2 is the symmetric univariate Cauchy distribution. Let λ_1 be any distribution on $\mathbb{R}^2$ which is concentrated on V_1 such that $P_1\lambda_1$ is a symmetric univariate distribution with a variance of 2. Also, let λ_2 be the distribution on $\mathbb{R}^2$, concentrated on V_2, given by

$$\lambda_2(\{(0, y): y \le t\}) := \begin{cases} \dfrac{2}{\pi|t|} & \text{if } t < -\dfrac{4}{\pi}, \\ \dfrac{1}{2} & \text{if } -\dfrac{4}{\pi} \le t < \dfrac{4}{\pi}, \\ 1 - \dfrac{2}{\pi t} & \text{if } \dfrac{4}{\pi} \le t. \end{cases}$$

Now, set $\lambda := (\lambda_1 + \lambda_2)/2$. Then supp $\lambda = V_1 \cup V_2$, so λ cannot be the convolution of some ν_1 and ν_2 with supp $\nu_i \subseteq V_i$, $i = 1, 2$. However, $P_1\lambda = (\lambda_1 + \delta(0))/2$ has variance one, so it is in DONA(μ_1), and $P_2\lambda = (\delta(0) + \lambda_2)/2$ is in DONA(μ_2) because its tail behavior is correct as seen by Theorem 4.11.5 with $d = 1$. Therefore, by Theorem 4.11.4, $\lambda \in$ DONA(μ).

4.12 MOMENTS

By the *p*th *moment* of γ is meant the integral $\int \|x\|^p \gamma(dx)$. In this section, we show that for full operator-stable distributions the finiteness of the *p*th moment depends only on the maximum of the real parts of the eigenvalues of an exponent. This is analogous to the one-dimensional case where the eigenvalue is the reciprocal of the exponent, so one would expect p to be less than the minimum exponent, that is, to be less than the reciprocal of the maximum eigenvalue. For ν in GDOA, a somewhat less precise result is obtained. The first result concerns the *p*th moment of any distribution.

Lemma 4.12.1. *For any $\gamma \in \mathscr{P}(\mathbb{R}^d)$,*

(a) *if $0 < p$ and $\int \|x\|^p \gamma(dx) < \infty$, then for every $y \in \mathbb{R}^d$ we have $\int |\langle x, y \rangle|^p \gamma(dx) < \infty$;*

(b) *if $0 < p \le 2$ and $\int |\langle x, y \rangle|^p \gamma(dx) < \infty$ for $y \in \{e_1, \ldots, e_d\}$, where $\{e_1, \ldots, e_d\}$ is an orthonormal basis for $\mathbb{R}^d$, then $\int |\langle x, y \rangle|^p \gamma(dx) < \infty$ for every $y \in \mathbb{R}^d$ and $\int \|x\|^p \gamma(dx) < \infty$.*

Proof. For (a), simply apply the Cauchy–Schwarz inequality. For (b), since $p \leq 2$,

$$\|x\|^p = \left(\sum \langle e_i, x\rangle^2\right)^{p/2} \leq \sum |\langle e_i, x\rangle|^p.$$

Since each summand is integrable, so is $\|x\|^p$. By (a), $\int |\langle x, y\rangle|^p \gamma(dx) < \infty$ for every $y \in \mathbb{R}^d$. Q.E.D.

Throughout this section, μ denotes a full operator-stable distribution on $\mathbb{R}^d$ with exponent B, and, for $\nu \in \text{GDOA}(\mu)$, we have $A_n \in \text{Aut}$ and $a_n \in \mathbb{R}^d$ such that $A_n \nu^n * \delta(a_n) \Rightarrow \mu$. Also, let

$$\lambda := \min\{\text{Re}(x): x \text{ is an eigenvalue of } B\},$$
$$\Lambda := \max\{\text{Re}(x): x \text{ is an eigenvalue of } B\}.$$

Then $1/2 \leq \lambda \leq \Lambda$ (cf. Theorem 4.6.5). Note that the set $\{\text{Re}(x): x \text{ is an eigenvalue of } B\}$ is the same for all exponents of μ (cf. Sections 4.2 and 4.6).

Theorem 4.12.2. *Let $\nu \in \text{GDOA}(\mu)$ and let $p \in (0, 1/\Lambda)$. Then*

$$\int \|x\|^p \nu(ds) < \infty,$$

$$\int \|x\|^p (A_n \nu^n * \delta(a_n))(dx) \to \int \|x\|^p \mu(dx) < \infty.$$

Proof. Let $\xi, \xi_1, \xi_2, \ldots$ be i.i.d. r.v.'s with common distribution ν and let $S_n := \sum_{i=1}^n \xi_i$ with $S_0 := 0$. For now, assume ν is symmetric and each $a_n = 0$. We first show that

$$\sup_n E(\|A_n S_n\|^p) < \infty. \tag{4.12.1}$$

Fix $s \in (p, 1/\Lambda)$. By Corollary 4.5.4, there are constants $c > 0$ and N_1 such that for every $n \geq N_1$ and for every $m \geq 1$ we have

$$\|A_n A_{nm}^{-1}\| < cm^{1/s}. \tag{4.12.2}$$

Let $\varepsilon > 0$ be small and, by tightness, select a so large that $2P[\|A_n S_n\| > a] < \varepsilon$ for all n. Using Lévy's inequality for symmetric summands, we have

$$\left(P[\|A_{mn} S_n\| \leq a]\right)^m = P\left[\max_{k \leq m} \|A_{mn}(S_{nk} - S_{n(k-1)})\| \leq a\right]$$
$$\geq 1 - 2P[\|A_{mn} S_{mn}\| > a] > 1 - \varepsilon$$

for every n and m. Hence,

$$P[\|A_{mn}S_n\| > a] < 1 - (1-\varepsilon)^{1/m}.$$

For $\varepsilon \in (0,1)$, the quantity $m(1-(1-\varepsilon)^{1/m})$ is bounded for all $m \geq 1$. Therefore,

$$\sup_{n,m} mP[\|A_{mn}S_n\| > a] < \infty. \qquad (4.12.3)$$

Since $P[\|A_{mn}S_n\| > a] \geq P[\|A_n S_n\| > a\|A_n A_{nm}^{-1}\|]$, using (4.12.2) and (4.12.3), we have

$$\sup_{n \geq N_1} \sup_{m \geq 1} mP[\|A_n S_n\| > acm^{1/s}] < \infty.$$

Hence, there is t_0 such that

$$N_2 := \sup_{n \geq N_1} \sup_{t \geq t_0} t^{s/p} P[\|A_n S_n\| \geq t^{1/p}] < \infty.$$

This implies that

$$\sup_{n \geq N_1} E(\|A_n S_n\|^p) = \sup_{n \geq N_1} \int_0^\infty P[\|A_n S_n\| \geq t^{1/p}]\, dt$$

$$\leq t_0 + \frac{N_2 t_0^{1-s/p}}{s/p - 1} < \infty.$$

Therefore, (4.12.1) holds for symmetric ν.

For general $\nu \in \mathrm{GDOA}(\mu)$, let $\{U_n\}$ and $\{V_n\}$ be independent sequences of r.v.'s such that, for each $n \geq 1$, the distribution of U_n and the distribution of V_n is $A_n \nu^n * \delta(a_n)$. Then, by (4.12.1),

$$D := \sup_n E(\|U_n - V_n\|^p) < \infty.$$

Let $g_n(y) := E(\|U_n - y\|^p)$. Then by the independence of U_n and V_n we have

$$Eg_n(V_n) = E(\|U_n - V_n\|^p) \leq D.$$

Let b be so large that $P[\|U_n\| > b] < 1/2$ for all n. Also, let $B_b := \{x \in \mathbb{R}^d : \|x\| \leq b\}$ and $G_n := \{y \in \mathbb{R}^d : g_n(y) \leq 3D\}$. We see that B_b

is not contained in $\mathbb{R}^d \setminus G_n$, because if it were we would have

$$D \geq E\big(g_n(V_n) I_{B_b}(V_n)\big) \geq \frac{3D}{2},$$

which is impossible. Hence, $B_b \cap G_n \neq \emptyset$. Let $y_n \in B_b \cap G_n$ for $n \geq 1$. Then, for all $n \geq 1$,

$$E(\|U_n\|^p) \leq E(\|U_n - y_n\| + \|y_n\|)^p$$
$$\leq c_p\big(g_n(y_n) + \|y_n\|^p\big) \leq c_p(3D + b^p) < \infty,$$

where c_p is 1 when $p \leq 1$ and is 2^{p-1} when $p > 1$. Therefore, $\sup_n E(\|A_n S_n + a_n\|^p) < \infty$.

Hence, $E(\|A_n S_n\|^p) < \infty$ for each n. Since $\|A_n S_n\| \geq \|S_n\|/\|A_n^{-1}\|$, we have $E(\|S_n\|^p) < \infty$, so $\int \|x\|^p \nu(dx) < \infty$.

Since $\mu \in \text{GDOA}(\mu)$, we also have $\int \|x\|^p \mu(dx) < \infty$.

Finally, by the uniform integrability, we have the convergence of moments as stated. Q.E.D.

Corollary 4.12.3. *If $\nu \in \text{GDOA}(\mathcal{N}_d)$ and $0 < p < 2$, then the conclusion of Theorem 4.12.2 holds. If $\nu \in \text{DONA}(\mathcal{N}_d)$, then the covariance operator of ν exists and is equal to the covariance operator of $\mathcal{N}_d$; in particular,*

$$\int \|x\|^2 \nu(dx) < \infty.$$

Proof. For the first statement, $\Lambda = \lambda = 1/2$; and for the second statement, apply Theorem 4.11.6. Q.E.D.

Theorem 4.12.4. *Let $\nu \in \text{GDOA}(\mu)$ with $\mu := [a, R, M]$. Assume $M(\mathbb{R}^d) > 0$. If $p > 1/\lambda$, then*

$$\int \|x\|^p \nu(dx) = \infty.$$

Proof. Let $C := \{x \in \mathbb{R}^d : \|x\|_B \geq a\}$, where B is an exponent of μ [cf. (3.4.3)] and $a > 0$ is selected so that $M(C^c) > 0$ and $M(\partial C) = 0$

(cf. Corollary 4.3.6). Then

$$\int \|x\|_B^p \nu(dx) \geq \int_C \|A_n^{-1}x\|_B^p (A_n \nu)(dx)$$
$$\geq a^p \|A_n\|_B^{-p}(A_n \nu)(C).$$

Since $n(A_n \nu)(C) \to M(C) > 0$, it suffices to show that $n^{-1}\|A_n\|_B^{-p} \to \infty$. Let $\varepsilon \in (1/p, \lambda)$. By Theorem 4.5.3(a), $\|A_n\|_B \leq cn^{-\varepsilon}$ for some constant c. Therefore, $n^{-1}\|A_n\|_B^{-p} \geq c^{-p} n^{p\varepsilon - 1} \to \infty$. Since the conclusion holds for the norm $\|\cdot\|_B$, it holds for any norm. Q.E.D.

Theorem 4.12.5. *Let* $\nu \in \mathrm{DONA}(\mu)$ *with* $\mu := [a, R, M]$. *Assume* $M(\mathbb{R}^d) > 0$. *Then*

(a) *for* $p > 1/\Lambda$, $\int \|x\|^p \nu(dx) = \infty$;
(b) *for* $p < 1/\Lambda$, $\int \|x\|^p \nu(dx) < \infty$.

Proof. Recall the primary decomposition of $\mathbb{R}^d$ given by the linear operator B (cf. Theorem 4.6.5 and its proof). Let P be the canonical projection of $\mathbb{R}^d$ onto the subspaces corresponding to those eigenvalues of B having real parts equal to Λ. Then P is a polynomial in B, so P commutes with B. Since $\nu \in \mathrm{DONA}(\mu)$, we have $n^{-B}\nu^n * \delta(a_n) \Rightarrow \mu$. Hence,

$$n^{-PB}(P\nu)^n * \delta(Pa_n) = P(n^{-B}\nu^n * \delta(a_n)) \Rightarrow P\mu.$$

By Theorem 4.6.5, $P\mu$ is full and operator-stable on $P\mathbb{R}^d$ with exponent PB. Note that every eigenvalue of PB has a real part equal to Λ. Apply Theorem 4.12.5 to $P\nu$ and $P\mu$ to obtain

$$\int \|x\|^p \nu(dx) \geq \int \|x\|^p (P\nu)(dx) = \infty. \qquad \text{Q.E.D.}$$

Summarizing these results for θ, we have the following theorem.

Theorem 4.12.6. *Let* $\mu := [a, R, M]$ *be full and operator-stable. Then*

(a) $M(\mathbb{R}^d) = 0$ *implies* $\int \|x\|^p \mu(dx) < \infty$ *for all* $p > 0$;
(b) $M(\mathbb{R}^d) > 0$ *implies*

  (i) $\int \|x\|^p \mu(dx) < \infty$ *for* $0 < p < 1/\Lambda$;
  (ii) $\int \|x\|^p \mu(dx) = \infty$ *for* $p > 1/\Lambda$.

4.13 INDEPENDENT MARGINALS

We consider the question of when does an operator-stable measure μ have independent marginals. The first results concern infinitely divisible measures in general, which are then specialized to operator-stable measures. To fix the notation and terminology for this section, $\mu := [a, R, M]$ is infinitely divisible on $\mathbb{R}^d$, $J\mu$ is a *marginal* of μ provided $J \in \text{End}(\mathbb{R}^d)$ is such that $J^2 = J$, that is, J is an idempotent, and $J\mu$ is called a *k-dimensional* marginal provided $\dim J(\mathbb{R}^d) = k$ and is called a *univariate* marginal when $k = 1$. Note that marginals of infinitely divisible measures are again infinitely divisible. Also, if $J \in \mathbf{D}(\mu)$, then $\mu = J\mu * (I - J)\mu$, so they are independent marginals (cf. Section 2.3).

A set $\{J_1\mu, \ldots, J_r\mu\}$ of marginals is called *complete* if $I = \sum_{i=1}^r J_i$, and the set of marginals is called *orthogonal* if $J_i J_j = 0 = J_j J_i$ for all $i \neq j$. Let ξ be an $\mathbb{R}^d$-valued random vector whose distribution is μ. A set $\{J_1\mu, \ldots, J_r\mu\}$ of marginals is said to be an *independent* set of marginals if the random vectors $J_1\xi, \ldots, J_r\xi$ are independent. Note that two marginals $J_1\mu$ and $J_2\mu$ may be independent, but $J_1(\mathbb{R}^d)$ and $J_2(\mathbb{R}^d)$ may not be orthogonal subspaces of $\mathbb{R}^d$ (cf. Theorem 4.13.2); also, note that $J_1\mu$ and $J_2\mu$ may be orthogonal marginals, but $J_1(\mathbb{R}^d)$ and $J_2(\mathbb{R}^d)$ may not be orthogonal subspaces.

Let $\mathscr{F}$ and ID denote the classes of full measures and of infinitely divisible measures on $\mathbb{R}^d$, respectively. For $\mu := [a, R, M] \in \text{ID}$,

$$\hat{\mu}(y) = \exp\left\{i\langle y, a\rangle - \tfrac{1}{2}\langle Ry, y\rangle + \int_{\mathbb{R}^d \setminus \{0\}} \mathscr{I}_1(y; x) M(dx)\right\}.$$

Here and below, for $y_1, \ldots, y_k, x_1, \ldots, x_k \in \mathbb{R}^d$,

$$\mathscr{I}_k(y_1, \ldots, y_k; x_1, \ldots, x_k)$$
$$:= \exp\left(i\sum_{j=1}^k \langle y_j, x_j\rangle - 1 - \frac{i\sum_{j=1}^k \langle y_j, x_j\rangle}{1 + \sum_{j=1}^k \langle x_j, x_j\rangle}\right).$$

The function $\mathscr{I}_k$ arises in the representation of k measures on $\mathbb{R}^d$ (cf. the following lemma).

Lemma 4.13.1. *Let $\mu = [a, R, M] \in \text{ID}(\mathbb{R}^d)$ and let $J_1\mu, \ldots, J_r\mu$ be marginals. Then the joint distribution, ν, of $J_1\mu, \ldots, J_r\mu$ is $\text{ID}(\mathbb{R}^d)$*

and, for $y_1, \ldots, y_r \in \mathbb{R}^d$,

$$\hat{\nu}(y_1, \ldots, y_r) = \exp\left\{ i \sum_{j=1}^{r} \langle y_j, \tilde{a}_j \rangle - \frac{1}{2} \sum_{j,k=1}^{r} \langle J_j R J_k^* y_k, y_j \rangle \right.$$
$$\left. + \int_{\mathbb{R}^{rd} \setminus \{0\}} \mathscr{G}_r(y_1, \ldots, y_r; x_1, \ldots, x_r)(AM)(dx_1, \ldots, dx_r) \right\}$$

for some $\tilde{a}_j \in \mathbb{R}^d$, where $A: \mathbb{R}^d \to \mathbb{R}^{rd}$ is given by $A(x) := (J_1 x, \ldots, J_r x)$.

Proof. By adding and subtracting

$$\frac{i \sum_{j=1}^{r} \langle y_j, J_j x \rangle}{1 + \sum_{j=1}^{r} \langle J_j x, J_j x \rangle}$$

in the appropriate place in the characteristic function of ν, this is an easy calculation. Q.E.D.

The following result on independent marginals applies to any infinitely divisible measure, not necessarily operator-stable.

Theorem 4.13.2. *Let $\mu = [a, R, M]$ be full and let $J_1 \mu, \ldots, J_r \mu$ be marginals. Then $J_1 \mu, \ldots, J_r \mu$ are independent if and only if for all $j \neq k$ we have both*

$$J_j R J_k^* = 0, \qquad (4.13.1)$$

$$M\big((\ker J_j)^c \cap (\ker J_k)^c\big) = 0. \qquad (4.13.2)$$

Proof. Since $J_1 \mu, \ldots, J_r \mu$ are independent if and only if their joint characteristic function is the product of their individual characteristic functions, we see from Lemma 4.13.1 that they are independent if and only if, for all $y_1, \ldots, y_r \in \mathbb{R}^d$,

$$\sum_{j,k=1}^{r} \langle J_j R J_k^* y_k, y_j \rangle = \sum_{i=1}^{r} \langle J_i R J_i^* y_i, y_i \rangle, \qquad (4.13.3)$$

$$\int_{\mathbb{R}^{rd} \setminus \{0\}} \mathscr{G}_r(y_1, \ldots, y_r; x_1, \ldots, x_r)(AM)(dx_1, \ldots, dx_r)$$
$$= \sum_{j=1}^{r} \int_{\mathbb{R}^d \setminus \{0\}} \mathscr{G}_1(y_j; x_j)(J_j M)(dx_j). \qquad (4.13.4)$$

Obviously, (4.13.1) and (4.13.3) are equivalent. We wish to rewrite (4.13.4). Let $\delta_0 := \delta(0)$ and for $B_1, \ldots, B_r \in \mathscr{B}(\mathbb{R}^d)$ with $B_1 \times \cdots \times B_r \in \mathscr{B}(\mathbb{R}^{rd} \setminus \{0\})$ define the measure $\mathbb{M}$ on $\mathscr{B}(\mathbb{R}^{rd} \setminus \{0\})$ as a product measure by

$$\mathbb{M}(B_1 \times \cdots \times B_r)$$
$$:= \sum_{j=1}^{r} \delta_0(B_1) \cdots \delta_0(B_{j-1})(J_j M)(B_j)\delta_0(B_{j+1}) \cdots \delta_0(B_r).$$

Then, for an $\mathbb{M}$-integrable function $f: \mathbb{R}^{rd} \to \mathbb{R}^1$, we have

$$\int_{\mathbb{R}^{rd} \setminus \{0\}} f(x_1, \ldots, x_r)\mathbb{M}(dx_1, \ldots, dx_r)$$
$$= \sum_{j=1}^{r} \int_{\mathbb{R}^d \setminus \{0\}} f(0, \ldots, 0, x_j, 0, \ldots, 0)(J_j M)(dx_j).$$

Hence, $\mathbb{M}$ is a Lévy measure and (4.13.4) is equivalent to

$$(A\mathbb{M})(B_1 \times \cdots \times B_r) = \mathbb{M}(B_1 \times \cdots \times B_r). \quad (4.13.5)$$

Therefore, it suffices to show that (4.13.2) and (4.13.5) are equivalent. Assume (4.13.5) holds. For fixed $j \neq k$, let $B_j = B_k = \mathbb{R}^d \setminus \{0\}$ and let the other B_i's be $\mathbb{R}^d$. Then $\mathbb{M}(B_1 \times \cdots \times B_r) = 0$ and $(A\mathbb{M})(B_1 \times \cdots \times B_r) = M((\ker J_j)^c \cap (\ker J_k)^c)$, so (4.13.2) holds. Now, assume (4.13.2) holds. If there are $j \neq k$ with $0 \notin B_j$ and $0 \notin B_k$, then $\mathbb{M}(B_1 \times \cdots \times B_r) = 0$ and $(A\mathbb{M})(B_1 \times \cdots \times B_r) \leq M((\ker J_j)^c \cap (\ker J_k)^c) = 0$, so (4.13.5) holds in this case. If $0 \notin B_j$ for some j, but $0 \in B_i$ for all $i \neq j$, then $J_j^{-1}(B_j) \subset (\ker J_j)^c$, $J_i^{-1}(B_i^c) \subset (\ker J_i)^c$ for $i \neq j$, and

$$(A\mathbb{M})(B_1 \times \cdots \times B_r) = M\left(\bigcap_{i \neq j} (J_i^{-1}(\mathbb{R}^d) \setminus J_i^{-1}(B_i^c))\right)$$
$$= M\left(J_j^{-1}(B_j) \cap \left(\bigcap_{i \neq j} J_i^{-1}(\mathbb{R}^d)\right)\right)$$
$$= (J_j M)(B_j) = \mathbb{M}(B_1 \times \cdots \times B_r),$$

where the first equality uses (4.13.2), so (4.13.5) holds in this case. (A slight change in the argument is needed when $r = 2$.) Since 0 cannot be in every B_i, we have (4.13.2) implies (4.13.5). Q.E.D.

This theorem shows that for marginals of an infinitely divisible law to be independent it is necessary and sufficient that pairs behave properly.

Corollary 4.13.3. *Let μ be full and infinitely divisible on $\mathbb{R}^d$. The marginals $J_1\mu, \ldots, J_r\mu$ are independent if and only if they are pairwise independent.*

Remark 4.13.4. It is not necessary for the marginals to be on orthogonal subspaces in order for them to be independent, because (4.12.2) may hold without the subspaces being orthogonal subspaces.

Lemma 4.13.5. *Let $\mu = [a, R, M]$ and let $J_1\mu, \ldots, J_r\mu$ be marginals. Then the condition (4.13.2) implies*

$$\operatorname{supp} M \subset \bigcup_{j=1}^{r} \ker J_j. \qquad (4.13.6)$$

When $J_1\mu, \ldots, J_r\mu$ are a complete set of orthogonal marginals, then (4.13.2) is equivalent to

$$\operatorname{supp} M \subset \bigcup_{j=1}^{r} J_j(V). \qquad (4.13.7)$$

Proof. If $(\operatorname{supp} M) \cap (\bigcup_{j=1}^{r} \ker J_j)^c \neq \emptyset$, then $(\operatorname{supp} M) \cap (\ker J_1)^c \cap (\ker J_2)^c \neq \emptyset$, which contradicts (4.13.2). Therefore, (4.13.2) implies (4.13.6). When we have a complete set of orthogonal marginals, $I = \sum_{j=1}^{r} J_j$ and $J_j J_k = 0 = J_k J_j$ for all $j \neq k$. This implies that

$$\bigcup_{j=1}^{r} J_j(V) = \bigcap_{j \neq k} \left((\ker J_j) \cup (\ker J_k) \right).$$

Clearly, this equality shows that (4.13.2) and (4.13.7) are equivalent.
Q.E.D.

One might wonder whether there are different choices for independent univariate marginals, possibly only one choice if we also require them to be complete and orthogonal. However, if μ is Gaussian and full on $\mathbb{R}^d$ with $d \geq 2$, there are obviously an infinite number of such univariate marginals, since marginals of a Gaussian are Gaussians and they are independent when on orthogonal subspaces. But, if μ has a "significant" Poisson component, we see in the next theorem that μ has at most one choice for such marginals.

Proposition 4.13.6. *Let $\mu = [a, R, M]$ and assume $[a, 0, M]$ is full. If μ has a set of complete independent orthogonal univariate marginals, then μ has exactly one such set of marginals.*

Proof. Suppose $J_1\mu, \ldots, J_d\mu$ and $K_1\mu, \ldots, K_d\mu$ are sets of complete independent orthogonal univariate marginals. By Theorem 4.13.2 and Remark 4.13.5,

$$\text{supp } M \subset \left(\bigcup_{i=1}^{d} J_i(\mathbb{R}^d)\right) \cap \left(\bigcup_{j=1}^{d} K_j(\mathbb{R}^d)\right).$$

Fix i. Since $[a, 0, M]$ is full, $\text{lin}(\text{supp } M) = \mathbb{R}^d$ (cf. Chapter 1). Hence,

$$\bigcup_{j=1}^{d} \left(\text{supp}_0 M \cap J_i(\mathbb{R}^d) \cap K_j(\mathbb{R}^d)\right) = \text{supp}_0 M \cap J_i(\mathbb{R}^d) \neq \emptyset,$$

where $\text{supp}_0 M := (\text{supp } M) \setminus \{0\}$. Hence, there is $j(i)$ such that $\text{supp}_0 M \cap J_i(\mathbb{R}^d) \cap K_{j(i)}(\mathbb{R}^d) \neq \emptyset$. Since $J_i\mu$ and $K_{j(i)}\mu$ are univariate, we have $J_i(\mathbb{R}^d) = K_{j(i)}(\mathbb{R}^d)$. By orthogonality,

$$J_k K_{j(i)} = 0 = K_{j(i)} J_k, \qquad i \neq k.$$

Using completeness,

$$J_i = \sum_{k=1}^{d} J_i K_{j(k)} = J_i K_{j(i)}$$

$$= \sum_{k=1}^{d} J_k K_{j(i)} = K_{j(i)}. \qquad \text{Q.E.D.}$$

We now specialize to full t^B-stable measures $\mu = [a, R, M]$. The decomposition of such μ given in Theorem 4.6.5 and the information about B given in Theorem 4.6.12 will be used in stating and obtaining the results. To fix the notation in the remainder of this section,

$$\mu = \mu_1 * \mu_2 \quad \text{and} \quad \mathbb{R}^d = V_1 \oplus V_2,$$

μ_1 is full Gaussian on V_1, μ_2 has no Gaussian component and is full on V_2, and the subspaces V_1 and V_2 correspond to those eigenvalues of B with real parts equal to $1/2$ and greater than $1/2$, respectively. Let L_1 and L_2 be the projections of $\mathbb{R}^d$ onto V_1 and V_2, respectively, with $L_1 L_2 = 0 = L_2 L_1$. Then $L_i \mu = \mu_i$. Also, μ_i is $t^{L_i B}$-stable. Note that V_1 and V_2 need not be orthogonal. The following is a main result concerning operator-stable measures.

Theorem 4.13.7. *Let μ be full and operator-stable on $\mathbb{R}^d$. If $\dim V_1 \leq 1$ and if $J_1 \mu, \ldots, J_d \mu$ and $K_1 \mu, \ldots, K_d \mu$ are two sets of complete orthogonal independent univariate marginals, then $\{J_1, \ldots, J_d\} = \{K_1, \ldots, K_d\}$.*

Proof. When $\dim V_1 < 1$, $\mu = [a, 0, M]$, so apply Proposition 4.13.6. Now, assume $\dim V_1 = 1$. By Lemma 4.13.5, $\operatorname{supp} M \subset \bigcup_{i=1}^d J_i(\mathbb{R}^d)$. Hence, $J_i(\mathbb{R}^d) \subset V_2$ for i such that $\operatorname{supp}_0 M \cap J_i(\mathbb{R}^d) \neq \emptyset$. Since $J_1 \mu, \ldots, J_d \mu$ is a set of complete orthogonal univariate marginals, there is a unique j such that $V_1 = J_j(\mathbb{R}^d)$ and $V_1 \cap J_i(\mathbb{R}^d) = \{0\}$ for $i \neq j$. Hence, we may assume $J_1 = L_1 = K_1$. For $i > 1$, $J_i \mu = J_i \mu_2$ and $K_i \mu = K_i \mu_2$. Considering μ_2, J_i, and K_i for $2 \leq i \leq d$ on V_2, by Proposition 4.13.6 we have $\{J_2, \ldots, J_d\} = \{K_2, \ldots, K_d\}$. Q.E.D.

Remark 4.13.8. When $\dim V_1 > 1$, Theorem 4.13.7 is false, because the projections into V_1 are not unique, whereas the projections into V_2 are still unique.

For $\mu = [a, R, M] \in \mathscr{F} \cap \mathbb{OS}$ with univariate marginals $J_1 \mu, \ldots, J_d \mu$, let

$$T_1 := \{j : \dim J_j(V_1) = 1\}, \qquad T_2 := \{j : \dim J_j(V_2) = 1\}.$$

Theorem 4.13.9. *Let $\mu = [a, R, M]$ be full and operator-stable on $\mathbb{R}^d$ and let $J_1 \mu, \ldots, J_d \mu$ be a set of complete orthogonal univariate marginals. Then they are independent if and only if*

$$R = \sum_{j \in T_1} J_j R J_j^*, \qquad (4.13.8)$$

$$\operatorname{supp} M \subset \bigcup_{j \in T_2} J_j(\mathbb{R}^d). \qquad (4.13.9)$$

Proof. Let $T := \{j : \operatorname{supp}_0 M \cap J_j(\mathbb{R}^d) \neq \emptyset\}$. Assume the marginals are independent. They by Theorem 4.13.2 and Lemma 4.13.5 we have supp $M \subset \bigcup_{j=1}^{d} J_j(\mathbb{R}^d)$. Hence, supp $M \subset \bigcup_{j \in T} J_j(\mathbb{R}^d)$. But, lin(supp M) = V_2, so obviously $T = T_2$; hence, (4.13.9) holds. Since μ_2 has no Gaussian component, $(L_2 R^{1/2})(L_2 R^{1/2})^* = L_2 R L_2^* = 0$, so $L_2 R = 0$. Hence, for $j \in T_2$, $J_j R = 0$. By Theorem 4.13.2, $J_j R J_k^* = 0$ for $j \neq k$. Therefore,

$$R = \sum_{j,k} J_j R J_k^* = \sum_{j \in T_1} J_j R J_j^*.$$

The converse part of this theorem follows by noting that (4.13.9) implies (4.13.6), and that (4.13.8) with the first equality above implies (4.13.1). Q.E.D.

Theorem 4.13.10. *Let $\mu = [a, R, M]$ be full and operator-stable on $\mathbb{R}^d$ and let $J_1\mu, \ldots, J_d\mu$ be a set of complete orthogonal independent univariate marginals. Then*

(a) *the marginals $J_1\mu_1, \ldots, J_d\mu_1$ are independent;*
(b) *the marginals $J_1\mu_2, \ldots, J_d\mu_2$ are independent;*
(c) *for $j \in T_1$, $J_j\mu$ is univariate Gaussian;*
(d) *for $j \in T_2$, $J_j\mu$ is univariate stable non-Gaussian;*
(e) *there are $\alpha_j > 1/2$ for $j \in T_2$ such that the linear operator*

$$\frac{1}{2}L_1 + \sum_{j \in T_2} \alpha_j J_j$$

is an exponent for μ.

Proof. As in Lemma 4.13.1, define $A: \mathbb{R}^d \setminus \{0\} \to \mathbb{R}^{d^2} \setminus \{0\}$ by $A(x) := (J_1 x, \ldots, J_d x)$. By independence, for $y_1, \ldots, y_d \in \mathbb{R}^d$,

$$(A\mu)\hat{\,}(y_1, \ldots, y_d) = \prod_{j=1}^{d} (J_j\mu)\hat{\,}(y_j).$$

Since $\mu = \mu_1 * \mu_2$,

$$(A\mu)\hat{\,}(y_1, \ldots, y_d) = (A\mu_1)\hat{\,}(y_1, \ldots, y_d) \cdot (A\mu_2)\hat{\,}(y_1, \ldots, y_d).$$

For fixed j, set $y_i = 0$ for $i \neq j$ to obtain

$$(J_j\mu)\hat{\,}(y_j) = (J_j\mu_1)\hat{\,}(y_j) \cdot (J_j\mu_2)\hat{\,}(y_j).$$

Hence,

$$(A\mu_1)\hat{}(y_1,\ldots,y_d) \cdot (A\mu_2)\hat{}(y_1,\ldots,y_d)$$
$$= \prod_{j=1}^{d}(J_j\mu_1)\hat{}(y_j) \cdot \prod_{i=1}^{d}(J_i\mu_2)\hat{}(y_i).$$

This proves (a) and (b) by the uniqueness of the decomposition of an infinitely divisible law into its Gaussian and non-Gaussian components.

For (c) and (d), $J_j\mu = J_j\mu_1 * J_j\mu_2$. Then $J_j\mu$ is Gaussian if and only if $J_j\mu_2 = 0$. For $j \in T_1$, $J_jV_2 = 0$, so $J_j\mu_2 = 0$. Also, $J_j\mu$ has no Gaussian component if and only if $J_j\mu_1 = 0$. For $j \in T_2$, $J_jV_1 = 0$, so $J_j\mu_1 = 0$.

Clearly, (e) follows from (c) and (d). Q.E.D.

Corollary 4.13.11. *Under the assumptions in Theorem* 4.13.10, μ *has exactly one exponent if and only if* $\dim V_1 \leq 1$.

Proof. If $\dim V_1 > 1$, the Gaussian measure μ_1 on V_1 has many exponents. This implies that μ also does. If $\dim V_1 \leq 1$, the uniqueness of the exponent of μ follows from the uniqueness of the characteristic index of any univariate stable measure. Q.E.D.

Corollary 4.13.12. *Let* μ *be full and operator-stable on* $\mathbb{R}^d$. *Then* μ *has a set of complete orthogonal independent univariate marginals if and only if there is an exponent of* μ *which is diagonalizable.*

4.14 MULTIVARIATE STABLE MEASURES

We say that $\mu \in \mathcal{P}(\mathbb{R}^d)$ is *multivariate stable* if there are $a_n > 0$, $v_n \in \mathbb{R}^d$, and $\nu \in \mathcal{P}(\mathbb{R}^d)$ such that

$$T_{a_n}\nu^n * \delta(v_n) \Rightarrow \mu. \qquad (4.14.1)$$

In particular, when $v_n = 0$ in (4.14.1), we say that μ is *strictly multivariate stable*. For brevity, we use the terms *stable* and *strictly stable*. Recall that, for $a \in \mathbb{R}^1$, $T_ax := ax$ for all $x \in \mathbb{R}^d$. Thus, stable measures are a special subclass of the operator-stable measures obtained by requiring the *norming sequence* $\{A_n\}$ *to be of the restrictive form* $A_n = a_nI$ with $a_n > 0$. Therefore, the results in the earlier sections of this chapter can be specialized to this setting. However, the assumption that μ be full can be weakened to assuming that μ is nondegenerate, that is, $\mu \neq \delta(a)$ for any $a \in \mathbb{R}^d$. Also, some proofs are much simpler and straightforward than their counterparts in the general framework of operator-stable measures. Note that $\{T_{a_n}\nu: n \geq 1\}$ is an infinitesimal array (cf. Section 4.1), and that when μ is nondegenerate, we have $a_n \to 0$ and $a_{n+1}/a_n \to 1$ as $n \to \infty$ (cf. Proposition 3.9.1).

Proposition 4.14.1. *A measure $\mu \in \mathcal{P}(\mathbb{R}^d)$ is stable if and only if for every $n \geq 1$ there are $b_n > 0$ and $x_n \in \mathbb{R}^d$ such that*

$$\mu^n = T_{b_n}\mu * \delta(x_n). \tag{4.14.2}$$

In particular, μ is strictly stable if and only if $x_n = 0$ in (4.14.2).

Proof. When μ is degenerate, (4.14.2) is obvious. Assume μ is nondegenerate and stable. Fix $n \geq 1$. Then $T_{a_k}\nu^{kn} * \delta(n\nu_k) \Rightarrow \mu^n$ as $k \to \infty$. Let $\mu_k := T_{a_{kn}}\nu^{kn} * \delta(\nu_{kn})$. Then $\mu_k \Rightarrow \mu$ and $T_{a_k a_{kn}^{-1}}\mu_k * \delta(n\nu_k - a_k a_{kn}^{-1}\nu_{kn}) \Rightarrow \mu^n$ as $k \to \infty$. By Proposition 2.2.11, there are $b_n > 0$ and $x_n \in \mathbb{R}^d$ such that (4.14.2) holds. The remainder of the proof is obvious. Q.E.D.

We now obtain a particular form of (4.2.1) when μ is stable, but not necessarily full. We cannot simply apply Theorem 4.2.12 because its proof used the symmetry group of operators which is neither a group nor compact when μ is not full. However, since for μ stable we are limited to multiplying by constants, the analogous symmetry group is a compact group.

Theorem 4.14.2. *Let $\mu \in \mathcal{P}(\mathbb{R}^d)$ be nondegenerate. Then μ is stable if and only if there is $\alpha > 0$ such that for all $t > 0$ there is $x_t \in \mathbb{R}^d$ with*

$$\mu^t = T_{t^{1/\alpha}}\mu * \delta(x_t). \tag{4.14.3}$$

Proof. When (4.14.3) holds, μ is stable by Proposition 4.14.1. Now assume μ is nondegenerate and stable. For $t > 0$, let $G_t := \{b \neq 0: \mu^t = T_b\mu * \delta(x) \text{ for some } x \in \mathbb{R}^d\}$ and let $G := \bigcup_{t>0} G_t$. Using Proposition 4.14.1 and making slight changes in the proofs of Lemmas 4.2.3 to 4.2.7, we see that G is a closed subgroup of the group $(\mathbb{R}^1 \setminus \{0\}, \cdot)$ and that G_1 is a compact subgroup of G. So, select $c > 0$

such that $e^c \notin G_1$. Recall that the function $l\colon G \to \mathbb{R}_+$ given by $l(b) := t$ if $b \in G_t$ is a continuous homeomorphism (cf. Lemma 4.2.8). Since $h\colon \mathbb{R}^1 \to \mathbb{R}^1$ given by $h(s) := \log l(e^{sc})$ is a continuous homeomorphism (cf. Lemma 4.2.9), there is $a \neq 0$ such that $h(s) = as$ for all s. Let $\alpha := a/c$. Then $\log l(e^{s/\alpha}) = s$, so $\log l(t^{1/\alpha}) = \log t$ for all $t > 0$, that is, $t^{1/\alpha} \in G_t$. Therefore, (4.14.3) holds. Q.E.D.

Using this theorem and the proof of Theorem 4.6.5, we obtain the following corollaries.

Corollary 4.14.3. *Let $\mu \in \mathscr{P}(\mathbb{R}^d)$ be nondegenerate. Then μ is stable if and only if (4.14.3) holds for some $\alpha \in (0,2]$. Furthermore, when $\alpha = 2$, μ is Gaussian, and when $\alpha < 2$, μ has no Gaussian component.*

Corollary 4.14.4. *In Proposition 4.14.1, we may take b_n to be $n^{1/\alpha}$ with $0 < \alpha \leq 2$.*

Using the ideas in the proofs of Proposition 4.3.4 and Corollary 4.3.5, we obtain the following result.

Theorem 4.14.5. *Let $\mu := [a, R, M]$ be nondegenerate and stable on $\mathbb{R}^d$. Then exactly one of the following is true.*

(a) *$M = 0$ and μ is Gaussian.*
(b) *$R = 0$ and, for $A \in \mathscr{B}(\mathbb{R}^d \setminus \{0\})$,*

$$M(A) = \int_{S^{d-1}} \int_0^\infty I_A(tx) t^{-(1+\alpha)} \, dt \, m(dx), \quad (4.14.4)$$

where $0 < \alpha < 2$ and m is a Borel measure on the unit sphere S^{d-1} in $\mathbb{R}^d$ given by

$$m(F) := M([F;[1,\infty)])$$

for $F \in \mathscr{B}(S^{d-1})$.

Furthermore, when (b) holds

$$\operatorname{supp} M = \left(\bigcup_{t>0} t(\operatorname{supp} m) \right) \cup \{0\}.$$

Proof. Let $S_t := \{x \in \mathbb{R}^d : \|x\| = t\}$ for $t > 0$. Then (cf. Proposition 4.3.4)

$$M(A) = \int_{S_{1/\alpha}} \int_0^\infty I_A(s^{1/\alpha} x) s^{-2} \, ds \, m_1(dx).$$

By a change of variables, this becomes (4.14.4). Q.E.D.

Corollary 4.14.6. *The measure $\mu := [a, 0, M]$ is stable if and only if*

$$\hat{\mu}(y) = \exp\left\{i\langle a, y\rangle + \int_{S^{d-1}} \int_0^\infty \left(e^{it\langle x, y\rangle} - 1 - \frac{it\langle x, y\rangle}{1 + \|x\|^2}\right) t^{-(1+\alpha)} \, dt \, m(dx)\right\}.$$

It is obvious that if μ is stable on $\mathbb{R}^d$ with exponent α, then, for each $x \in \mathbb{R}^d$, $\Pi_x \mu$ is univariate stable with the same exponent α, where $\Pi_x(y) := \langle x, y\rangle$. Is the converse true? That is, do all the univariate marginals of $\mu \in \mathscr{P}(\mathbb{R}^d)$ being stable imply μ is stable on $\mathbb{R}^d$? In this generality, the answer is no! We shall show that with the added assumption that μ is infinitely divisible the answer is yes.

Lemma 4.14.7. *Let $\mu \in \mathscr{P}(\mathbb{R}^d)$. If for all $x \in \mathbb{R}^d$ the univariate distribution $\Pi_x \mu$ is strictly stable, then each distribution in the set $\{\Pi_x \mu : x \in \mathbb{R}^d$ and $\Pi_x \mu \neq \delta(0)\}$ has the same exponent $\alpha \in (0, 2]$.*

Proof. Let $\alpha(x)$ denote the exponent of $\Pi_x \mu$ with the convention that $\alpha(x) = 2$ when $\Pi_x \mu = \delta(0)$. For each $\alpha \in (0, 2]$, let $F(\alpha) := \{x \in \mathbb{R}^d : \alpha(x) \geq \alpha\}$. We first show that each $F(\alpha)$ is a subspace of $\mathbb{R}^d$. For $t \neq 0$, $\Pi_{tx}\mu$ and $\Pi_x\mu$ have the same exponent, because

$$(\Pi_{tx}\mu)\hat{\,}(s) = \hat{\mu}(stx) = (\Pi_x\mu)\hat{\,}(st).$$

Hence, $tx \in F(\alpha)$ for all t and for all $x \in F(\alpha)$. Now, let $x, y \in F(\alpha)$ and let $\{\xi_k\}$ be a sequence of independent identically distributed random vectors with common distribution μ. Then, for each $n \geq 1$, $n^{-1/\alpha(x)}\Pi_x(\xi_1 + \cdots + \xi_n)$ and $\Pi_x \xi_1$ have the same distribution, and $n^{-1/\alpha(y)}\Pi_y(\xi_1 + \cdots + \xi_n)$ and $\Pi_y \xi_1$ also have the same distribution. Hence, for any $\varepsilon \in (0, \alpha)$, $\mathscr{L}(n^{-1/\varepsilon}\Pi_{x+y}(\xi_1 + \cdots + \xi_n)) \Rightarrow \delta(0)$.

Consequently, $\varepsilon < \alpha(x+y)$ for every $\varepsilon \in (0, \alpha)$, so $\alpha \leq \alpha(x+y)$. Therefore, $x + y \in F(\alpha)$, and $F(\alpha)$ is a subspace.

Second, let $\alpha_F := \{\alpha(x): x \in F\}$, where F is any subspace of $\mathbb{R}^d$. Then, if dim $F = 2$, α_F cannot have more than two elements. To see this, suppose to the contrary that there is a subspace F of dimension 2 with $x, y, z \in F$ for which $\alpha(x) < \alpha(y) < \alpha(z)$. Then no two of x, y, z can be colinear since for colinear vectors ω_1, ω_2 we have $\alpha(\omega_1) = \alpha(\omega_2)$. Since F has dimension 2, $x = t_1 y + t_2 z$ for some nonzero t_1, t_2. Since $F(\alpha(y))$ is a subspace and $y, z \in F(\alpha(y))$, we have $x = t_1 y + t_2 z \in F(\alpha(y))$, that is, $\alpha(x) \geq \alpha(y)$, a contradiction.

Third, assume F is a two-dimensional subspace such that there are $\alpha_1, \alpha_2 \in \alpha_F$ with $\alpha_1 < \alpha_2$. Then $\alpha_F = \{\alpha_1, \alpha_2\}$. Let $F_2 := F \cap F(\alpha_2)$. Then F_2 is a subspace of F and dim $F_2 < 2$ since $\alpha_1 < \alpha_2$. For $x \in F_2$, select $x_n \in F \setminus F_2$ such that $x_n \to x$. Then $\Pi_{x_n}\mu \Rightarrow \Pi_x\mu$ (cf. Corollary 1.7.3), and the exponent of $\Pi_x\mu$ is α_2, whereas each $\Pi_{x_n}\mu$ has exponent α_1. This is not possible, unless $\alpha_2 = 2$ and $\Pi_x\mu = \delta(0)$.

Finally, since any two points in $\mathbb{R}^d$ lie in some two-dimensional subspace, we obtain the lemma. Q.E.D.

Theorem 4.14.8. *Let $\mu \in \mathscr{P}(\mathbb{R}^d)$. Then μ is strictly stable on $\mathbb{R}^d$ if and only if $\Pi_x\mu$ is strictly stable on $\mathbb{R}^1$ for each $x \in \mathbb{R}^d$. In which case, μ and $\Pi_x\mu$ have the same exponent α whenever $\Pi_x\mu \neq \delta(0)$.*

Proof. The converse part is immediate, so assume that each $\Pi_x\mu$ is strictly stable. By Lemma 4.14.7, the exponent of each $\Pi_x\mu$ is the same α for each $\Pi_x\mu \neq \delta(0)$. When $\Pi_x\mu = \delta(0)$ for every x, the theorem is obvious. Hence, assume $\Pi_x\mu \neq \delta(0)$ for at least one x. We will show that given $a > 0$ and $b > 0$, there is $c > 0$ such that $T_a\mu * T_b\mu = T_c\mu$. For x with $\Pi_x\mu \neq \delta(0)$, select $c > 0$ so that $T_a\Pi_x\mu * T_b\Pi_x\mu = T_c\Pi_x\mu$. Since $\Pi_x\mu$ is strictly stable with exponent not depending on x, such c exists and does not depend on x. Set $\nu := T_{a/c}\mu * T_{b/c}\mu$. Since T and Π commute, $\Pi_x\nu = \Pi_x\mu$, so $\Pi_x\nu$ is strictly stable. Hence, $\hat{\nu} = \hat{\mu}$, so $T_a\mu * T_b\mu = T_c\mu$. Now, by induction, for every $n \geq 1$, $\mu^n = T_{b_n}\mu$ for some $b_n > 0$. By Proposition 4.14.1, μ is strictly stable. Q.E.D.

Now, we have an obvious corollary.

Corollary 4.14.9. *Let $\mu \in \mathscr{P}(\mathbb{R}^d)$. If each $\Pi_x\mu$ is stable on $\mathbb{R}^1$, then μ^0 is strictly stable on $\mathbb{R}^d$.*

This brings us to a main result.

Theorem 4.14.10. *Assume μ is infinitely divisible on $\mathbb{R}^d$. Then μ is stable on $\mathbb{R}^d$ if and only if $\Pi_x \mu$ is stable on $\mathbb{R}^1$ for all $x \in \mathbb{R}^d$.*

Proof. When μ is stable on $\mathbb{R}^d$, it is obvious that each $\Pi_x \mu$ is stable on $\mathbb{R}^1$. So, we now assume that each $\Pi_x \mu$ is stable on $\mathbb{R}^1$ with exponent $\alpha \in (0, 2]$.

First, assume $1 < \alpha \leq 2$. Let m denote the mean of μ and let $\nu := \mu * \delta(-m)$. Then each $\Pi_x \nu$ is strictly stable on $\mathbb{R}^1$. By Theorem 4.14.8, ν is strictly stable on $\mathbb{R}^d$, so μ is stable on $\mathbb{R}^d$.

Second, assume $0 < \alpha < 1$. Since $\Pi_x \mu$ is stable, $\Pi_x \mu^0$ is strictly stable. Hence, μ^0 is strictly stable on $\mathbb{R}^d$. Consequently, μ has no Gaussian component. Let $\mu = [a, 0, M]$. Then $M + M^-$ is given by (4.14.4), where $M^-(A) := M(-A)$. Hence,

$$\int_{\|x\| \leq 1} \|x\| M(dx) = \frac{1}{2} \int_{\|x\| \leq 1} \|x\| (M + M^-)(dx)$$
$$= \frac{m(S^{d-1})}{2(1 - \alpha)} < \infty.$$

Therefore, $\mu = \nu * \delta(b)$ for some $b \in \mathbb{R}^d$, where

$$\hat{\nu}(y) := \exp\left\{ \int (e^{i\langle y, v \rangle} - 1) M(dv) \right\}.$$

Since $\Pi_x \mu$ is stable with exponent $\alpha < 1$, we have $\Pi_x \mu = \rho_x * \delta(a_x)$, for some $a_x \in \mathbb{R}^1$ with ρ_x strictly stable on $\mathbb{R}^1$ with exponent α. Hence,

$$T_{n^{-1/\alpha}} \Pi_x \mu^n = \rho_x * \delta(n^{1 - 1/\alpha} a_x) \Rightarrow \rho_x$$

as $n \to \infty$ since $\alpha < 1$. Let ρ' be any limiting measure, not necessarily a probability measure, of the set $\{T_{n^{-1/\alpha}} \mu^n : n \geq 1\}$. Then $\Pi_x \rho' = \rho_x$ for all $x \in \mathbb{R}^d$. Therefore, there is $\rho \in \mathscr{P}(\mathbb{R}^d)$ such that $\Pi_x \rho = \rho_x$ for all x and $T_{n^{-1/\alpha}} \mu^n \Rightarrow \rho$. Since each ρ_x is strictly stable with exponent α, ρ is strictly stable with the same exponent (cf. Theorem 4.14.8). Consequently,

$$\hat{\rho}(y) = \exp\left\{ \int (e^{i\langle y, v \rangle} - 1) N(dv) \right\}$$

for some Lévy measure N (cf. Theorem 4.14.5). Hence, for all x, $\Pi_x \nu * \delta(\langle x, b \rangle) = \Pi_x \mu = \Pi_x \rho * \delta(a_x)$. By the uniqueness of the representation of an infinitely divisible measure, $a_x = \langle x, b \rangle$ and $\nu = \rho$. Therefore, $\mu = \rho * \delta(b)$, so μ is stable on $\mathbb{R}^d$.

Third, assume $\alpha = 1$. As before, μ^0 is strictly stable with exponent 1. Hence, μ has no Gaussian component, so we can write, for some $a \in \mathbb{R}^d$, $\mu = \nu * \delta(a)$, with

$$\hat{\nu}(y) := \exp\left\{ \int \left(e^{i \langle y, v \rangle} - 1 - i \langle y, v \rangle I_{[\|v\| \leq 1]} \right) M(dv) \right\}. \quad (4.14.5)$$

For notational convenience, denote any measure ν whose characteristic function $\hat{\nu}$ is of this form by $c_{1,d}(M)$, for d-dimensional and centered at 1 with Lévy measure M. Then, for all $x \in \mathbb{R}^d$, there are $a_x \in \mathbb{R}^1$ and $c_{1,1}(M_x)$ such that $\Pi_x \mu = c_{1,1}(M_x) * \delta(a_x)$, with $c_{1,1}(M_x)$ being stable with exponent 1.

Since μ^0 is strictly stable with exponent 1, $T_{n^{-1}}(\mu^n * \bar{\mu}^n) = \mu^0$. Hence, by Theorem 1.7.1(b), the sequence $\{T_{n^{-1}} \mu^n : n \geq 1\}$ is shift tight. Therefore, letting $m_n := \int_{[\|x\| \leq 1]} x \mu(dx)$, the sequence $\{T_{n^{-1}} \mu^n * \delta(-m_n) : n \geq 1\}$ is tight, and any measure η which is the limit of some subsequence of it is of the form $\eta = c_{1,d}(N)$, with N possibly depending on the subsequence. Select such as subsequence so that

$$T_{n^{-1}} \mu^n * \delta(-m_n) \Rightarrow c_{1,d}(N) \quad (4.14.6)$$

along that subsequence.

Since $\Pi_x \mu$ is stable with exponent 1, for each $n \geq 1$ and each $x \in \mathbb{R}^d$, there is $j(n, x) \in \mathbb{R}^1$ such that

$$T_{n^{-1}}(\Pi_x \mu)^n * \delta(j(n, x) \log n) = \Pi_x \mu.$$

Hence,

$$T_{n^{-1}}(\Pi_x \mu)^n * \delta(-m_n) * \delta(m_n + j(n, x) \log n) = \Pi_x \mu. \quad (4.14.7)$$

From (4.14.6) and (4.14.7), there is $h_x \in \mathbb{R}^1$ such that $m_n + j(n, x) \log n \to h_x$ along that subsequence. Consequently, from (4.14.5) and (4.14.6),

$$\Pi_x c_{1,d}(N) * \delta(h_x) = \Pi_x \mu$$
$$= \Pi_x c_{1,d}(M) * \delta(\langle x, a \rangle). \quad (4.14.8)$$

For any Lévy spectral measure G of a stable measure with exponent $\alpha = 1$ and any $x \in \mathbb{R}^d$, let

$$c(G, x) := \int \left(I_{[|\langle x,v\rangle| \le 1]} - I_{[\|v\| \le 1]} \right) \langle x, v \rangle G(dv).$$

Note that $c(G, x)$ exists by Theorem 4.14.5. Then

$$\Pi_x c_{1,d}(G) = c_{1,1}(\Pi_x G) * \delta(c(G, x)).$$

Hence, (4.14.8) becomes

$$c_{1,1}(\Pi_x N) * \delta(h_x + c(N, x)) = c_{1,1}(\Pi_x M) * \delta(\langle x, a \rangle + c(M, x)).$$

Thus, $\Pi_x N = \Pi_x M$ for all x. In particular, we have $c(M, x) - c(N, x) = \langle x, b \rangle$, for some $b \in \mathbb{R}^d$. Consequently,

$$\Pi_x c_{1,d}(M) = c_{1,1}(\Pi_x N) * \delta(c(M, x))$$
$$= \Pi_x c_{1,d}(N) * \delta(\Pi_x b).$$

Therefore, $c_{1,d}(M) = c_{1,d}(N) * \delta(b)$, so $b = 0$ and $M = N$. From (4.14.5), $\mu = c_{1,d}(N) * \delta(a)$ with $c_{1,d}(N)$ being stable. Therefore, μ is stable on $\mathbb{R}^d$. Q.E.D.

Example 4.14.11. It can be shown that the complex-valued function g defined on $\mathbb{R}^2$ by

$$g(x, y) := e^{i\gamma r \cos 3\theta} - r^\alpha,$$

where (r, θ) are the polar coordinates of (x, y), $0 < \alpha < 1$, and $\gamma > 0$ is sufficiently small, is the characteristic function corresponding to a nonstable measure on $\mathbb{R}^2$. However, all its univariate marginals are stable with exponent α.

What is surprising is that Example 4.14.11 essentially captures the necessity to assume that μ is infinitely divisible in Theorem 4.14.10. This assumption can be weakened.

Theorem 4.14.12. *Let $\mu \in \mathscr{P}(\mathbb{R}^d)$ be such that all of its two-dimensional marginals are infinitely divisible. Then μ is stable on $\mathbb{R}^d$ if and only if each $\Pi_x \mu$ is stable on $\mathbb{R}^1$.*

Proof. Assume each $\Pi_x \mu$ is stable on $\mathbb{R}^1$. Let $u_1, u_2 \in \mathbb{R}^d$ be linearly independent, let $V := \text{lin}\{u_1, u_2\}$, and let J be the orthogonal projection of $\mathbb{R}^d$ onto V. for $x \in \mathbb{R}^d$, let $x' := Jx$. Then $\Pi_x(J\mu) = \Pi_{x'}\mu$. Since each $\Pi_{x'}\mu$ is stable on $\mathbb{R}^1$ and $J\mu$ is infinitely divisible by assumption, by Theorem 4.14.10 we have that $J\mu$ is stable on V. Therefore, given $a > 0$ and $b > 0$, there are $c > 0$ and $f(u_1, u_2) \in V$ such that

$$T_{ac}(J\mu) * T_{bc}(J\mu) * \delta(f(u_1, u_2))$$
$$= J(T_{ac}\mu * T_{bc}\mu) * \delta(f(u_1, u_2)) = J\mu. \quad (4.14.9)$$

For $x \in \mathbb{R}^d$, let $g(x) \in \mathbb{R}$ be such that

$$\Pi_{x'}(T_{ac}\mu * T_{bc}\mu) * \delta(g(x)) = \Pi_{x'}\mu. \quad (4.14.10)$$

Using $\Pi_x(J\mu) = \Pi_{x'}\mu$, (4.14.9), and (4.14.10), we see that g is linear on $\mathbb{R}^d$. Also, if $x_n \to 0$, then $g(x_n) \to 0$. Therefore, there is z such that $g(x) = \langle x, z \rangle$. Hence, $\Pi_{x'}(T_{ac}\mu * T_{bc}\mu * \delta(r)) = \Pi_{x'}\mu$, where $r \in \mathbb{R}^d$ is selected so $Jr = z$. This implies that *all* univariate marginals of $T_{ac}\mu * T_{bc}\mu * \delta(r)$ and of μ agree. By the Cramér–Wold device, the two measures agree, so μ is stable. Q.E.D.

Example 4.14.13 (Lévy). Let X and Y be independent real-valued random variables, each $\mathcal{N}(0, 1/4)$, and let μ be the distribution of the random vector $(X^2, 2XY, Y^2)$. Then μ is not infinitely divisible, but the joint distribution of any two of X^2, $2XY$, and Y^2 is infinitely divisible.

Example 4.14.14. There are noninfinitely divisible measures on $\mathbb{R}^d$ with $d > 2$ which have all of their two-dimensional marginals being infinitely divisible. For $y \in \mathbb{R}^d$, $d > 2$, define

$$\varphi(y) := \exp\left\{ \int (e^{i\langle y, x \rangle} - 1) Z_\varepsilon(dx) \right\},$$

where $Z_\varepsilon(F) := \lambda(F \cap D_1) - \varepsilon \lambda(F \cap D_2)$ for $0 < \varepsilon < 1$, λ is a Lebesgue measure on $\mathbb{R}^d$, $D_1 := \{x \in \mathbb{R}^d : 1 < \|x\| < 2\}$, and $D_2 := \{x \in \mathbb{R}^d : \|x\| < \delta\}$, with $\delta > 0$ small. Then φ is a characteristic function for each ε and δ. Also, by selecting $\delta > 0$ sufficiently small, the two-dimensional marginals of Z_ε are positive measures. Consequently, these marginals are infinitely divisible.

Let us now examine the domain of attraction of a stable measure on $\mathbb{R}^d$. For a stable measure μ with exponent $\alpha \in (0,2]$, its *domain of normal attraction*, DONA, consists of those $\nu \in \mathscr{P}(\mathbb{R}^d)$ such that

$$n^{-(1/\alpha)I}\nu^n * \delta(v_n) \Rightarrow \mu$$

for some sequence $\{v_n\}$ in $\mathbb{R}^d$.

Theorem 4.14.15. *Let μ be stable on $\mathbb{R}^d$ with exponent α. When $\alpha = 2$, so μ is Gaussian, $\nu \in \text{DONA}(\mu)$ if and only if the covariance operators of μ and of ν are the same. When $\alpha < 2$, $\nu \in \text{DONA}(\mu)$ if and only if*

$$\lim_{t \to \infty} t \cdot \nu(\{sx : x \in E, s > t^{1/\alpha}\}) = m(E)$$

for every $E \in \mathscr{B}(S^{d-1})$, where m is given in Theorem 4.14.5.

This theorem is a consequence of Theorem 4.14.5.

Next, we consider the finiteness of moments. When μ is Gaussian, the results are contained in Theorem 4.12.2, Corollary 4.12.3, and Proposition 4.12.4. For μ non-Gaussian, from Theorems 4.12.5 and 4.12.6 we obtain the following result.

Theorem 4.14.16. *Let μ be stable on $\mathbb{R}^d$ with exponent $\alpha < 2$. Then $\int \|x\|^p \mu(dx) < \infty$ if and only if $p < \alpha$.*

Finally, we consider the problem of the existence of a set of complete orthogonal independent univariate marginals (cf. Section 4.13). The Gaussian case is immediately obvious, so we address the non-Gaussian case.

Theorem 4.14.17. *Let μ be stable on $\mathbb{R}^d$ with exponent $\alpha < 2$. Then there is exactly one set of complete orthogonal independent univariate marginals. Furthermore, $J_1\mu, \ldots, J_d\mu$ is such a set if and only if the Lévy measure M of μ satisfies*

$$\text{supp } M \subset \bigcup_{j=1}^{d} J_j(\mathbb{R}^d).$$

Proof. By Corollary 4.13.12, such a set of univariate marginals exists, and, by Theorem 4.13.6, there is only one such set. The final statement follows from Theorem 4.13.9. Q.E.D.

4.15 THE CASE WHEN $d = 3$

Let μ be full and operator-stable on $\mathbb{R}^3$. We know that its symmetry group, $\mathbf{A}(\mu)$, is conjugate to a closed subgroup, $\mathscr{O}_0$, of the full orthogonal group $\mathscr{O}$ on $\mathbb{R}^3$ (cf. Theorem 2.4.1).

Lemma 4.15.1. *The dimension of the tangent space $\mathscr{T}(\mathscr{O}_0)$ is either 0, 1, or 3.*

Proof. We know that $\mathscr{T}(\mathscr{O})$ is $\mathscr{K}$, the set of all skew-symmetric operators on $\mathbb{R}^3$ (cf. Example 1.5.7). Since $\mathscr{O}_0$ is a closed subgroup in $\mathscr{O}$, we have that $\mathscr{T}(\mathscr{O}_0) \subset \mathscr{T}(\mathscr{O})$ and that $\mathscr{T}(\mathscr{O}_0)$ is closed under the Lie bracket $[\cdot, \cdot]$ (cf. Lemma 1.5.3). Define the linear transform f from $\mathbb{R}^3$ onto $\mathscr{K}$ by

$$f(a, b, c) := \begin{bmatrix} 0 & -a & -b \\ a & 0 & -c \\ b & c & 0 \end{bmatrix}.$$

Note that, for $x, y \in \mathbb{R}^3$, $[f(x), f(y)] = f(x \times y)$, where $\times$ is the usual cross product on $\mathbb{R}^3$. Hence, $(\mathbb{R}^3, \times)$ is isomorphic to $\mathscr{K}$. Therefore, either $\mathscr{T}(\mathscr{O}_0) = \{0\}$, $\dim \mathscr{T}(\mathscr{O}_0) = 1$, or $\mathscr{O}_0 = \mathscr{O}$. Q.E.D.

When $\mathscr{T}(\mathscr{O}_0) = \{0\}$, μ has an unique exponent. When $\mathbf{A}(\mu)$ is conjugate to $\mathscr{O}$, the class of exponents $\mathscr{E}_s(\mu)$ is given by $cI + W^{-1}\mathscr{K}W$ for some $c \geq 1/2$ and for some $W \in \mathrm{Aut}(\mathbb{R}^3)$ (cf. Theorem 4.8.1), and the characteristic function $\hat{\mu}$ is given in Theorem 4.8.4. We now consider the case when $\dim \mathscr{T}(\mathscr{O}_0) = 1$.

Lemma 4.15.2. *Let $\mathscr{O}_0$ be a closed subgroup of $\mathscr{O}$ with $\dim \mathscr{O}_0 = 1$. Then*

$$\mathscr{T}(\mathscr{O}_0) = \left\{ \begin{bmatrix} 0 & -c & 0 \\ c & 0 & 0 \\ 0 & 0 & 0 \end{bmatrix} : c \in \mathbb{R}^1 \right\},$$

with respect to some orthonormal basis for $\mathbb{R}^3$.

Proof. We know $\dim \mathscr{T}(\mathscr{O}_0) = 1$, so let $\{Q\}$ be a basis for $\mathscr{T}(\mathscr{O}_0)$. Then $Q \in \mathscr{K}$. Since $\det Q = \det(-Q^*) = (-1)^3 \det Q$, $\det Q = 0$, that is, Q is singular. Let $u \in \mathbb{R}^3$ be such that $|u| = 1$ and $Qu = 0$. Let this u be a third member of an orthonormal basis for $\mathbb{R}^3$. This is the basis needed in the lemma. By skew-symmetry the third row and third column are all zeros, and by skew-symmetry the diagonal is all zeros and the (1, 2)-element is the negative of the (2, 1)-element. Since $\{Q\}$ is a basis for $\mathcal{T}(\mathcal{O}_0)$, we have the stated result. Q.E.D.

Theorem 4.15.3. *Let μ be full operator-stable on $\mathbb{R}^3$. If $\mathcal{T}(\mathbf{A}(\mu))$ is neither $\{0\}$ nor conjugate to $\mathcal{K}$, then there is a positive-definite self-adjoint linear operator W on $\mathbb{R}^3$ and there are real numbers $a, b \geq 1/2$ such that, with respect to some orthonormal basis for $\mathbb{R}^3$,*

$$\mathcal{E}_s(W^{-1}\mu) = \left\{ \begin{bmatrix} a & -c & 0 \\ c & a & 0 \\ 0 & 0 & b \end{bmatrix} : c \in \mathbb{R}^1 \right\}.$$

Proof. Since $\mathbf{A}(\mu)$ is conjugate to a closed subgroup $\mathcal{O}_0$ of $\mathcal{O}$, select the W so that $\mathbf{A}(W^{-1}\mu) = \mathcal{O}_0$. By Lemma 4.15.2,

$$\mathcal{T}(\mathbf{A}(W^{-1}\mu)) = \left\{ \begin{bmatrix} 0 & -c & 0 \\ c & 0 & 0 \\ 0 & 0 & 0 \end{bmatrix} : c \in \mathbb{R}^1 \right\},$$

with respect to some orthonormal basis. Use this basis for all matrix representations. Let $B \in \mathcal{E}_s(W^{-1}\mu)$ with the representation (b_{ij}). By Lemma 1.5.3 and the proof of Theorem 4.6.7, $BQ - QB \in \mathcal{T}(\mathbf{A}(W^{-1}\mu))$ for all $Q \in \mathcal{T}(\mathbf{A}(W^{-1}\mu))$, that is,

$$B \begin{pmatrix} J & 0 \\ 0 & 0 \end{pmatrix} - \begin{pmatrix} J & 0 \\ 0 & 0 \end{pmatrix} B = \begin{pmatrix} 0 & -c & 0 \\ c & 0 & 0 \\ 0 & 0 & 0 \end{pmatrix}$$

for some $c \in \mathbb{R}^1$, where

$$J := \begin{pmatrix} 0 & -1 \\ 1 & 0 \end{pmatrix}.$$

This implies that $b_{31} = b_{32} = b_{12} = b_{23} = 0$ and $b_{12} = b_{21}$. It only remains to show that $b_{11} = b_{22}$.

We now know that $B' := \mathrm{diag}(b_{11}, b_{22}, b_{33})$ is in $\mathcal{E}_s(W^{-1}\mu)$. Since B' commutes with every Q in $\mathcal{T}(\mathbf{A}(W^{-1}\mu))$, we have that the $\mathrm{diag}(b_{11}, b_{22})$ commutes with every 2×2 skew-symmetric matrix. Using Schur's lemma, this implies that $\mathrm{diag}(b_{11}, b_{22})$ is a multiple of the identity. Hence, $b_{11} = b_{22}$. Q.E.D.

THE CASE WHEN $d = 3$

Theorem 4.15.4. *Assume* $\dim(\mathcal{T}(\mathbf{A}(\mu))) = 1$ *and*

$$\mathcal{T}(\mathbf{A}(\mu)) = \left\{ \begin{bmatrix} 0 & -c & 0 \\ c & 0 & 0 \\ 0 & 0 & 0 \end{bmatrix} := c \in \mathbb{R}^1 \right\},$$

μ has an exponent B whose matrix representation is $\mathrm{diag}(a, a, b)$ with respect to the usual basis, where a and b are both greater than $1/2$, and M is the Lévy measure of μ. Then we have that $\mathbf{A}(\mu) = \mathcal{O}'$, where $\mathcal{O}'$ is the subgroup of $\mathcal{O}$ generated by all orthogonal transformations which leave the z-axis invariant. Furthermore, we have that there is a finite Borel measure ν on $[-\pi/2, \pi/2]$ such that, for Borel $A \subset \mathbb{R}^3 \setminus \{0\}$,

$$M(A) = \int_{-\pi/2}^{\pi/2} \int_0^{2\pi} \int_0^\infty I_A(t^B x(\theta, \varphi)) t^{-2} \, dt \, d\theta \, \nu(d\varphi),$$

where $x(\theta, \varphi) = (\cos\theta \cos\varphi, \sin\theta \cos\varphi, \sin\varphi)$.

Conversely, if ν is a finite Borel measure on $[-\pi/2, \pi/2]$ and if for Borel $A \subset \mathbb{R}^3 \setminus \{0\}$, $\overline{M}(A)$ is defined by the previous triple integral with $x(\theta, \varphi)$ as before and B is $\mathrm{diag}(a, a, b)$, with a and b greater than $1/2$, then $\overline{M}$ is a Lévy measure, $t^B \overline{M} = t \cdot \overline{M}$ for all $t > 0$, and $\mathbf{A}(\overline{M}) \subset \mathcal{O}'$.

[In the converse part of this theorem, we *cannot* conclude that $B \in \mathcal{E}_s(\mu)$ implies $\mathbf{A}(\mu) = \mathcal{O}'$. That is, μ may have an exponent which is diagonalizable even though $\mathbf{A}(\mu)$ is discrete.]

Proof. By the assumed form of $\mathcal{T}(\mathbf{A}(\mu))$, we have $\mathbf{A}(\mu) = \mathcal{O}'$. The first task is to construct ν. For A a Borel subset of S_B (cf. Theorem 4.3.7), let $m(A) := M\{t^B x : x \in A, t > 1\}$. Let $= \Phi \colon S_B \times \mathbb{R}^+ \to \mathbb{R}^3 \setminus \{0\}$ be given by

$$\Phi(x, t) := t^B x$$

(cf. Proposition 4.3.4). Let $\tilde{M}$ be the measure on $S_B \times \mathbb{R}^+$ given by $\tilde{M} := m \times \gamma$, where $d\gamma := t^{-2} dt$ on $\mathbb{R}^+$. Clearly, $M = \Phi \tilde{M}$.

Next, we show that there is a finite Borel measure ν on $[-\pi/2, \pi/2]$ such that $m = \lambda \times \nu$, where λ is a Lebesgue measure on $[0, 2\pi)$. Let $R(\theta)$ denote the counterclockwise rotation about the z-axis through an angle of θ radians. Let D and E be Borel subsets of $[0, 2\pi)$ and $[-\pi/2, \pi/2]$, respectively. Since $\mathbf{A}(M) = \mathcal{O}'$, $R(\theta) m(D \times E) = m(D \times E)$. Set $\alpha_E(D) := m(D \times E)$, for E fixed. Then α_E is a finite Borel measure on $[0, 2\pi)$ which is invariant under rotations, so $\alpha_E =$

$a(E) \cdot \lambda$, where $a(E)$ is a constant depending on E. But, $a(\cdot)$ as a function of E is a finite Borel measure on $[-\pi/2, \pi/2]$. Set $\nu(E) = a(E)$. Clearly, $m = \lambda \times \nu$. This establishes the stated representation for M.

Now, we prove the converse of the theorem. We first show that $\overline{M}$ is a Lévy measure. Clearly, $\overline{M}$ is a measure on $\mathbb{R}^3 \setminus \{0\}$, so it remains to show that $\int (|x|^2 \wedge 1) \overline{M}(dx) < \infty$, where $c \wedge d$ means the minimum of c and d. Clearly, for $0 < t \leq 1$, $\|t^B\| \leq 3r^\alpha$, where $\alpha = a \wedge b$. Thus,

$$\int (|x|^2 \wedge 1) \overline{M}(dx)$$

$$\leq k_1 + k_2 \int_{-\pi/2}^{\pi/2} \int_0^{2\pi} \int_0^1 \left(|t^B x(\theta, \varphi)|^2 \wedge 1 \right) t^{-2} \, dt \, d\theta \, \nu(d\varphi)$$

for some constants k_1 and k_2. But,

$$\int_0^1 \left(|t^B x(\theta, \varphi)|^2 \wedge 1 \right) t^{-2} \, dt \leq 9 \int_0^1 t^{2(\alpha-1)} \, dt < \infty,$$

since $\alpha > 1/2$. Hence $\overline{M}$ is a Lévy measure.

The rest of the theorem follows easily. Q.E.D.

4.16 BIBLIOGRAPHIC COMMENTS

The beginning of the study of the operator-stable laws presented in Section 4.2 is due independently to Sharpe (1969) and Sakovic (1961, 1964). Our presentation follows Sharpe (1969) and Hudson and Mason (1981). The material in Section 4.3 partially follows Hudson and Mason (1981), Jurek (1982c, 1984, 1989a), Krakowiak (1979), and Sharpe (1969). Section 4.4 is based on Jurek (1982c, 1984). The ideas in Section 4.5 are due to Urbanik (1975, 1977, 1978). The structural results and results on the exponents of an operator-stable law in Section 4.6 follow Holmes, Hudson, and Mason (1982), Jurek (1979a), and Sharpe (1969). Hudson, Mason, and Veeh (1983) is the basis for Section 4.7, whereas Section 4.8 is from Holmes, Hudson, and Mason (1982) and Luczak (1984). Material on the centering function is due to Sharpe (1969). Also concerning $b(t)$, Sato (1987) finds condition to eliminate it, that is, $\mu^t = t^B \mu$, when 1 is an eigenvalue of the exponent B. For Section 4.10 on the absolute continuity, see Hudson (1980).

The domain of normal attraction material is from several sources, Hudson, Mason and Veeh (1983), Jurek (1980), and Meerschaert (1990). Material concerning moments is from Hudson, Jurek, and Veeh (1986) and Meerschaert (1990). Section 4.13 on independent marginals follows Hudson (1980), Hudson, Mason, and Tucker (1981), and Veeh (1982). Also of interest is that there are *nonstable* laws on $\mathbb{R}^2$ with *all* univariate marginals being stable [cf. Marcus (1983)]. However, if all two-dimensional marginals are infinitely divisible, then μ is stable if and only if all of its one-dimensional marginals are stable [cf. Giné and Hahn (1983)]. The material on multivariate stable measures is well known [e.g., Lévy (1937)], but also see Giné and Hahn (1983) and Dudley and Kanter (1974). The example following Lévy's example is due to Linnik and Ostrovskii (private communication). Finally, when $d = 3$ is from Holmes, Hudson, and Mason (1982).

Epilogue

The selection of topics in this monograph was biased by our point of view. Nevertheless, we want to indicate some areas of research which are close to the theory of operator-limit distributions and are not mentioned here.

Operator-Semistability

Let $\xi_1, \xi_2, \ldots$ be independent identically distributed random vectors, either $\mathbb{R}^d$-valued or Banach space valued, let $A_1, A_2, \ldots$ be bounded linear operators, and let $a_1, a_2, \ldots$ be vectors, as in Section 4.1. We say that μ is *operator-semistable* if there exists a sequence of natural numbers $1 \leq k_1 < k_2 < \cdots$ such that

$$A_n(\xi_1 + \cdots + \xi_{k_n}) + a_n \Rightarrow \mu \quad \text{and} \quad \frac{k_{n+1}}{k_n} \to r \geq 1$$

as $n \to \infty$. Obviously, this yields the class of operator-stable laws whenever $k_n = n$, or for more general sequences with $r = 1$. The following is the main characterization of such μ.

A full probability μ is operator-semistable if and only if there exist $c \in (0, 1)$, an operator B, and a vector b such that

$$\mu^c = B\mu * \delta(b)$$

[cf. Jajte (1977), Krakowiak (1979, 1980), Kruglov (1972), Luczak (1981, 1984), and Siebert (1986)]. In fact, the above equation was studied in Lévy (1937) for the real line. Similar to the spectral characterization of exponents of operator-stable probabilities (cf. Theorem 4.6.5), there is a characterization of the operator B which appears in the above factorization of operator-semistable probabilities [cf. Jajte (1977)].

$\mathscr{G}$-Stability

One way of looking at the Sakovic–Sharpe characterization is the following. Since $t^B\mu = \mu^t * \delta(-b_t)$ for all $t > 0$, we have

$$t^B\mu * s^B\mu = (s+t)^B\mu * \delta(b_{t+s} - b_t - b_s)$$

for all $t, s > 0$, where $\{t^B: t > 0\}$ is a group of linear operators. Thus, for an arbitrary group $\mathscr{G}$ of bounded linear operators, we say that μ is $\mathscr{G}$-stable if for each $G_1, G_2 \in \mathscr{G}$, there exist $G_3 \in \mathscr{G}$ and a vector v such that

$$G_1\mu * G_2\mu = G_3\mu * \delta(v).$$

Taking $\mathscr{G} = \{T_a: a > 0\}$, where $T_a v = av$, this becomes the classical notion of stability of measure (cf. Section 4.14). Parthasarathy and Schmidt (1975) and Schmidt (1975) contain most of what is known in the case $\mathscr{G} \subset \text{Aut}(\mathbb{R}^d)$. Obviously, μ essentially depends on how large $\mathscr{G}$ is. More precisely, μ is called *completely stable* when μ is $\text{Aut}(\mathbb{R}^d)$-stable. Parthasarathy (1973) showed that μ is completely stable, $d \geq 2$, if and only if μ is Gaussian; also, see Mincer (1982a). In the case of an infinite-dimensional Hilbert space H and the group $\text{Aut}(H)$, the situation is very different. The class of completely stable measures includes Gaussian measures and some non-Gaussian ones [cf. Mincer and Urbanik (1979)].

One-Parameter Group Normalization

In Chapters 3 and 4, we dealt with the limit distributions of sequences

$$A_n(\xi_1 + \cdots + \xi_n) + a_n,$$

with sequences $\{A_n\}$ from the full group Aut. However, the final characterizations of limiting μ were

$$\mu = e^{-tQ}\mu * \nu_t, \qquad \mu^t = t^B\mu * \delta(b_t), \qquad t > 0,$$

for operator-self decomposable and operator-stable measures, respectively (cf. Theorems 3.5.5 and 4.2.12). Hence, one sees that we could take A_n from a one-parameter group $\{t^C: t > 0\} = \{e^{-sC}: s \in \mathbb{R}\}$ with a fixed operator C. In infinite-dimensional linear spaces not all one-parameter groups are of exponential form (cf. Section 1.4). This leads to the following concept. For a given one-parameter strongly continuous group $\mathbb{U}$ of bounded linear operators on a Banach space X, we say that $\mu \in L(\mathbb{U})$ if there exist $U_{t_n} \in \mathbb{U}$ and X-valued independent random variables ξ_n such that

(i) $U_{t_n}\xi_j$, $1 \le j \le n$, $n \ge 1$, are uniformly infinitesimal;
(ii) $U_{t_n}(\xi_1 + \cdots + \xi_n) + x_n \Rightarrow \mu$

for some $x_n \in X$. Similarly, $\mu \in \mathscr{S}(\mathbb{U})$ if the sequence $\{\xi_n\}$ is also identically distributed (cf. Sections 3.4 to 3.6, 4.3, and 4.4). The concept of normalization by elements from a group $\mathbb{U}$ was investigated in Jurek (1983a) and Kehrer (1983).

Bibliography

Araujo, A. and Giné E. (1980). *The Central Limit Theorem for Real and Banach Valued Random Variables.* Wiley.

Billingsley, P. (1966). Convergence of types in k-spaces. *Z. Wahrsch. Verw. Gebiete* **5** 175–179.

Billingsley, P. (1968). *Convergence of Probability Measures.* Wiley.

Billingsley, P. (1986). *Probability and Measure.* Second Edition, Wiley.

Chorny, V. (1986). Operator-semistable distributions on $\mathbb{R}^d$. *Theory Probab. Appl.* **31** 703–705.

Cuppens, R. (1975). *Decomposition of Multivariate Probability.* Academic.

de Acosta, A. (1980). Exponential moments of vector valued random series and triangular arrays. *Ann. Probab.* **8** 381–389.

de Acosta, A. (1982). Invariance principles in probability for triangular arrays of B-valued random vectors and some applications. *Ann. Probab.* **10** 346–373.

deConinck, J., (1984). Infinitely divisible distribution functions of class L and the Lee–Yang Theorem. *Comm. Math. Phys.* **96** 373–385.

Dudley, R. M., and Kanter, M. (1974). Zero–one laws for stable measures. *Proc. Amer. Math. Soc.* **148** 623–624.

Dynkin, E. B. (1965). *Markov Processes* **1**. Springer-Verlag.

Fenchal, W. (1936). *Ueber beschraenkte lineare Gruppen. Mat. Tidsshar.* **B** 10.

Fisz, M. (1954). A generalization of a theorem of Khintchine *Studia Math.* **14** 310–313.

Freedman, D. (1971). *Brownian Motion.* Holden-Day.

Gikhman, I. I., and Skorohod, A. V. (1969). *Introduction to the Theory of Random Processes.* W. B. Saunders.

Giné, E. and Hahn, M. G. (1983). On stability of probability laws with univariate stable marginals. *Z. Wahrsch. Verw. Gebiete* **64** 157–165.

Gnedenko, B. V., and Kolmogorov, A. N. (1954). *Limit Distributions for Sums of Independent Random Variables.* Addison-Wesley.

Griffin, P. (1986). Matrix normalized sums of independent identically distributed random vectors. *Ann. Probab.* **14** 224–246.

Hahn, M. G., and Klass, M. J. (1980). Matrix normalization of sums of random vectors in the domain of attraction of the multivariate normal. *Ann. Probab.* **8** 262–280.

Hahn, M. G., and Klass, M. J. (1981). The multidimensional central limit theorem for arrays normed by affine transformations. *Ann. Probab.* **9** 611–623.

Hahn, M. G. and Klass, M. J. (1985). Affine normability of partial sums of I.I.D. random vectors: a characterization. *Z. Wahrsch. Verw. Gebiete* **69** 479–505.

Hazod, W. (1985). Stable probability measures on groups and on vector spaces: a survey. *Probability measures on Groups. VIII. Proceedings, Oberwofach.* Springer-Verlag.

Hazod, W., and Nobel, S. (1990). Convergence of types theorem for simply connected nilpotent Lie groups. Preprint.

Hille, E., and Phillips, R. S. (1957). *Functional Analysis and Semi-Groups.* American Mathematical Society, Colloq. Publication **31**.

Hoffman, K. and Kunze, R. (1971). *Linear Algebra 2nd ed.* Prentice-Hall.

Holmes, J. P., Hudson, W. N., and Mason, J. D. (1982). Operator-stable laws: multiple exponents and elliptical symmetry. *Ann. Probab.* **10** 602–612.

Hudson, W. N. (1980). Operator-stable distributions and stable marginals. *J. Multivariate Anal.* **10** 26–37.

Hudson, W. N., and Mason, J. D. (1981). Operator-stable laws. *J. Multivariate Anal.* **11** 434–447.

Hudson, W. N., and Mason, J. D. (1981a). Operator-stable distribution on $\mathbb{R}^2$ with multiple exponents. *Ann. Probab.* **9** 482–489.

Hudson, W. N., and Mason, J. D. (1981b). Exponents of operator-stable laws. *Lecture Notes in Math.* **860** 291–298.

Hudson, W. N., and Mason, J. D. (1982). Operator-self-similar processes in a finite-dimensional space. *Trans. Amer. Math. Soc.* **273** 281–297.

Hudson, W. N., Jurek, Z. J., and Veeh, J. A., (1986). The symmetry group and exponents of operator stable probability measures. *Ann. Probab.* **14** 1014–1023.

Hudson, W. N., Mason, J. D., and Tucker, H. G. (1981). Operator-stable distributions with independent marginals. *Z. Wahrsch. Verw. Gebiete* **58** 285–297.

Hudson, W. N., Mason, J. D., and Veeh, J. A. (1983). The domain of normal attraction of an operator-stable law. *Ann. Probab.* **11** 178–184.

Ilinskii, A. I. (1978). On c-decomposability of characteristic functions. *Litovsk. Mat. Sb.* **18** 45–50 (in Russian).

Jajte, R. (1964). On stable distributions in Hilbert space. *Studia Math.* **30** 63–71.

Jajte, R. (1977). Semi-stable probability measures on $\mathbb{R}^N$. *Studia Math.* **61** 29–39.

Jurek, Z. J. (1979a). Remarks on operator-stable probability measures. *Ann. Soc. Math. Polon. Ser. I Comment. Math. Prace Mat.* **21** 71–74.

Jurek, Z. J. (1979b). On stability of probability measures in Euclidean spaces. *Lecture Noters in Math.* **828** 17–26. Springer-Verlag.

Jurek, Z. J. (1980). Domains of normal attraction of operator-stable measures on Euclidean spaces. *Bull. Acad. Polon. Sci. Sér. Sci. Math.* **28** 397–409.

Jurek, Z. J. (1981). Convergence of types, selfdecomposability and stability of measures on linear spaces. *Lecture Notes in Math.* **860** 257–267. Springer-Verlag.

Jurek, Z. J. (1982a). An integral representation of operator-selfdecomposable random variables. *Bull. Acac. Polon. Sci.* **30** 385–393.

Jurek, Z. J. (1982b). Structure of a class of operator-selfdecomposable probability measures. *Ann. Probab.* **10**, 849–856.

Jurek, Z. J. (1982c). How to solve the inequality: $U_t m \leq m$ for every $0 < t < 1$? II. *Bull. Acad. Polon. Sci.* **30** 476–483.

Jurek, Z. J. (1982d). Remarks on V-decomposable measures. *Bull. Acad. Polon. Sci.*, **30** 395–401.

Jurek, Z. J. (1983a). Limit distributions and one-parameter groups of linear operators on Banach spaces. *J. Multivariate Anal.* **13** 578–604.

Jurek, Z. J. (1983b). The classes $L_m(Q)$ of probability measures on Banach spaces. *Bull. Acad. Polon. Sci. Sér. Sci. Math.* **31** 51–62.

Jurek, Z. J. (1984). On polar coordinates in Banach spaces. *Bull. Polon. Acad. Math.* **3** 61–66.

Jurek, Z. J. (1985a). Relations between the s-selfdecomposable and selfdecomposable measures. *Ann. Probab.*, **13** 592–608.

Jurek, Z. J. (1985b). Random integral representation for another class of limit laws. *Lecture Notes in Math.* **1153** 297–309. Springer-Verlag.

Jurek, Z. J. (1988a). Random integral representations for classes of limit distributions similar to Lévy class L_0. *Probab. Theory Related Fields* **78** 473–490.

Jurek, Z. J. (1988b). Remarks concerning the theory of operator-limit distributions. *Bull. Acac. Polon. Sci*, **36** 307–313.

Jurek, Z. J. (1989a). Linear support and absolute continuity. *Lecture Notes in Math.* **1391** 140–147. Springer-Verlag.

Jurek, Z. J. (1989b). Random integral representations for classes of limit distributions similar to Lévy class L_0. II. *Nagoya Math. J.* **114** 53–64.

Jurek, Z. J. (1992). Operator exponents of probability measures and Lie semigroups. *Ann. Probab.* **20** 1053–1062.

Jurek, Z. J. and Rosinski, J. (1988). Continuity of certain random integral mappings and the uniform integrability of infinitely divisible measures. *Theory Probab. Appl.* **33** 560–572 (Russian edition).

Jurek, Z. J. and Smalara, J. (1981). On integrability with respect to infinitely divisible measures. *Bull. Acad. Polon. Sci.* **29** 179–185.

Jurek, Z. J. and Vervaat, W. (1983). An integral representation for self-decomposable Banach space valued random variables. *Z. Wahrsch. Verw. Gebiete* **62** 247–262.

Kehrer, E. (1983). Stabilkitat von Wahrscheinlichkeitsmassen unter Operatorgruppen auf Banachraumen, Ph.D. dissertation, Universität Tübingen, Germany.

Klosowska, M. (1980). Domain of operator attraction of a Gaussian measure in $\mathbb{R}^N$. *Ann. Soc. Math. Polon. Ser. I Comment. Math. Prace Mat.* **22** 73–80.

Krakowiak, W. (1979). Operator-stable probability measures on Banach spaces. *Colloq. Math.* **41** 313–326.

Krakowiak, W. (1980). Operator semi-stable probability measures on Banach spaces. *Colloq. Math.* **43** 351–363.

Kruglov, V. N. (1972). On the extension of the class of stable distributions. *Theory Probab. Appl.* **17** 685–694.

Kumar, A. and Mandrekar, V. (1972). Stable probability measures on Banach space. *Studia Math.* **42** 133–144.

Kumar, A. and Schreiber, B. (1975). Self-decomposable probability measures on Banach space. *Studia Math.* **53** 55–71.

Kucharczak, J. (1975). Remarks on operator-stable measures. *Colloq. Math.* **34** 109–119.

Laha, R. G. and Rohatgi, V. K. (1980). Semistable measures on a Hilbert space, *J. Multivariate Anal.* **10** 88–94.

Lamperti, J. W. (1962). Semi-stable stochastic processes. *Trans. Amer. Math. Soc.* **104** 62–78.

Lamperti, J. W. (1972). Semi-stable Markov processes. *Z. Wahrsch. Verw. Gebiete* **22** 205–225.

Lange, S. (1969). *Analysis II*. Addison-Wesley.

Lange, S. (1975). $SI_2(R)$. Addison-Wesley.

Lévy, P. (1937). *Theorie de l'Addition des Variables Aleatoires*. Gauthier-Villars.

Linde, W. (1986). *Probability in Banach Spaces—Stable and Infinitely Divisible Distributions*. Wiley-Interscience.

Lindvall, T. (1973). Weak convergence of probability measures and random functions in the function space $D[0, \infty)$. *J. Appl. Probab.* **10** 109–121.

Luczak, A. (1981). Operator semi-stable probability measures on $\mathbb{R}^N$. *Colloq. Math.*, **45** 287–300.

Luczak, A. (1984). Elliptical symmetry and characterization of operator-stable and operator semi-stable measures. *Ann. Probab.* **12** 1217–1223.

Lukacs, E. (1960). *Characteristic Functions*. Griffin and Co.

Marcus, D. (1983). Non-stable laws with all projections stable. *Z. Wahrsch. Verw. Gebiete* **64** 139–156.

Marczewski, E., and Ryll-Nardzewski, C. (1953). Remarks on the compactness and nondirect products of measures. *Fund. Math.* **40** 165–170.

Mason, J. D. (1982). A comparison of the properties of operator-stable distributions and operator-self-similar processes. *Colloq. Math. Soc. Janos Bolyai* **36** 751–760.

Meerschaert, M. M. (1990). Moments of random vectors which belong to some domain of normal attraction. *Ann. Probab.* **18** 870–876.

Michalicek, J. (1972). Der Anziehungsbereich von operator-stabilen Verteilungen im $\mathbb{R}^2$. *Z. Wahrsch. Verw. Gebiete* **25** 57–70.

Michalicek, J. (1972). Die Randverteilungen der operator-stabilen Masse im 2-dimensionalen Raum. *Z. Wahrsch. Verw. Gebiete* **21** 135–146.

Mincer, B. (1982a). On $GL(n, R)$-stable measures. *Probab. Math. Statist* **2** 193–195.

Mincer, B. (1982b). $U(X)$-stable measures on Banach spaces. *Probab. Math. Statist* **3** 310–315.

Mincer, B., and Urbanik, K. (1979). Completely stable measures on Hilbert spaces. *Colloq. Math.* **42** 301–307.

Niedbalsha, T. (1979). An example of the decomposability semigroup, *Colloq. Math.* **39** 137–139.

Niedbalsha-Rajba, T. (1981). On decomposability semigroups on the real line. *Colloq. Math.* **44** 347–358.

Numakura, K. (1952). On bicompact semigroups. *Math. J. Okayama Univ.* **1** 99–108.

Paalman, A. B., and de Miranda, D. (1964). *Topological Semigroups*. Math. Centre Tracts 11. Mathematisch Centrum, Amsterdam.

Parthasarathy, K. R. (1967). *Probability Measures on Metric Spaces*. Academic.

Parthasarathy, K. R. (1973). Every completely stable distribution is normal. *Sankhyā Ser. A* **35** 35–38.

Parthasarathy, K. R. and Schmidt, K. (1975). Stable positive definite functions. *Trans. Amer. Math. Soc.* **203** 161–174.

Paulauskas, V. J. (1984). Some remarks on multivariate stable distributions, *J. Multivariate Anal.* **6** 356–368.

Pollard, D. (1984). *Convergence of Stochastic Processes*. Springer Series in Statistics. Springer-Verlag.

Sakovic, G. N. (1961). Solution of a multivariate functional equation. *Ukrain. Mat. Zh.*, **13** 173–189 (in Russian).

Sakovic, G. N. (1965). Multivariate stable distributions. Ph.D. dissertation, Kiev (in Russian).

Sato, K.-I. (1973). A note on infinitely divisible distributions and their Lévy measures. *Science Reports of the Tokyo Kyoiku Daigaku, Section A* **12** 101–109.

Sato, K.-I. (1985). Lectures on multivariate infinitely divisible distributions and operator-stable processes. Technical Report 54, Laboratory for Research in Statistics and Probability, Carleton University, Ottawa, Canada.

Sato, K.-I. (1987). Strictly operator-stable distributions. *J. Multivariate Anal.* **22** 278–295.

Sato, K.-I. (1991). Self-similar processes with independent increments. *Probab. Theory Related Fields* **89** 285–300.

Sato, K.-I. and Yamazato, M. (1984a). Stationary processes of Ornstein–Uhlenbeck type. *Lecture Notes in Math.* **1021**, *Proceedings of the Japan–USSR Symposium on Probability Theory*. Springer-Verlag.

Sato, K.-I. and Yamazato, M. (1984b). Operator-selfdecomposable distributions as limit distributions of processes of Ornstein–Uhlenbeck type. *Stochastic Process. Appl.* **17** 73–100.

Schmidt, K. (1975). Stable probability measures on $\mathbb{R}^v$. *Z. Wahrsch. Verw. Gebiete* **33** 19–31.

Semovskii, S. V. (1984). The Bernstein–Feller central limit theorem in $\mathbb{R}^p$. *Theory Probab. Appl.* **29** 586–591.

Semovskii, S. V. (1988). Operator-normalized sums of random vectors: convergence to the normal law together with convergence of moments. *Theory Probab. Appl.* **32** 748–749.

Sharpe, M. (1969). Operator-stable probability distributions on vector groups. *Trans. Amer. Math. Soc.* **136** 51–65.

Siebert, E. (1986). Supplements to operator-stable and operator-semistable laws on Euclidean spaces. *J. Multivariate Anal.* **19** 329–341.

Skitovich, W. P. (1954). Linear forms of independent random variables and normal distribution. *Izv. Akad. Nauk SSSR* **15** 185–200 (in Russian).

Tucker, H. G. (1967). *A Graduate Course in Probability Theory*. Academic.

Urbanik, K. (1972). Lévy's probability measures on Euclidean spaces. *Studia Math.* **44** 119–148.

Urbanik, K. (1975). Decomposability properties of probability measures. *Sankhyā Ser. A* **37** 530–537.

Urbanik, K. (1976). Some examples of decomposability semigroups. *Bull. Acad. Polon. Sci.* **10** 915–918.

Urbanik, K. (1977). A characterization of Gaussian measures on Banach spaces. *Studia Math.* **59** 275–281.

Urbanik, K. (1978). Lévy's probability measures on Banach spaces. *Studia Math.* **63** 283–308.

Urbanik, K. (1979). Geometric decomposability properties of probability measures. *Banach Center Publication* **5** 249–254.

Veeh, J. A. (1982). Infinitely divisible measures with independent marginals. *Z. Wahrsch. Verw. Gebiete* **61** 303–308.

Watanabe, S. (1968). A limit theorem of branching processes and continuous state branching processes. *J. Math. Kyoto Univ.* **8** 381–396.

Weissman, I. (1976). On convergence of types and processes in Euclidean space. *Z. Wahrsch. Verw. Gebiete* **37** 35–41.

Williams, D. (1979). *Diffusions, Markov Processes and Martingale Foundations* **1**. Wiley.

Wolfe, S. J. (1980). A characterization of Lévy probability distribution functions on Euclidean spaces. *J. Multivariate Anal.* **10** 379–384.

Wolfe, S. J. (1982a). On a continuous analogue of the stochastic difference equation $X_n = \rho X_{n-1} + B_n$. *Stochastic Process. Appl.* **12** 301–312.

Wolfe, S. J. (1982b). A characterization of certain stochastic integrals. *Stochastic Process. Appl.* **12** 136.

Yamazato, M. (1983). Absolute continuity of operator-selfdecomposable distributions on $\mathbb{R}^2$. *J. Multivariate Anal.* **13** 550–560.

Zolotarev, V. M. (1986). One-dimensional stable distributions. *Trans. Math. Monographs* **65**. Amer. Math. Soc.

INDEX

Absolute continuity, 162, 163
　of exp($-tQ$)-decomposable, 164, 173, 175
　of infinitely divisible, 164, 165
　of operator-stable, 237
Affine transformation, 24, 48, 59, 192
Araujó and Giné, xiii, 44, 281
Aut(X), 4

Banach:
　dual, 3
　space, 3, 67
Banach–Steinhaus theorem, 24
Billingsley, 44, 70, 281
Boundary, 20

Center of a group, 221
Centering function, 192, 236
Characteristic functional, 25, 32, 33, 57, 62, 89, 101, 103, 119, 197, 264
Chorny, 281
Compact:
　conditional, 22, 25, 49, 50
　sequential, 22
　shift conditionally, 23, 25, 29
Continuity theorem, 26, 62
Continuous mapping theorem, *see* Weak convergence, mapping theorem
Convergence:
　of operator types, 52, 187, 202
　of types, 45, 48, 64, 65, 67, 68
　weakly, 20
Convergence of operator types, 52, 187, 202

Convergence of types, 45, 48, 64, 65, 67, 68
Convolution, 23
Coordinate mapping, 41
Cramér theorem, 38, 58, 89
Cramér–Wold device, 26, 269
Cuppens, 281

de Acosta, 44, 281
deConinck, x, 281
de Miranda, 44, 285
Determinant, 6
Differentiable, 8
　continuously, 8
Distribution, 22
　converge in distribution, 22
Domain of attraction, 239
Domain of normal attraction, 239, 243, 247, 252
DONA, 240
Dudley, 275, 281
Dynkin, 44, 281

Elliptically symmetric, 180, 229, 234
End (X), 4
Exponent:
　commuting, 219, 221
　exp($-tQ$)-decomposable, 89, 96
　operator-selfdecomposable, 76
　operator-stable, 185, 186, 192, 205, 211, 212, 215, 272
Exponential:
　exp($-tQ$)-, 89, 101, 161, 162, 173
　U-, 85, 179

289

Fenchal, 70, 281
Fisz, x, 70, 281
Freedman, 183, 281
Functional, 24, 25

Gaussian, 31, 58, 88, 90, 127, 142, 162, 213, 214, 246, 270
   centered, 32, 33, 38, 39
   moment, 252
GDOA, 240
Generalized domain of attraction, 250
Gikhman and Skorohod, 44, 281
Giné, 275, 281
Gnedenko and Kolmogorov, x, 281
Griffin, xii, 282

Haar measure, 181, 220, 223, 231
Hahn, xii, xiii, 275, 281, 282
Hausdorff measure, 230
Hazod, xiii, 282
Hilbert space, 3, 26, 62
Hille and Phillips, 24, 44, 82, 282
Hoffman and Kunze, 44, 282
Holmes, 274, 275, 282
Hudson, viii, xi, 274, 275, 282

Ideal, 1
Idempotent, 1, 2, 55, 77, 78
ID(X), 28, 34, 35
ID$_{\log}$(X), 36, 124, 127, 128, 132, 149, 168, 173, 198, 199, 201
Ilinskii, 67, 282
Infinitely divisible, 28, 29, 240
   absolute continuity, 163, 165
Infinitesimal system, 35, 38, 72, 84
Integrable Bochner, 8
Integral:
   random, 117, 120
   representation, 116
Invariant:
   semigroup, 50, 65
   space, 6, 63
Inverse function theorem, 9

J_1 metric, 41
Jajte, xiii, 277, 278, 283
Jurek, 44, 70, 182, 183, 274, 275, 279, 282, 283

Kanter, 275, 281

Kehrer, 279, 284
Klass, xii, xiii, 282
Klosowska, 284
Krakowiak, xiii, 274, 277, 284
Kruglov, 277, 284
Kucharczak, xiii, 284
Kumar, 183, 284

Laha, 284
Lamperti, x, 284
Lange, 180, 284
Lévy, 269, 275, 278, 284
   class, 71
   –Kintchin representation, 33, 101
   spectral measure, 31, 33, 34, 95
Lie:
   bracket, 15, 17
   tangent space, 15, 85, 189, 212, 271
   theory, 15
lin(A), 105
lin$_Q$(A), 105
Linde, x, 284
Lindvall, 284
Luczak, 274, 277, 284, 285
Lukacs, 237, 285

Mandrekar, 284
Marcus, 275, 285
Marczewski and Ryll-Nardzewski, 285
Marginal, 254
   complete, 254
   independent, 254, 255, 259
   k-dimensional, 254
   orthogonal, 254
Markov process, 144, 150
Mason, x, 274, 275, 282, 285
Measurable function, 7
Measure:
   decomposable, 72
   equivalent, 162
   regular, 20
Meerschaert, 275, 285
Michalicek, 285
Mincer, xiii, 278, 285
Minimal polynomial, 6, 205
Moment, 249, 253, 270
   exponential, 32, 35
   Gaussian, 252
   logarithmic, 96, 103, 122, 124, 127, 130, 161, 179, 198

INDEX

Monothetic semigroup, 2, 3, 55

Niedbalsha-Rajba, xiii, 67, 285
Nobel, 282
Norm:
  new, 92
  operator, 4
Norming sequence, 72, 74, 201
Numakura, 44, 285
  theorem, 1, 2, 55, 76, 77, 80, 81

One-parameter semigroup, 10, 71, 76, 122, 124
  contraction, 10, 13, 147, 152
  operator, 144
  strongly continuous, 10, 12, 13, 14, 78, 83, 279
Operator:
  bounded, 4
  closed, 5, 11
  contraction, 10
  convergence of types, 52
  core, 14
  covariance, 33
  extension, 5
  generator, 10, 11, 12, 13, 14, 108, 116, 142, 144, 149, 152
  graph, 4
  log of, 17, 82
  S-, 33, 37, 88, 90
  -selfdecomposable, 71, 72, 74, 76, 78, 83
  -semistability, 277
  -stable, 185, 191
  topology, 24
  trace, 6
  type, 51
Ornstein–Uhlenbeck, x, 149, 150, 161, 183
Orthogonal transformation, 18, 59, 60, 61

Paalman, 44, 285
Parthasarathy, 44, 278, 285
Paulauskas, 285
Poisson measure, 30
Polar coordinate, 92, 110
Pollard, 44, 285
Portmanteau theorem, 20
Primary decomposition theorem, 6, 205, 206, 210, 215, 253
Prohorov:
  metric, 22
  theorem, 22, 23

Random integral representation, 71, 178
Resolvent equation, 145
Riesz representation theorem, 4
Rohatgi, 284
Rolle's theorem, 111
Rosinski, 183, 284

Sakovic, x, 186, 274, 285, 286
Sakovic–Sharpe characterization, 186
Sato, x, xiii, 183, 274, 286
Schmidt, 278, 285, 286
Schreiber, 183, 284
Schur's lemma, 180, 181, 230, 272
Selfdecomposable, 177
sem(a), 2, 55, 75, 76, 77
Semigroup:
  decomposable, 45, 53
  Feller–Dynkin, 152
  full, 46, 47, 63
  invariant, 50, 65
Semovskii, 286
Separable, 23, 39
Sharpe, xi, 70, 186, 274, 286
Siebert, 277, 286
Skew-symmetric transformation, 18, 88, 213, 229, 271
Skitovich theorem, 38, 58, 286
Skorohod:
  metric, 41, 42, 43
  space, 39, 42, 43, 117
Smalara, 44, 284
Spectral:
  function, 94, 109, 112
  measure, 31
Stable:
  multivariate, 261, 263, 265, 266, 268
  strictly, 261
  t^B, 193
Stationary, 161
Stochastically equivalent, 40
Subadditive, 35, 36
Submultiplicative, 35, 36
Support, 26, 27, 28, 38, 105, 106, 197, 205, 270

Symmetric:
  elliptically, 234
  group, 45, 53, 59
  semigroup, 53
Symmetrization, 26, 29

Tangent space, 15, 85, 189, 212, 271
Tight, 20, 22, 49
Topological:
  semigroup, 1
  subsemigroup, 1
Trace, 6
Tucker, xiii, 275, 282, 286

U-exponent, 85
  commuting, 88
Uniformly asymptotically negligible, 35
Urbanik, xiii, 44, 45, 67, 70, 178, 182, 274, 278, 285, 286, 287
  decomposability semigroup, 45, 53, 59, 66, 67, 71, 76, 83, 108

Veeh, viii, 274, 275, 282, 287
Vervaat, 183, 284

Watanabe, 44, 287
  lemma, 14, 160
Weak convergence, 20, 38
  mapping theorem, 21, 24, 86
Weissman, 70, 287
Williams, 183, 287
Wolfe, 182, 183, 287

Yamazato, xiii, 183, 286, 287

Zolotarev, x, 287

Applied Probability and Statistics (Continued)
  HOEL · Elementary Statistics, *Fifth Edition*
  HOGG and KLUGMAN · Loss Distributions
  HOLLANDER and WOLFE · Nonparametric Statistical Methods
  HOSMER and LEMESHOW · Applied Logistic Regression
  IMAN and CONOVER · Modern Business Statistics
  JACKSON · A User's Guide to Principle Components
  JOHN · Statistical Methods in Engineering and Quality Assurance
  JOHNSON · Multivariate Statistical Simulation
  JOHNSON and KOTZ · Distributions in Statistics
    Discrete Distributions
    Continuous Univariate Distributions—1
    Continuous Univariate Distributions—2
    Continuous Multivariate Distributions
  JUDGE, GRIFFITHS, HILL, LÜTKEPOHL, and LEE · The Theory and Practice of Econometrics, *Second Edition*
  JUDGE, HILL, GRIFFITHS, LÜTKEPOHL, and LEE · Introduction to the Theory and Practice of Econometrics, *Second Edition*
  KALBFLEISCH and PRENTICE · The Statistical Analysis of Failure Time Data
  KASPRZYK, DUNCAN, KALTON, and SINGH · Panel Surveys
  KISH · Statistical Design for Research
  KISH · Survey Sampling
  LAWLESS · Statistical Models and Methods for Lifetime Data
  LEBART, MORINEAU, and WARWICK · Multivariate Descriptive Statistical Analysis: Correspondence Analysis and Related Techniques for Large Matrices
  LEE · Statistical Methods for Survival Data Analysis, *Second Edition*
  LePAGE and BILLARD · Exploring the Limits of Bootstrap
  LEVY and LEMESHOW · Sampling of Populations: Methods and Applications
  LINHART and ZUCCHINI · Model Selection
  LITTLE and RUBIN · Statistical Analysis with Missing Data
  MAGNUS and NEUDECKER · Matrix Differential Calculus with Applications in Statistics and Econometrics
  MAINDONALD · Statistical Computation
  MALLOWS · Design, Data, and Analysis by Some Friends of Cuthbert Daniel
  MANN, SCHAFER, and SINGPURWALLA · Methods for Statistical Analysis of Reliability and Life Data
  MASON, GUNST, and HESS · Statistical Design and Analysis of Experiments with Applications to Engineering and Science
  McLACHLAN · Discriminant Analysis and Statistical Pattern Recognition
  MILLER · Survival Analysis
  MONTGOMERY and PECK · Introduction to Linear Regression Analysis, *Second Edition*
  NELSON · Accelerated Testing, Statistical Models, Test Plans, and Data Analyses
  NELSON · Applied Life Data Analysis
  OCHI · Applied Probability and Stochastic Processes in Engineering and Physical Sciences
  OKABE, BOOTS, and SUGIHARA · Spatial Tesselations: Concepts and Applications of Voronoi Diagrams
  OSBORNE · Finite Algorithms in Optimization and Data Analysis
  PANKRATZ · Forecasting with Dynamic Regression Models
  PANKRATZ · Forecasting with Univariate Box-Jenkins Models: Concepts and Cases
  RACHEV · Probability Metrics and the Stability of Stochastic Models
  RÉNYI · A Diary on Information Theory
  RIPLEY · Spatial Statistics
  RIPLEY · Stochastic Simulation
  ROSS · Introduction to Probability and Statistics for Engineers and Scientists
  ROUSSEEUW and LEROY · Robust Regression and Outlier Detection
  RUBIN · Multiple Imputation for Nonresponse in Surveys
  RYAN · Statistical Methods for Quality Improvement
  SCHUSS · Theory and Applications of Stochastic Differential Equations

Applied Probability and Statistics (Continued)

BUCKLEW · Large Deviation Techniques in Decision, Simulation, and Estimation
BUNKE and BUNKE · Nonlinear Regression, Functional Relations and Robust Methods: Statistical Methods of Model Building
CHATTERJEE and HADI · Sensitivity Analysis in Linear Regression
CHATTERJEE and PRICE · Regression Analysis by Example, *Second Edition*
CLARKE and DISNEY · Probability and Random Processes: A First Course with Applications, *Second Edition*
COCHRAN · Sampling Techniques, *Third Edition*
*COCHRAN and COX · Experimental Designs, *Second Edition*
CONOVER · Practical Nonparametric Statistics, *Second Edition*
CONOVER and IMAN · Introduction to Modern Business Statistics
CORNELL · Experiments with Mixtures, Designs, Models, and the Analysis of Mixture Data, *Second Edition*
COX · A Handbook of Introductory Statistical Methods
*COX · Planning of Experiments
CRESSIE · Statistics for Spatial Data
DANIEL · Applications of Statistics to Industrial Experimentation
DANIEL · Biostatistics: A Foundation for Analysis in the Health Sciences, *Fifth Edition*
DAVID · Order Statistics, *Second Edition*
DEGROOT, FIENBERG, and KADANE · Statistics and the Law
*DEMING · Sample Design in Business Research
DILLON and GOLDSTEIN · Multivariate Analysis: Methods and Applications
DODGE and ROMIG · Sampling Inspection Tables, *Second Edition*
DOWDY and WEARDEN · Statistics for Research, *Second Edition*
DRAPER and SMITH · Applied Regression Analysis, *Second Edition*
DUNN · Basic Statistics: A Primer for the Biomedical Sciences, *Second Edition*
DUNN and CLARK · Applied Statistics: Analysis of Variance and Regression, *Second Edition*
ELANDT-JOHNSON and JOHNSON · Survival Models and Data Analysis
EVANS, PEACOCK, and HASTINGS · Statistical Distributions, *Second Edition*
FISHER and VAN BELLE · Biostatistics: A Methodology for the Health Sciences
FLEISS · The Design and Analysis of Clinical Experiments
FLEISS · Statistical Methods for Rates and Proportions, *Second Edition*
FLEMING and HARRINGTON · Counting Processes and Survival Analysis
FLURY · Common Principal Components and Related Multivariate Models
GALLANT · Nonlinear Statistical Models
GROSS and HARRIS · Fundamentals of Queueing Theory, *Second Edition*
GROVES · Survey Errors and Survey Costs
GROVES, BIEMER, LYBERG, MASSEY, NICHOLLS, and WAKSBERG · Telephone Survey Methodology
HAHN and MEEKER · Statistical Intervals: A Guide for Practitioners
HAND · Discrimination and Classification
*HANSEN, HURWITZ, and MADOW · Sample Survey Methods and Theory, Volume I: Methods and Applications
*HANSEN, HURWITZ, and MADOW · Sample Survey Methods and Theory, Volume II: Theory
HEIBERGER · Computation for the Analysis of Designed Experiments
HELLER · MACSYMA for Statisticians
HOAGLIN, MOSTELLER, and TUKEY · Exploratory Approach to Analysis of Variance
HOAGLIN, MOSTELLER, and TUKEY · Exploring Data Tables, Trends and Shapes
HOAGLIN, MOSTELLER, and TUKEY · Understanding Robust and Exploratory Data Analysis
HOCHBERG and TAMHANE · Multiple Comparison Procedures

*Now available in a lower priced paperback edition in the Wiley Classics Library.